Reduction and Emergence in Science and Philosophy

Grand debates over reduction and emergence are playing out across the sciences, from superconductors to slime mold, but these discussions have reached a stalemate with both sides declaring victory on empirical grounds. This book provides new theoretical frameworks which illuminate the novel positions of scientific reductionists and emergentists, the recent empirical advances that drive these new views, and the deeper issues in their ongoing battles. Carl Gillett highlights the flaws in existing philosophical frameworks and shows how philosophical discussions can be reoriented to reflect recent scientific advances and the new set of issues which they raise – including collectives and their behavior, the nature of "parts" and "wholes," the character of aggregation, and hence the continuity of nature itself. His book also shows how the ongoing disputes about concrete scientific cases are empirically resolvable, and thus how we can break the scientific stalemate in future work.

CARL GILLETT is Professor of Philosophy at Northern Illinois University. He is the editor of *Physicalism and its Discontents* (with Barry Loewer, Cambridge, 2001) and the author of numerous journal articles.

Reduction and Emergence in Science and Philosophy

CARL GILLETT
Northern Illinois University

CAMBRIDGE
UNIVERSITY PRESS

CAMBRIDGE
UNIVERSITY PRESS

University Printing House, Cambridge CB2 8BS, United Kingdom

One Liberty Plaza, 20th Floor, New York, NY 10006, USA

477 Williamstown Road, Port Melbourne, VIC 3207, Australia

314-321, 3rd Floor, Plot 3, Splendor Forum, Jasola District Centre, New Delhi - 110025, India

79 Anson Road, #06-04/06, Singapore 079906

Cambridge University Press is part of the University of Cambridge.

It furthers the University's mission by disseminating knowledge in the pursuit of education, learning and research at the highest international levels of excellence.

www.cambridge.org
Information on this title: www.cambridge.org/9781107428072

First published 2016
First paperback edition 2018

A catalogue record for this publication is available from the British Library

Library of Congress Cataloging in Publication data
Names: Gillett, Carl, 1967– author.
Title: Reduction and emergence in science and philosophy /
Carl Gillett, Northern Illinois University.
Description: New York : Cambridge University Press, 2016. |
Includes bibliographical references and index.
Identifiers: LCCN 2016023275| ISBN 9781107075351 (hardback: alk. paper) |
ISBN 9781107428072 (pbk. : alk. paper)
Subjects: LCSH: Philosophy and science. | Science – Philosophy. |
Reductionism. | Emergence (Philosophy)
Classification: LCC B67.G54 2016 | DDC 110–dc23
LC record available at https://lccn.loc.gov/2016023275

ISBN 978-1-107-07535-1 Hardback
ISBN 978-1-107-42807-2 Paperback

Contents

Preface

The challenge and the crackling of thin ice are what give science its metaphysical excitement.

– E. O. Wilson[1]

Our opening passage from the prominent scientific reductionist E. O. Wilson highlights how some of the most exciting debates in the sciences are metaphysical, ongoing, and of real import – or so I seek to show in this book. These scientific battles over reduction, emergence, and the structure of nature are what drew me into philosophy as a student from a planned career in the sciences. Philosophers of science had long discussed "reduction," and to a lesser extent "emergence," so it seemed obvious that philosophy was the place to look for insight. However, over the years since my switch into philosophy it has slowly dawned on me that philosophical frameworks are actually now part of the problem in our stalemated discussions about reduction and emergence.

One of my main contentions in this book is that philosophical discussions need to be refreshed by a re-engagement with the sciences, but my main work is to illuminate what is at issue in the scientific discussions. Rather than providing a comprehensive treatment of every researcher in the voluminous literature on reduction and emergence, spanning science and philosophy, I consequently focus on providing a story about the main scientific positions and I therefore apologize to writers whose work I have not critically engaged in depth or sometimes at all. Instead, I pursue the positive project of laying out better theoretical frameworks for understanding the scientific debates, their live positions, and deeper issues to highlight how empirical evidence can move these discussions forward. Along the way, I also detail why, and how, philosophical frameworks need to be refreshed and reoriented given my findings about the scientific debates.

[1] Wilson (1998), p. 55.

Although I wrote a doctoral thesis in the philosophy of mind, I was especially lucky to work in graduate school at Rutgers with Jerry Fodor, Barry Loewer, and Brian McLaughlin, who are all central figures in philosophical debates over reduction and emergence. In particular, Brian's seminal graduate course on the British emergentists ignited my interest in scientific positions and my concerns about existing philosophical frameworks. I have subsequently discussed the issues of the book with many people over the years and I apologize to anyone I have forgotten. But I recall productive discussions with: Ken Aizawa, Colin Allen, Mark Bedau, Charlotte Brown, Craig Callender, John Churchill, Lenny Clapp, Carl Craver, Seth Crook, Michael Esfeld, Jerry Fodor, Jeff Goldstein, Cleo Kearns, Jaegwon Kim, Paul Lodge, Barry Loewer, Jonathan Lowe, Brian McLaughlin, Andrew Melnyk, Ted Morris, Tim O'Connor, David Papineau, Tom Polger, Bill Ramsey, Bob Richardson, Brad Rives, Christian Sachse, J. R. Schrader, Larry Shapiro, Sydney Shoemaker, Mark Siderits, Michael Silberstein, Jim Simeone, Achim Stephan, Leopold von Steubenberg, Jackie Sullivan, Sven Walter, Fritz Warfield, Jessica Wilson, Gene Witmer, and Julie Yoo. I owe special thanks for the comments of my NIU colleagues Hal Brown and Valia Allori, who were members of a reading group that worked through a very rough, early draft of the book. And I am also grateful for discussion with the many wonderful NIU graduate students in that group, and in classes where I presented related material, including: Matt Babb, Rene Bolinger, Brian Chambliss, Dean Da Vee, Louis Gularte, Travis Hobbs, Mike McCourt, Brock Rough, Amy Seymour, Jeff Snapper, Peter Van Elswyk, and Jason Winning, amongst others. I am also grateful for the comments of three reviewers of the manuscript, an anonymous referee, and especially Andrew Melnyk and William Hasker, who generously followed up their reports with still further productive comments and conversations.

A lot of other people have supported me over the years, starting with two wonderful colleagues who were my department chairs, and I am thankful to Charlotte Brown, at IWU, and David Buller, at NIU, for all their help. A very early draft of the book was completed in a semester's research leave from teaching supported by a grant from the Templeton Foundation, for which I am grateful. Thanks are also due to my ever-accommodating Cambridge University Press editor Hilary Gaskin. And I owe thanks to Jonathan Lowe for his help placing the book with the Press and I am very sad that he is not with us to see it come out. I am also especially grateful to Barry Loewer and Jaegwon Kim for their intellectual and personal support over the years, which helped me in ways great and small.

Finally, I owe very special thanks to Sarah McKibben and Alasdair Gillett-McKibben for their love, support, and inevitable sacrifice through my work on the book.

Introduction

I think a good case can be made that science has now moved from an Age of Reductionism to an Age of Emergence, a time when the search for ultimate causes of things shifts from the behavior of parts to the behavior of the collective.

– Robert Laughlin[1]

When you are criticizing the philosophy of an epoch, do not chiefly direct your attention to those intellectual positions which its exponents feel it necessary to defend. There will be some fundamental assumptions which ... the epoch unconsciously presuppose[s]. Such assumptions appear so obvious that people do not know what they are assuming ... With these assumptions a certain limited number of types of philosophic system are possible, this group constitutes the philosophy of the epoch.

– Alfred North Whitehead[2]

As we start the new century, we find ourselves in the midst of a new cycle of scientific debates over "reduction" and "emergence" or what we may colorfully term, following Robert Laughlin the Nobel Prize-winning physicist and scientific emergentist, a new "Battle of the Ages." For defenders of the Age of Reductionism, like the other Nobel Prize-winning physicist Steven Weinberg, biologists like Francis Crick or E. O. Wilson, and many others, are not giving up without a fight. We find clashes between the two sides across bitter fights in physics over the supercollider or high-energy superconductors, biological debates over cells and their molecular components, disputes over neurons and neuronal populations, and many other concrete cases at various levels of the sciences.[3]

---

[1] Laughlin (2005), p. 208.
[2] Whitehead (1925), p. 71.
[3] See Scott (2007) for a more comprehensive survey of such examples.

It is hard to come by clear formulations of the opposing hypotheses pressed by scientific reductionists and emergentists about such examples, but each side does supply pithy slogans to publicize their claims. Thus scientific reductionists continue to press increasingly sophisticated views under the banner that "Wholes are nothing but their parts." On the other side, scientific emergentists of this generation, like Laughlin or the other condensed matter physicist Philip Anderson, chemists such as Ilya Prigogine, neuroscientists like Walter Freeman, and many in systems biology or the sciences of complexity, still defend ontological positions claiming that "Wholes are more than the sum of their parts," but the new wave of emergentists now make the latter claim alongside views that I contend can be sloganized as "Parts behave differently in wholes."[4]

Superficially, one might therefore think that very little has changed in the sciences. At the beginning of the *last* century, one of the great scientific disputes also raged between reductionists and emergentists of various kinds.[5] However, closer inspection reveals important differences with these older debates whose central question was whether all natural phenomena are *composed* and, in particular, the chemical and biological entities whose composed status had, at that point, remained inscrutable for decades. Putting the issues in a different way, the earlier battles focused on whether all levels of nature, including chemical or biological phenomena, were amenable to what I shall term "compositional explanation" – that is, to explanations of higher-level entities built around lower-level entities taken to *compose* them. Famously, however, during the course of the twentieth century the rise of quantum mechanics and molecular biology finally provided compositional explanations in chemical and biological cases. And such headline-making advances occurred against the backdrop of continuing waves of compositional explanation in the full range of other sciences. The earlier debates were consequently settled in the sciences, since these explanations provided *qualitative* accounts of the components of the entities found at many levels of nature, including chemical and biological phenomena.

---

[4] The second slogan, as I show in Chapters 6 and 7, expresses a central claim of contemporary scientific emergentists once we dig into their ontological commitments.

[5] For example, see Haraway (1976) for a survey of these earlier battles in biology.

But scientific innovation never ceases and amongst recent advances, as Laughlin notes, is our improved understanding of the components of complex aggregations, or "collectives" to use Laughlin's term, from high-energy superconductors, to eukaryotic cells, and on to slime mold, neural populations, or eusocial insect colonies. Often using new techniques, for the first time we now also often have *quantitative* accounts of the components found in such complex collectives. And these discoveries apparently fuel our new cycle of debates. Thus contemporary scientific reductionists and emergentists each apparently endorse the ubiquity of compositional explanation and their disputes no longer concern the *existence* of composition at all levels in nature, but instead focus upon opposing accounts of its *character* and *implications*. The two sides thus clash over an array of what Whitehead would class as diverging *epochal* ontological commitments implicitly assumed by either side in contrasting views of the very nature of composition (i.e. "parts" and "wholes"), the form of the aggregation that always accompanies such composition, the varieties of determination we find in the universe, and the character of the fundamental laws, amongst a variety of other issues.[6]

These exciting debates focus on Big Questions of interest to anyone concerned with what the sciences tell us about the structure of the universe or the best ways for understanding it. Unsurprisingly, the scientific disputes have consequently played out not just across academic journals and monographs in the sciences, but also across a dizzying array of best-selling books, newspaper and magazine articles, and even congressional hearings.[7] However, in contrast to this popular engagement, philosophers have shown comparatively little interest in such scientific discussions.

It is striking that philosophers, at least in any numbers, have not engaged the scientific debates, though there have been notable

---

[6] On one plausible story, Anderson (1972) arguably fired the first salvo in the present cycle of contemporary debates and four decades later the scientific disputes only burn more brightly.

[7] The congressional testimony was focused on the funding of the so-called "supercollider" that was taken to be central to the scientific reductionist program. For a matched pair of books setting out the two sides of the ensuing debates, see Weinberg (1992), written as a defense of the supercollider, and Laughlin (2005), which provides the contrasting emergentist perspective. For just a few of the other books on these debates, again from both sides, see Holland (1999), Lewin (1992), and Wilson (1998).

exceptions.[8] And some philosophers have even taken the time to dismiss scientific discussions of reduction and emergence as mere empty rhetoric used to support funding grabs.[9] This situation is not so surprising, since philosophers have erected their own proprietary views of the nature of scientific composition, reduction, and emergence diverging from the accounts apparently used in the sciences. Applying their different theoretical frameworks, the reigning view in mainstream philosophy is consequently that reductionism is basically a *dead*, and perhaps even somewhat *distasteful*, position. And many (most?) philosophers dismiss discussions of emergence as, at best, *kooky* and, at worst, *incoherent* (to use far more polite terms than are usual in such dismissals).

Scientists in recent debates are well aware of these dislocations with philosophical discussions. *Both* scientific reductionists *and* emergentists take a jaundiced view of philosophical frameworks that do not comfortably fit their views or what they see as the deeper issues. Thus one finds Weinberg explicitly spurning philosophical models of reduction, repeatedly pressing the different nature of his scientific reductionist position, and penning a chapter entitled "Against Philosophy."[10] And the emergentist biologist Ernst Mayr counsels avoiding the received philosophical frameworks for reduction and emergence, since he concludes that such frameworks distort the issues in damaging ways.[11]

It is obviously troubling to find theoretical differences between philosophical and scientific debates on such important topics. And it leaves us with a pressing question: Are philosophical or scientific

---

[8] I locate the work of these writers in the frameworks of coming chapters. For example, Jaegwon Kim, and writers such as Alexander Rosenberg, John Heil, Andrew Melnyk, and Barry Loewer, press ontological accounts of reductionism with affinities to scientific reductionism. And other writers such as John Bickle and Kenneth Schaffner have directly engaged the kinds of reduction espoused by scientists in specific scientific cases. With regard to emergence, Mark Bedau, John Dupré, Robin Hendry, Alicia Juarrero, Sandra Mitchell, Robert Richardson, and Achim Stephan, amongst others, have all begun to engage the claims of various species of scientific emergentist.

[9] The suggestion that philosophers of science should not co-facilitate such rhetorically based funding grabs, and hence should abandon the terms "reduction" and "emergence," was forcefully pressed by the philosopher of physics John Norton at the 2009 Paris–Pittsburgh Conference on Reduction and Emergence held at the Pittsburgh HPS Department.

[10] Weinberg (1992), Chapter 7. See his (2001) for similar sentiments.

[11] See Mayr (2002).

approaches to reduction and emergence, as well as connected foundational issues such as the nature of scientific composition, more likely to be closer to the truth about the deeper issues and live positions of our times? One of my main negative conclusions in this book is that the scientists are right to be wary, for I show at length that the received philosophical wisdom has gotten things dead wrong about the nature of scientific composition, the varieties and implications of reduction and emergence, the viable positions, and the deeper issues between them. And I show that even those philosophical pioneers who have hit on one of the live views, embracing positions akin to either scientific reductionism or emergentism, have still overlooked the strongest opposing views, missed the deeper issues, and formed flawed conclusions about what our empirical evidence has shown. I consequently also establish that philosophers on both sides of the debates have misstepped in their arguments and assessments of our empirical evidence based upon them.

Perhaps more importantly, on the positive side, I highlight how reconnecting with, and following, the scientific debates allows us to correct these deficiencies and finally appreciate the very different intellectual landscape in which our recent empirical advances have left us. However, I also illuminate the ways in which the scientific debates themselves need better theoretical frameworks to move beyond their present stalemate by articulating their arguments and positions in order to properly appreciate the new debates and move them forward. I subsequently show that scientific emergentists and reductionists have also overlooked the strongest opposing views, and the key issues with these rivals, thus also leading these scientific researchers to flawed conclusions about what our empirical evidence has shown so far.

Some Big Claims about Still Bigger Debates

Much of the work of later chapters is devoted to providing detailed arguments to support these conclusions, but even at the outset I suggest we have some compelling reasons at this point to favor the sciences as the place to look for insight in order to move the debates over reduction and emergence forward.

First, compositional explanations, and connected empirical findings, plausibly provide our core empirical evidence about the structure of nature both locally and globally. Scientific reductionism and emergent-ism are each focused directly on such explanations, so engaging these views provides a way to reconnect with our core empirical evidence and more carefully assess where it leaves us, thus refreshing and reori-enting our discussions. Although philosophical debates have often focused on "global" positions or theses, about the whole of nature, my primary focus throughout the book therefore follows the sciences in being "local" and trained on specific examples of compositional explanation in the sciences and their implications.

Second, the concepts of composition used in the sciences, the notions of "parts" and "wholes" that underpin both scientific reductionism and emergentism, are used in our awesomely successful composi-tional explanations. Such explanations are basically the most success-ful explanatory applications of compositional notions to entities in nature. In contrast, putting it politely, it is far harder to see what the explanatory successes are for the philosophical frameworks recently used to understand composition in nature. To the degree to which we take explanatory success to be a marker of truth, we should thus favor scientific notions of composition, rather than philosophical frame-works, as reflecting the structure of nature. And we therefore have clear reasons to take the scientific positions directly focused upon the implications of such concepts to be more likely to engage the deeper issues.

Third, I have already marked how debates over reduction and emer-gence have gone through a continuing evolution because they have an important empirical component. And empirical evidence can plausibly be expected to play a key role in transforming, or having transformed, contemporary debates. Given their closer contact with scientific inno-vations, scientists plausibly react more swiftly to empirical advances than philosophers. So we have another reason to favor engaging the scientific positions because scientific debates often more swiftly reflect important empirical advances than philosophical discussions. I detail in later chapters how this is just the situation that has come to pass in the distinct sets of debates over reduction and emergence in science and philosophy.

Fourth, and finally, by carefully following the scientific debates, and focusing on specific examples of compositional explanation, we can get a better, clearer grip on our present epistemic situation to discern which positions are alive or dead and why. Consequently, we have the chance to discern to what degree the debates are empirically resolvable and what types of evidence are relevant to the key disputes. Building on such a platform, we can therefore refresh and reorient the debates by illuminating what has really been established so far from empirical evidence in various cases, or even globally, and hence also clarify the future work that can address the open questions.

My starting hypothesis is therefore that scientific concepts and positions come closer to the deeper issues, and live positions, of our day than those of philosophy. And, as we begin to explore the claims of scientific reductionists and emergentists, it is important to be neutral between the various sciences, since one of the hotly contested questions between the two sides in the sciences concerns the relative status of various scientific disciplines. As a working assumption, I thus also endorse what I term "Inclusivism" in the position that we need to consider evidence from a *range* of sciences, and not just fundamental physics, in addressing issues over reduction and emergence.[12] Overall, in contrast to much recent philosophy, *I therefore side with the sciences, but I do not take sides between the higher and lower sciences except where I identify good reasons.*[13]

Though following the sciences, I also show that many of the central notions, theses, arguments, and commitments of the battling positions in the sciences have remained largely unarticulated, just as Whitehead would predict, since they are, or are based upon, unarticulated epochal ontological assumptions implicitly assumed by the protagonists on either side of the debates. Given the unfortunate state of the theoretical frameworks in the scientific discussions, my primary goal is therefore to provide more adequate, positive theoretical accounts of

[12] The alternative position is what I dub "Exclusivism," which takes fundamental physics to be the only science whose evidence is required to resolve debates over reduction and emergence or the structure of nature. Consider, for example, the explicit, Exclusivist position in Ladyman and Ross (2007). Many other philosophers have often recently fallen into Exclusivist positions, but I show that scientific reductionism, when properly understood, endorses Inclusivism.

[13] My stance thus deliberately mirrors a maxim of David Lewis, but then tweaks the resulting claim to avoid either the obvious reductionist or Exclusivist bias it contains. Famously, Lewis suggested: "Materialist metaphysicians want to side with physics, but not to take sides within physics" (Lewis (1983), pp. 37–8). But this makes the blatantly reductionist or Exclusivist assumption that there is only one science of interest, physics, in debates over reduction and emergence.

the scientific notions of composition deployed in compositional explanations and then for the positions of scientific reductionists and emergentists that build upon them.

As a way to move the scientific discussions forward, my theoretical focus obviously contrasts with a "More Data" strategy that simply seeks to pile on more, and more, empirical findings to resolve the scientific disputes. However, the More Data approach now looks ineffective. As we shall see, we do not lack for data and both sides have declared victory because they each claim our empirical evidence has settled the issues – but in favor of opposing views! The scientific debates have thus bogged down into a de facto stalemate. In this situation, the old scientific saw that "Theory without data is useless, data without theory is blind" appears to offer sound guidance. And I therefore contend that our most pressing need is presently for better theory to understand the scientific positions and finally clarify the empirical evidence relevant to confirming, or disconfirming, their opposing claims.

Given the shared focus of scientific reductionists and emergentists on the notions of composition deployed in compositional explanations, i.e. the "parts" and "wholes" in their slogans, I begin by constructing a theoretical framework for these foundational phenomena. My approach, in its first step, is to *start* with the scientific explanations and their notions, describe the features of these concepts, and *then*, in its second step, to construct a theoretical framework that captures the highlighted features of these scientific notions of composition. My methodology thus contrasts with recent philosophical approaches that pay little detailed attention to the nature of scientific notions of composition and simply seek to shoehorn such concepts into pre-existing philosophical machinery developed for other purposes. Using real scientific examples, I confirm that existing philosophical frameworks provide inadequate accounts of scientific composition because they fail to accommodate key features of such notions.

Taking my better account of "parts" and "wholes" in the sciences as a platform, I then repeat this procedure for scientific reductionism and emergentism in turn. Once more, I start with each scientific position to describe its commitments and arguments in the first step, and then in the second step construct a theoretical framework for the view, rather than simply seeking to jam the scientific position into pre-existing philosophical accounts of either reduction or emergence.

Again I establish that this approach is warranted. I show that scientific reductionists and emergentists both defend novel positions largely overlooked by philosophers, and that each of these views highlights significant flaws in the dominant philosophical accounts of reduction and emergence. Still more importantly, and contrary to the received wisdom in philosophy, I ultimately establish that the kinds of position articulated in the scientific debates exhaust the two kinds of live positions in contemporary debates, whether in science, philosophy, or more widely.

I therefore pursue this *three-step methodology* throughout the book. First, I pursue a descriptive project of articulating the features of a scientific concept or position. Then, second, using my descriptive account, I construct a theoretical framework for the concept or position that allows me to assess the arguments built upon the concept or position. But my work is thus not merely descriptive, although its initial phase does focus on articulating and theoretically reconstructing scientific concepts and positions. For, third, I then also prosecute the *prescriptive* project of assessing both philosophical, but also scientific, positions and arguments about scientific composition, reduction, and emergence.[14]

For instance, using my better theoretical frameworks, I outline why the most widely endorsed views in philosophy and the sciences are not amongst the viable positions about the structure of cases of compositional explanation. And I illuminate the false dichotomies philosophers have endorsed about both reduction and emergence. However, I also detail a number of places where both scientific reductionists and emergentists have made bad arguments. For example, I detail how the most common argument offered by scientific reductionists for their claim that "Wholes are nothing but their parts" is actually *invalid*. And, contrary to the claims of scientific emergentists, I outline why the existence of multiple realization, feedback loops, the indispensability of higher sciences, and/or the necessity of using non-linear dynamics or explanation by simulation do *not* alone provide good arguments against scientific reductionism once it is properly understood.

Appreciating the flawed arguments recently offered in the sciences, and in philosophy, I show that both sides in scientific debates, although each hitting on one of the live positions, have plausibly failed

---

[14] For a more detailed discussion of this methodology see Gillett (2016).

to appreciate the strongest *opposing* views. And I show that similar problems affect even those philosophers pursuing pioneering work on the scientific views. Given the comparative nature of scientific theory appraisal, where a theory is assessed by comparison to the strongest relevant rivals, I consequently show that extant defenses using empirical evidence from both scientific reductionists and emergentists, as well as their allies in philosophy, are presently all plausibly unsuccessful. My theoretical work thus illuminates how our Battle of the Ages is very much an *ongoing* one and rather different than even participants in the scientific debates, let alone philosophers, have supposed.

My final conclusion is that Laughlin is correct that we have indeed entered a very different era focused upon collective phenomena and their components. But I show how the protagonists on all sides of the debate, in philosophy and the sciences, including emergentists like Laughlin, need to make important adjustments to this new era. I highlight how we need to recalibrate philosophical discussions and replace existing philosophical accounts with more adequate theoretical frameworks for scientific composition, reduction, and emergence that highlight the very different nature of the live positions and underlying issues of our new debates. But I also move the scientific discussions forward by providing abstract theoretical frameworks that clarify the key scientific positions and their epochal claims, allow us to evaluate which of their arguments are good and bad, illuminate their differences about the structure of concrete scientific cases – and hence see how to empirically break the deadlock over such examples.

The Dislocation between Philosophy and the Sciences – and a Way Forward

Coming chapters range over often highly theoretical debates, in science and philosophy, and challenge a swathe of received philosophical wisdom. So to guide the reader I now want to provide overviews of the contents of the book's four parts in turn. Clearing space for later work is my focus in Part I, "Groundwork," where I provide brief overviews of the state of scientific and philosophical debates, diagnose the disconnect between them, and then construct a theoretical framework for scientific composition that helps to bridge the gap.

Chapter 1 provides surveys of scientific reductionism and emergentism, outlining how each position is pursuing what I term the "metaphysics *of* science" in the study of concepts and issues arising *from* or *within* the sciences. Both groups endorse what I term "everyday" reductionism, i.e. the pursuit of compositional explanations, and then explore the implications of the compositional notions of such explanations. In particular, my surveys highlight how each position defends integrated nomological, semantic, and methodological views apparently driven by their opposing ontological claims about the implications of compositional explanations and other empirical evidence. However, I also highlight how the two scientific groups overlap in their common failure to articulate their key ideas and arguments, leaving us with harmful theoretical deficits.

In contrast, my overview of various philosophical debates, whether about reduction or scientific composition, suggests philosophers have to different degrees fallen into what I term "metaphysics *for* science" in the practice of unwittingly shoehorning scientific concepts and positions into unsuitable theoretical machinery developed for other purposes. For example, I detail how philosophical debates over "reduction" have often followed the Positivists by focusing upon their semantic model of reduction that (i) uses an account of scientific explanation *excluding* compositional explanation, (ii) focuses upon the relations of semantic entities like statements, (iii) entails that higher sciences are dispensable and lower sciences are semantically omnipotent, and (iv) implies that there is no macro-world. Unfortunately, the semantic model of reduction, and the accounts of nature and the sciences it implies, fit very poorly with what we know of both nature and the sciences.

More recently, writers like Jaegwon Kim and others have pioneered philosophical approaches focused on ontological reduction, sidestepping the semantic model and its problems. However, even such pioneering accounts still use frameworks for scientific composition based around "functionalist" machinery developed for different phenomena. Once again, I show how questions consequently arise about whether even these pioneering accounts are examples of metaphysics for science's imposing alien frameworks upon both scientific composition and reduction in the sciences.[15]

---

[15] Throughout the book I keep scare-quotes on "functionalism" and "functionalist" to avoid serious confusions, since many writers wrongly assume such terms have a univocal meaning. As I highlight in Chapter 1, these terms refer to a variety of distinct, and often opposing, views and their proprietary theoretical notions.

Together my surveys confirm the dislocation between *both* the notions of reduction *and* the concepts of emergence used in philosophical and scientific debates.[16] And my overviews also flag potential dangers with popular philosophical treatments of scientific composition. My work points to a diagnosis of the genesis of this dislocation in the tumultuous turn against "metaphysics" in philosophy in the middle of the twentieth century. After this radical turn in philosophy, scientific debates over reduction and emergence have been left as the area most directly pursuing the metaphysics *of* science in their focus on the ontological implications of scientific concepts of composition, whilst philosophical debates have ended up pursing metaphysics *for* science, to lesser or greater degrees, in the application of scientifically alienated frameworks in understanding composition, reduction, and/ or emergence in the sciences.

Happily, however, my surveys also suggest an obvious remedy for the dislocation: Construct an adequate theoretical framework for the foundational scientific notions of composition, use this framework to reconstruct the positions and arguments of scientific reductionism and emergentism, and then use these reconstructed positions to refresh our understanding of debates over reduction and emergence. Adopting this strategy, I therefore look at concrete examples from the sciences, in Chapter 2, to provide a better theoretical account of scientific composition.

Looking at cases of compositional explanation using my methodology, I suggest that there are a number of relations of scientific composition holding between a variety of different ontological categories of entity, i.e. properties, individuals, and processes, but I outline fifteen common features distinctive of all of these scientific notions. Ultimately, I argue that we can best understand such scientific composition as involving what I term "working components" bearing relations of what I dub "joint role-filling." Put roughly, I highlight how we have compositional relations in the sciences when we have collectives of entities with certain roles whose relations to each other allow them to jointly fill the qualitatively distinct productive role individuative of the relevant composed entity. Using this approach, I offer a starting theoretical framework capturing the general nature of the scientific notions and I also consequently provide detailed frameworks for the specific relations of

---

[16] I offer my survey of notions of emergence in Part III in Chapter 5.

parthood between individuals and realization between property/relation instances.

Though focusing primarily on articulating the positive nature of scientific concepts, and supplying a theoretical framework to capture it, I also use my descriptive examination of concrete cases to highlight the deficiencies of extant philosophical treatments for scientific composition. Looking at standard "functionalist" accounts, and at work from the philosophy of science based on manipulability, I detail how each of these approaches fails to capture key features of scientific notions of composition and hence fails to provide an adequate account.

The Scientifically Manifest Image and the Challenge of Scientific Reductionism

Using my framework for scientific composition I reconstruct, and evaluate, *both* scientific reductionism *and* emergentism, since I ultimately show that we can properly understand, and assess, either of these opposing positions only when we appreciate its rival. Over the history of the sciences, a wide variety of different views have been given the name "reductionism," but my focus in Part II, "The Roots of Reduction," is on providing a better theoretical framework for the position presently defended by scientists like Weinberg, Crick, Wilson, and others. For simplicity of exposition, I refer to this view as "scientific reductionism" and we can start to appreciate this position by outlining the context in which it is defended.

Within the domains of particular sciences, we use *intra*-level productive explanations to understand specific phenomena using other entities at this level. But, as Figure IA illustrates, the sciences then *go on* to explain the entities used in such explanations through compositional explanations positing compositional relations to entities at lower levels. Putting together the apparent commitments of both intra-level productive, and inter-level compositional, explanations we have a picture of nature with many levels of determinative entities which are all *horizontally* determinative by producing various events at their own levels, but where we also have *vertical* relations of determination flowing upward in compositional relations from the lower- to higher-level entities posited in inter-level explanations.

Adapting Sellars's (1963) famous terminology, the result is what we may term the "Scientifically Manifest Image" because it is the picture

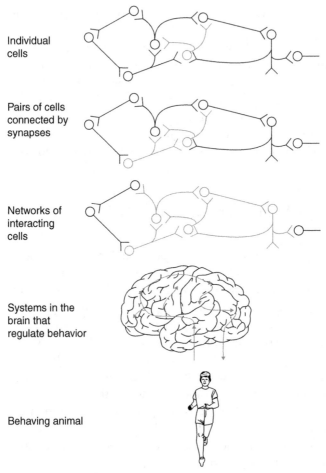

Individual
cells

Pairs of cells
connected by
synapses

Networks of
interacting
cells

Systems in the
brain that
regulate behavior

Behaving animal

Figure IA Scientific textbook diagram of levels of compositionally related entities.

of nature manifest in our scientific accounts and it is unsurprisingly widely endorsed in the sciences. The Scientifically Manifest Image also plausibly underlies the reigning "non-reductive physicalism" in philosophy, which is "physicalist" by taking all entities to be composed and "non-reductive" in accepting composed entities are nonetheless still determinative at their own levels.

Appreciating the Scientifically Manifest Image is important, I argue in Chapter 3, because it is the main target of scientific reductionism that claims to offer a *better* account of the import of compositional explanations and hence the *best* account of both the structure of nature and the sciences that study it. Crucially, the scientific reductionist *accepts* the descriptive account of compositional explanations, and compositional notions, outlined in Chapter 2, but then offers arguments that putatively show that these concepts lead to a starkly different picture of nature and the sciences than that offered by the Scientifically Manifest Image.

Utilizing my framework for composition, I reconstruct the ontological parsimony reasoning using compositional concepts that the scientific reductionist plausibly uses to defend their emblematic claim that "Wholes are nothing but their parts." Initially, I focus on the simplest of version of such reasoning, in what I term the "Argument from Composition," which is based solely upon the successful application of compositional concepts in some scientific case. Given the nature of composition, rather than explaining by positing determinative composed *and* component entities, the reductionist concludes we can instead explain everything at the higher and lower levels equally well using the existence of *component entities alone*. But, given this subconclusion, applying the Parsimony Principle the scientific reductionist further concludes that we should only accept that component entities are either the only determinative, or even existent, entities in a case of compositional explanation.

Although the scientific reductionist's picture of nature, which I term "Fundamentalism," takes there to only be one level of determinative entities in a case of compositional explanation, in the relevant components, I use my account of composition to highlight why Fundamentalism is not committed to the hackneyed "atoms-in-the-void" picture so often ascribed to reductionism. Crucially, I highlight why, given her focus on compositional explanation, the scientific reductionist is committed to what I term a "Collectivist Ontology" encompassing isolated components *and also* collective phenomena such as collectives of interrelated individuals. Most importantly, I outline how the Collectivist Ontology allows the scientific reductionist to endorse a sophisticated account of the macro-world, and its levels, encompassing multiple realization, Qualitative emergence, and the fact that "More is Different." I also begin to fill out the Fundamentalist's alternative, positive account of the nature of compositional explanations under which such explanations are true. And I consequently detail why the scientific reductionist's Fundamentalist position endorses both a macro-world and indispensable higher sciences that study it – just not the treatment of these phenomena implied by the Scientifically Manifest Image.

Building on this ontological account in Chapter 4, I then highlight a range of commitments of the Fundamentalist position. Most importantly, I lay out why scientific reductionism accepts that higher sciences are *in principle indispensable* because I show that under the reductionist's Collectivist Ontology the predicates of higher sciences are needed to express many of the truths about nature concerning collective phenomena. I thus conclude that Weinberg is right to press what he aptly terms the "compromising" nature of contemporary scientific reductionism (Weinberg (2001), p. 13), since my work establishes that this position is best understood as a combination of a thoroughgoing ontological reductionism and an equally robust semantic *anti*-reductionism that takes higher sciences to be significant and indispensable.

My theoretical framework for scientific reductionism highlights its many contrasts with the philosophically influential semantic model of reduction. For the Fundamentalist framework of scientific reductionism (i*) accepts the central role in the sciences of compositional explanations and their compositional concepts; (ii*) presses an ontological form of reduction; (iii*) accepts the in principle indispensability of higher sciences, and (iv*) embraces a complex macro-world where "More is Different." Each of these commitments is the contrary of the claims made by semantic reduction. And I consequently show that the famous philosophical objections to semantic reduction, utilizing the features of compositional explanations and their compositional concepts, all *fail* against scientific reductionism, which accepts the premises and conclusions of these critiques, albeit on its own terms.

My work thus establishes that, unlike the Positivist account of reduction, scientific reductionism seeks to provide a plausible account of all the sciences and all the levels of nature. Although endorsing priority for lower-level component entities, the lower-level laws about them, and the lower-level sciences that study such components, I show that the Fundamentalist account of scientific reductionism still accepts the existence, and importance, of the macro-world, its higher-level laws, and the higher-level sciences that study them. Given its nature, rather than being killed by arguments utilizing the features of compositional explanations in the sciences, I show that we actually need a reason to reject scientific reductionism *despite* these explanations that drive its core arguments.

Overall, I show that scientific reductionism is thus very much a live position and the pressing question that confronts us is one Weinberg has long emphasized: Why are we not *all* scientific reductionists now? Appreciating scientific reductionism thus highlights the need for philosophical discussions to undergo a transition from a largely *semantic* orientation to a more *ontologically* oriented approach. Philosophical pioneers such as Kim, Heil, Loewer, or Rosenberg have begun to make this change, but my work also shows that, in contrast to these pioneers, this new ontological focus must be founded in the concepts and evidence supplied by our best scientific theories, explanations, or models, rather than being driven or framed by alien ontological accounts designed for other purposes.

The (Not Too) Different World of Scientific Emergentism

There are now many distinct notions of emergence, so to start Part II, "The Fruits of Emergence," Chapter 5 provides a taxonomy of notions of "emergence" and assesses which legitimately pose a challenge to scientific reductionism. My taxonomic work highlights how philosophers have embraced a dichotomy of substantive options for "emergence" apparently resulting from something like the scientific reductionist's Argument from Composition. Philosophers and scientific reductionists alike thus reject the kind of emergence advocated by scientific emergentists, such as Anderson, Laughlin, Prigogine, Freeman, and others, in what I term "Strong emergence" in composed properties of higher-level individuals that are nonetheless determinative. However, despite their trenchant opposition, I note that philosophers have overlooked the ontological ideas that scientific emergentists articulate to make sense of Strong emergence. And my taxonomy suggests that such Strong emergence offers the only viable response to the arguments of scientific reductionism.

In Chapter 6, I therefore focus on reconstructing the core ontological claims about Strong emergence that scientific emergentists draw from concrete scientific examples where we not only have compositional explanations, but also now have quantitative accounts of the components in these complex collectives. I start with what I term the scientific emergentist's "Conditioned view of aggregation," which implies components have some powers in complex collectives that they would not have if the laws or principles applying in simpler collectives exhaustively applied in the complex aggregation. I term such powers of components "differential powers." In contrast, I show that scientific reductionists like Weinberg are committed to what I term the "Simple" view of aggregation" under which components aggregate in a continuous fashion across all simple and complex collectives, and where these components are determined only by other components, thus implying that components never have differential powers. I show that this difference in epochal commitments about the nature of aggregation is important because as a result of their Conditioned view of aggregation scientific emergentists can endorse a novel kind of "downward" determination from "wholes" to their "parts." Overlooked by philosophers, this is a species of non-causal, but also non-compositional, determination between composed and component entities focused upon role-shaping or role-constraining, rather than role-filling. I dub this novel relation "machresis" or "machretic determination."

My work shows that, basically, scientific reductionists and emergentists have a theoretical disagreement over what scientific composition allows. To resolve which side is correct, I therefore pursue a scientific thought experiment focusing on a scenario where scientific concepts of composition apply comprehensively alongside the novel commitments of scientific emergentism in Conditioned aggregation and machresis. My positive conclusion from this thought experiment is that the scientific emergentist is right that we can have Strong emergence in such situations where concepts of composition apply comprehensively. And my corresponding negative conclusions are that the scientific reductionist's Argument from Composition is actually *invalid* and that philosophers have endorsed a *false dichotomy* about the options for emergence and associated kinds of emergentism.

As well as illuminating the difficulties of scientific reductionism, I further show, in Chapter 7, that scientific emergentism offers an important positive position in its own right in a view of a comprehensive compositional hierarchy with *many* levels of determinative entities, but where we have *mutually determinative*, and *interdependent*, composed and component entities. I consequently dub this position "Mutualism" and I suggest it offers the only viable form of an ontologically non-reductive physicalism. Alongside its embrace of comprehensive composition, I illustrate how Mutualist scientific emergentism accepts the ubiquity of compositional explanation and various types of derivability and predictability, including for Strongly emergent entities. I also provide an account of the kinds of fundamental "organizational" or "emergent" laws that go along with this position.

Overall, scientific emergentism thus provides us with an alternative picture of the ontological structure of a compositional explanation that is compatible with its features, but with a structure radically different from that posited by the Fundamentalist account of scientific reductionism. And appreciating the Mutualist position of scientific emergentism again teaches us that an important transition is necessary, but this time in both philosophical and also scientific debates.

My work shows that scientific reductionists, and many philosophers, are fixated on the issues of the last generation and wrongly assume that the applicability of compositional concepts, and hence the existence of compositional explanation, *by themselves* suffice for an ontological reduction and the truth of Fundamentalism. But my theoretical frameworks establish that scientific emergentism, and Strong emergence, are compatible with both compositional explanation and the applicability of compositional concepts.

We therefore need to transition away from a view of our present debates that takes them to be like those of the last generation in being resolved by the existence of compositional explanations. Instead, we need to follow the lead of scientific emergentists in exploring the more subtle, and complex, ontological disputes thrown up by our new cutting-edge empirical accounts in cases where we not only have compositional explanations supplying qualitative accounts of compositional relations, but now also often have quantitative accounts of the components in the relevant complex collectives.

Understanding Our New Intellectual Landscape: The Live Positions and What Has Been Successfully Established from the Scientific Evidence (So Far)

My work reconstructing scientific reductionism and emergentism shows that we are indeed in a new era of debates centered upon the nature of collectives and their components, and hence views of aggregation. The concluding section of the book, Part IV, "New Landscapes, New Horizons," draws together the conclusions and findings of earlier chapters to highlight the live positions in our new debates and the deeper issues between them. Perhaps most importantly, the section clarifies the empirical evidence required to resolve the ongoing debates between the live positions in concrete scientific cases of compositional explanation and shows how to break the present stalemate in the sciences. Building on this work, an assessment is then offered of what has been successfully established, so far, from our empirical evidence.

To start Chapter 8, I articulate important arguments in the service of Fundamentalism and Mutualism. First, I return to my findings about scientific reductionism to illuminate a *valid* argument for compositional reduction in a case of compositional explanation. This reasoning is based around the applicability of compositional concepts *plus* what I call the "determinative completeness" of components, so I term this the "Argument from Composition and Completeness." I also highlight how this new reasoning has far higher evidential requirements than the invalid Argument from Composition.

Second, I use what I term the "Argument from Exhaustion" to formally establish that machretic determination is required, whether alone or in concert with other forms of determination, in order for a composed entity to be determinative. I consequently confirm that in order to exist a compositional relation must always be accompanied by a machretic relation. And I also thus establish that Mutualism is plausibly the only viable ontologically non-reductive position about the structure of a case of compositional explanation or about nature itself.

With these two arguments in hand, as well as earlier findings, I then offer a brief survey of which positions about a case of compositional explanation are live and which are dead. To begin, I show that the most popular positions in science and philosophy, in positions built around the Scientifically Manifest Image or standard forms of non-reductive physicalism, are dead positions that we can now see must collapse into other views. In contrast, I show that there are three candidates for live positions in Mutualism and two versions of Fundamentalism. (I term these "candidates for live views" since I can only resolve what the live views finally are after a review of what has been shown from our empirical evidence in Chapter 10.)

Appreciating the main arguments, and live positions, allows me to highlight the two broad layers of ontological disputes in contemporary debates. The first, and more familiar, set of differences concerns the range of determinative entities, the kinds of determination we find in nature, and hence the overall structure of cases of compositional explanation. On one side of these disputes, there is the Fundamentalism of scientific reductionism that subtracts compositional relations and levels to leave us with a universe that has *one layer* of determinative entities, but alongside a rich macro-world involving their collectives and higher sciences that study them. On the other side of these disputes, one can embrace the Mutualism of scientific emergentism where we do have walls of compositional relations rising vertically upward in the house of nature, but *only alongside the addition* of a solid roof of machretic relations pressing downward from composed entities to components. The Scientifically Manifest Image is not amongst the viable positions, nor are the standard versions of non-reductive physicalism defended in philosophy, since I show that the only live options involve either *adding to*, or *subtracting from*, the ontological accounts of these popular positions.

On top of this first set of disputes, my work also illuminates that in the new era focused on collectives we have a second layer of ontological differences that has been largely overlooked by philosophers and which has been left implicit even in scientific debates. This second set of disputes focus on differences over the continuity/discontinuity of nature and, in particular, the continuity/discontinuity of aggregation across simpler and more complex collectives. Central to these disagreements are differing epochal assumptions in the opposing Simple and Conditioned views of aggregation held by scientific reductionists and emergentists.

However, my work shows this second set of disputes does not simply line up with the first set of issues and leaves us with a still more complex set of live positions. I highlight this situation by illuminating a final overlooked option in a species of Fundamentalism that follows Mutualism in embracing discontinuities in aggregation, and hence the Conditioned view of aggregation and the existence of differential powers, whilst *still* endorsing the determinative completeness of components and pressing parsimony arguments to show that component entities are the only determinative entities. I thus distinguish this overlooked position, in what I term "*Conditioned* Fundamentalism," from the Fundamentalism commonly held by scientific reductionists like Weinberg and many in philosophy that I label "*Simple* Fundamentalism."

In this second set of disputes about concrete scientific cases, we thus have Simple Fundamentalists on one side pressing the continuity of nature, the non-existence of differential powers of components, and the truth of the Simple view of aggregation. However, on the other side of this dispute, we have two positions in Mutualism and also Conditioned Fundamentalism that endorse discontinuities in aggregation, the existence of differential powers of components, and hence the truth of the Conditioned view of aggregation. However, these two views then themselves diverge over other issues. Crucially, Mutualism and Conditioned Fundamentalism differ over what entities determine the novel behaviors and differential powers of components, where the Mutualist claims "emergent" composed entities machretically determine these contributions and the Conditioned Fundamentalist claims other components determine these contributions.

This abstract, preliminary survey begins to draw out the nature of the three candidates for live views and the more complex array of disputes between them over the nature of concrete scientific cases were we have compositional explanations. However, this treatment of our new debates works at a high level of abstraction and hence at some distance from scientific discussions. In Chapter 9, I therefore directly focus on the import for cases of compositional explanation of the three candidates for live views. For each of these three views, I consequently detail what the hypothesis claims about the structure of a concrete example of compositional explanation and also briefly sketch its claims about associated laws and explanations.

The negative conclusions of this more fine-grained work confirm that debates in the new era of collectives have moved on from the issues of the last generation of debates that too many scientists and philosophers often still focus upon. Writers on both sides, including those who have hit on one of the live views, continue to take their *opponents* to hold views from the last generation of debates. However, my work throughout the book, encapsulated in my survey of the three live views of cases of compositional explanation, shows that our present disputes do *not* concern any of the following: whether we can supply a compositional explanation in the case; whether compositional concepts apply comprehensively in the case; whether there is multiple composition in the case; whether we may need to posit higher-level scientific explanations to capture all the truths about the case; whether we may need to use non-linear dynamics or explanation by simulation; whether feedback loops exist, amongst other issues. My work shows that *all* the live views in the new debates, both reductionist and emergentist, are agreed about all of these phenomena and that our discussions have moved on.

In contrast, my positive work clarifies how the three live views do differ over a finer-grained array of ontological issues in any case of compositional explanation. Broadly put, the hypothesis of Simple Fundamentalism is that, underneath all the semantic and epistemic complexity, aggregation really is continuous and we have no differential powers of components in complex collectives, the Simple view of aggregation applies to the case, and we only have component entities and the determination between them. In contrast, the hypotheses of Mutualism and Conditioned Fundamentalism both assert that aggregation is discontinuous, that components have differential powers in the relevant complex collective and hence that the Conditioned view of aggregation applies in the case. However, Mutualism and Conditioned Fundamentalism differ over the entities that determine the differential powers of components, the nature of the determinative entities, and the species of determination we find in the case.

Building on this work, I then articulate particular ways to empirically confirm or disconfirm the three live hypotheses about the structure of a case of compositional explanation. I show that the hypothesis of Simple Fundamentalism may be confirmed or disconfirmed by resolving whether components have differential powers using the usual scientific methods by which we illuminate productive relations and hence powers. Assuming that we have found there are differential powers in some example, I outline how we may confirm or disconfirm the hypotheses of Mutualism and Conditioned Fundamentalism by resolving whether the contribution of differential powers by components is determined by composed entities or by other components. Once again, I note that common scientific methods plausibly suffice to supply such findings. Overall, I therefore outline how the issues of the new debates are empirically resolvable and my work clarifies how to break the present stalemate in these discussions using scientific methods.

A by-product of this work is to highlight the threshold of evidence required to make discussions of the competing views productive through being empirically resolvable. As well as the existence of a

successful compositional explanation supplying a qualitative account
of the components in some complex collective, my work establishes
that we can only resolve the disputes in the new debates when we also
have quantitative accounts of the relevant components in both simpler
collectives and also the relevant complex collective.

After clarifying the character of the evidence needed to resolve the
disputes between what I am terming "local" hypotheses, which make a
claim about a specific scientific example of compositional explanation,
and hence one or more pairs of levels, I then go on to clarify what it
takes to confirm the *global* forms of the three positions in hypotheses
making claims about all the levels of nature. I detail the slightly different kinds of evidence needed to establish global Mutualism, Simple
Fundamentalism, and Conditioned Fundamentalism. But I show that
each of these global views has in common that it can plausibly only be
established through the local hypotheses associated with the view being
established in concrete scientific cases of compositional explanations.

My conclusion is that the issue of the correct global view of nature
can only be resolved *after* we have resolved which of the viable local
theories offer the best accounts of concrete scientific examples usually from an array of different levels. And I therefore also conclude
that Inclusivism is the most plausible epistemic position, rather than
simply being a useful working principle, because the evidence of *many*
sciences is necessary for resolving the issues over reduction and emergence, and the structure of nature, at both the local and global scales.

Having clarified the empirical evidence that can resolve our debates
in the new era of collectives, I use my findings, in Chapter 10, to
provide an assessment of what has been established so far from our
empirical evidence. I therefore survey extant defenses using empirical
evidence from both sides of the debates, in philosophy and the sciences, and focusing on cases from across the higher and lower sciences
ranging from quantum mechanical derivations of chemical phenomena and the nature of superconductors, at one extreme, through to the
nature of eusocial insects and their colonies at the other.

On the positive side, my work shows that in some cases we have,
or are close to getting, the kind of empirical evidence that satisfies the
threshold of evidence I argue is needed to resolve the new debates.
However, on the negative side, I show that each of the extant defenses
is plagued by the theoretical problems I have identified and hence
plausibly fails due to one or both of two problems.

The first kind of difficulty I highlight in extant defenses is that, contrary to the claims of their protagonists, once we understand the empirical evidence relevant to our ongoing disputes, then we often simply find that we do not yet have the kinds of evidence in examples passing the necessary threshold. In addition, the second problem I identify with extant defenses is still more important because it bites even when we do have sufficiently detailed evidence of the right kind. Crucially, for the reasons highlighted in earlier chapters, I show that extant defenses fail to engage their strongest relevant rivals or the key issues with them. But scientific theory appraisal is comparative, so extant defenses are unsuccessful because they have so far failed to engage relevant rival hypotheses and the key disputes with them – even in cases where I suggest we may well already have evidence meeting the threshold for productively engaging these disputes.

My conclusion is consequently that, at least as yet, it is still very much an open question whether we should accept either version of Fundamentalism, or instead endorse Mutualism, about any of the particular scientific examples where the proponents of such positions are presently battling across physics, chemistry, biology, and the sciences of complexity. Given the dependence of global positions and theses upon such local cases for their justification, I conclude we are also yet to establish one of the live global accounts of the structure of nature.

My work therefore shows that, far from being mere rhetoric, recent scientific battles over reduction and emergence involve competing, substantive, scientific hypotheses that differ over the nature, and structure, of concrete cases of compositional explanation in the sciences in empirically resolvable ways. Our new debates are thus plausibly important scientific debates over substantive matters. My final conclusion is that, just as Laughlin claims, we have indeed moved into a new era, and one focused on collectives and their components, but that it still remains to be established whether the scientific reductionist or emergentist is correct about cases of compositional explanation across an array of levels and sciences. Still more exciting, my work shows how to resolve these presently stalemated disputes and that we may already have the evidence to hand that allows us to do so once our better theory allow us to marshal it properly. So, in the Conclusion, I use my findings to provide a brief road map of what we need to do next in order to best address our Battle of the Ages.

By supplying better theoretical frameworks for the recent scientific debates, and the Big Questions they engage, we thus put ourselves into what Whitehead would take to be an exceptional historical situation. By providing such theory, we can make explicit the epochal ontological assumptions and positions doing battle in our times. But such work also allows us to go one exciting step further to see how to empirically resolve these disputes, too. In our new era focused on collectives and their components, we can thus begin to clearly see how to decide whether we should embrace the coming of the Age of Emergence or allow the Age of Reductionism to march on.

PART I

Groundwork

1 | *Scientific Composition, the Universe, and Everything: Or, Reductionism and Emergentism in the Sciences and Philosophy*

Metaphysics of science – The abstract examination of ontological concepts, issues, arguments and positions as they arise *within* or *from* the sciences and their explanations, theories, models, findings, etc.

Scientists and philosophers have each engaged the Big Questions about the structure of nature and the sciences through their respective discussions of reduction and emergence. Each group also appears intent on pursuing what I term the "metaphysics of science" as defined in the epigraph to this chapter. For all sides agree that the sciences and their findings must drive answers to the Big Questions. Unfortunately, I am going to suggest that we have reasons to believe that only the scientific discussions have fully succeeded in pursuing the metaphysics of science, whilst philosophical work ends up either avoiding the metaphysics of science and/or pursuing, to varying degrees, what I term "metaphysics *for* science" by imposing alien frameworks onto scientific practice.

In order to support these contentions, and situate the work of later chapters, I provide brief surveys of both scientific and philosophical debates. My survey illustrates the centrality of compositional explanations for both scientific reductionism and emergentism. And I outline how each of these scientific views clearly pursues the metaphysics of science to defend opposing answers about a range of important ontological and methodological issues. In contrast, in philosophy I outline how the radical commitments of the Positivists resulted in their interconnected semantic frameworks for explanation and reduction, which are both plausibly distant from scientific practice. Oddly, I highlight how although the Positivist's account of explanation is long since abandoned by most philosophers, their connected account of reduction continues to dominate philosophical debates because it is central to the reigning *anti*-reductionist consensus in philosophy of science.

My survey highlights the importance of scientific notions of composition to both scientific and recent philosophical debates, so I conclude by looking at two philosophical frameworks commonly applied to understand scientific composition. But I highlight reasons to be suspicious that each of these frameworks plausibly mischaracterizes the scientific notions by shoehorning these concepts into alien machinery.

1.1 The Scientific Background: Compositional Explanation, Reductionism, and Emergentism in the Sciences

I briefly highlight the historical role and character of compositional explanations, in 1.1.1, and note an important global thesis about nature that these explanations support. I then provide an overview of the characteristic claims of scientific reductionism, in 1.1.2, and scientific emergentism, in 1.1.3, being careful to highlight their common focus on compositional explanations and the implications of their concepts.

1.1.1 Compositional Explanations and their Concepts

What common sense makes manifest to us is a world filled with familiar kinds of individuals, like dogs, icebergs, apples, etc., with the properties, such as hardness, texture, warmth, etc., and processes, such as scratching, melting, splitting, etc., revealed by our unaided senses. We also perceive the world as having rudimentary compositional relations, such as bricks composing a house, or wheels, bed, and axles composing a wagon, and so on. Both scientific reductionists like Weinberg, as well as their scientific emergentist opponents such as Anderson, are in agreement that the sciences pierce these surface appearances and show nature is very different from what Sellars (1963) termed the "Manifest Image" of common sense.[1]

Compositional explanations have been one of the main engines of this remarkable transformation of our understanding of nature. Consider just a few examples of compositional explanations to appreciate their features and breadth. We take the movement of the earth's surface, including the movement of land masses, to be explained by the movement of the continental plates and currents of magma that we take to compose the earth. We explain the motility of cells using the properties and relations of the molecules that we take to compose

---

[1] Weinberg (1992) and Laughlin (2005), pp. 2018–19.

them. We understand why crystals bend light in certain ways in terms of the properties and relations of the atoms or molecules taken to compose them. We understand the inheritance of traits of offspring from their parents using the molecules that compose parent and offspring. And we could easily go on, and on, through such explanations across the sciences.

Given their nature, such explanations are plausibly termed *compositional* explanations, since they are founded around showing how lower-level entities of one kind (whether individuals, properties, or processes) *compose* entities of very different kinds at higher levels. Philosophers of science have used a range of terms for compositional explanation, including "reductive explanation," "functional analysis," "functional explanation" or "mechanistic explanation," and there is a substantial body of work on its nature, including a recent burst of research.[2] To provide a backdrop for my surveys in this chapter, in this section I simply note a couple of broad features of such explanations, and some of their more important consequences, in a manner neutral between competing philosophical accounts.

First, compositional explanations allow us to explain one kind of entity, such as cells or their moving, in terms of the *qualitatively different* kinds of entity taken to compose it, like molecules or molecular processes of polymerization, and this hence results in what I term the "Piercing Explanatory Power," or "PEP", of compositional explanations. Thus we can explain the entities of the Manifest Image, like the hardness of a diamond, using utterly different kinds of entity, such as the covalent bonding of carbon atoms. But we have now gone on to iterate compositional explanations across all the levels of entities, whether in cytology or chemistry, that we have now discovered in addition to those of the Manifest Image.

Second, we should mark that once we have successfully supplied a compositional explanation of certain entities in terms of certain others that compose them, then we have established that these entities are in some sense the same. This is what I will term the "Ontologically Unifying Power," or "OUP," of compositional explanations. For

---

[2] Contemporary work on compositional explanation goes back at least to early work by Fodor (1968) and Dennett (1969), through Cummins (1983), down to more recent work such as Bechtel and Richardson (1993), Glennan (1996), Machamer, Darden, and Craver (2000), and Craver (2007), amongst many others.

example, at a certain point it was an open question how digestion and molecular processes were related. But once we compositionally explained the process of digestion in terms of molecular processes that compose it, then we established that these processes are in some sense the same. Most importantly, as the latter case illustrates, a successful compositional explanation shows that the mass-energy, or force, associated with a certain entity just is the mass-energy, or force, of certain other entities. Thus supplying a compositional explanation of digestion did away with any need to posit special vital energies or forces with digestion.

Our huge array of compositional explanations supports plausible arguments that everything in nature is composed by entities, which are composed by other entities, which are composed by still other entities, until we come to the most fundamental entities we know of in fundamental physics. Alongside this finding is the connected conclusion that the forces and mass-energies of certain entities just are the forces and mass-energies of the entities that compose them, which just are the forces and mass-energies of still other entities, which are in turn all just the forces and mass-energies of the entities of fundamental physics.

Although we have no well-confirmed cases of entities that are uncomposed, I should note that not everything has yet been included in the net of compositional explanations. And we have seemingly intractable difficulties with some phenomena at the bottom and top of the hierarchy of composition.[3] But the crucial point is that we plausibly lack any examples of entities in space-time, studied by the sciences, that we know cannot be composed, whilst we have a vast catalog of entities where we have qualitative accounts of their components as a result of successful compositional explanations. Unsurprisingly, scientists including both scientific reductionists and emergentists therefore now accept the existence of a comprehensive compositional hierarchy in nature as one of their working assumptions. Philosophers often use the term "physicalism" for such a view of nature as a universal

---

[3] The two most obvious problems concern quantum mechanical phenomena at the base, and phenomenal consciousness near the top, of the hierarchy. The jury is thus plausibly still out about whether compositional concepts apply to quantum mechanical phenomena or phenomenal consciousness, but neither example presently provides a case where it has been established that compositional concepts and explanations are known not to apply.

compositional hierarchy and that is how I shall use the term.[4] A strong case can indeed be made that we are now justified in accepting the truth of physicalism as a result of our core empirical evidence from compositional explanations.[5]

As their slogans make clear, both scientific reductionists and emergentists are focused on the nature and implications of scientific composition when they each talk about it in terms of "parts" and "wholes." But scientists also use a range of other terms like "composition" and "component," "constituent," "implementation," and more, to express such concepts. Unfortunately, the "part–whole" terminology suggests a focus solely upon the *individuals* that bear part–whole relations. However, I show at length in Chapter 2 that compositional explanations also posit compositional relations between higher- and lower-level *properties/relations* and *processes* as well as individuals. Throughout my discussion I therefore use the alternative scientific terminology of "component" and "composed" entities in order to cover the full range of entities involved in compositional relations. I also consequently use "composition" as a catch-all term to refer generically to the variety of compositional relations we shall see are deployed between a number of different categories of entity in the sciences.[6]

---

[4] The reader should note that under physicalism, given its empirical basis, it is left open whether there are still further compositional levels below the entities presently posited by fundamental physics. Many philosophers also often add the so-called "Completeness of Physics" to the definition of physicalism. However, I engage this more committed usage in Chapter 8 and show it is unsupported.

[5] For the best outline of the abductive argument supporting the existence of a universal compositional hierarchy in nature see Melnyk (2003), Part III. However, a common philosophical contention is that physicalism is justified using the Completeness of Physics (Loewer (2001), Papineau (2001)). As I detail in Chapter 10, there are good reasons to doubt this common position is tenable.

[6] Analytic metaphysicians sometimes complain that I am using "composition" in a way different from their usage, but it is important to emphasize that I am simply following a long-standing, and continuing, scientific usage of the terms "composition" and "compose" that stretches back at least to Sir Isaac Newton and continues down into writers like Weinberg, Anderson, Mayr, and many other working scientists. Throughout the book, my usage of "composition" therefore solely refers to the relations and concepts deployed in compositional explanations. If there are still further concepts of "composition" used in the sciences (Healey (2013)), then I do not discuss them here.

It is also important to clarify a second point about terminology to avoid begging substantive questions at the heart of the scientific discussions. In the sciences, but also amongst some philosophers, a common term for a compositional explanation is that it is a "reductionist" explanation and the term "reductionist" is thus often simply used to refer to someone who seeks compositional explanations (Wimsatt (2007)) who is taken to be pursuing a "reductionist methodology." Unfortunately, as the next two subsections make plain, on these mundane usages *both* scientific reductionists *and* scientific emergentists are "reductionists" who explicitly seek "reductionist explanations" and endorse "reductionist methodology'!

A number of writers have consequently suggested these mundane usages have been damaging to debates.[7] And alternative terminology has been broached. For instance, Ernst Mayr (2002) consequently suggests we ought to refer to "analysis" rather than "reductionist explanation." I agree the common usage is problematic and that we need to be careful not to beg any questions between the two sides, so I shall therefore continue to solely use the neutral term "compositional explanation" to refer to the explanations both sides acknowledge as important. When I need to refer to the weaker, and universally endorsed, position espousing the search for compositional explanations, I shall refer to "everyday" reductionism. And I only use the term "scientific reductionist" for the richer, more substantive positions endorsed by writers like Weinberg that I detail further in the next section and in later chapters.

1.1.2 Scientific Reductionism and Some of its Characteristic Claims

Though I am going to use Steven Weinberg (1992), (2001) as my exemplar of a scientist reductionist, I should mark that related positions are defended by other prominent scientists including the chemist Peter Atkins, biologists like Francis Crick, Richard Dawkins, George Williams, and E. O. Wilson, and the neuroscientist Eric Kandel, amongst others.[8] Although Weinberg focuses primarily upon

---

[7] For example, Mayr (1988b).
[8] Atkins (1995), Crick (1966), (1994), Dawkins (1982), (1987), Williams (1985), Wilson (1998), and Kandel (2006).

fundamental physics and a *global* version of scientific reductionism, I will be careful to frame scientific reductionism so it can also be applied *locally* to any pair of disciplines whose entities are related by compositional explanations.

Two superficial features of scientific reductionism are worth noting before turning to its substantive claims. First, many scientific reductionists now routinely complain about the negative perceptions of their views such as when we find Dawkins contending that in the sciences "Reductionism is a dirty word, and a kind of 'holistier than thou' self-righteousness has become fashionable" (Dawkins (1982), p. 113). Alongside such claims, scientific reductionists press the point that their views are misunderstood. Thus Dawkins, again in typically colorful fashion, claims that:

To call oneself a reductionist will sound, in some circles, a bit like admitting to eating babies. But, just as nobody actually eats babies, so nobody is really a reductionist in any sense worth being against. (Dawkins (1987), p. 13)

Dawkins is instructive in highlighting the puzzlement that scientific reductionists often express about resistance to their views and this brings us to the second point I want to mark. For scientific reductionists often claim that if their view were properly appreciated then all sides (at least in the sciences) would see that they actually *endorse* scientific reductionism. As well as Dawkins, we find Weinberg making this point in his debates with the emergentist Ernst Mayr.[9]

One might explain such claims of scientific reductionists as simply being a failure to properly distinguish everyday reductionism, which is indeed universally accepted in the sciences, from their more substantive position. However, a more charitable interpretation is that such writers endorse arguments that, given the nature of everyday reductionism, then we should all also adopt the more substantive claims of scientific reductionism as well. The widespread scientific endorsement of everyday reductionism thus leaves Weinberg and Dawkins puzzled about why other scientists do not embrace their views. Let us therefore examine what they say about their arguments and their overall positions.

[9] Weinberg (2001), p. 13.

Scientific reductionists make an array of integrated claims about a wide range of issues in the sciences. However, at the base of scientific reductionism is a contention about what we ought to infer about the structure of nature from compositional explanations and especially their compositional concepts. For example, in discussing a long string of such explanations about a piece of chalk, and its properties, from an array of levels, Weinberg tells us that:

... we have a sense of direction in science. There are arrows of scientific explanation ... Having discovered many of these arrows, we can now look at the pattern that has emerged, and we notice a remarkable thing: perhaps the greatest scientific discovery of all. These arrows seem to converge to a common source! Start anywhere in science and, like an unpleasant child, keep asking "Why?" You will eventually get down to the level of the very small. (Weinberg (2001), pp. 17–18)

The arrows of compositional explanations always draw us downward as the compositional relations they posit point upward. For such explanations explain higher-level entities by taking them to be composed, as we saw in the last section, by lower-level entities. The claim of scientific reductionists like Weinberg is consequently that these components are responsible for everything that happens in the relevant case in so far as anything is responsible at all.

Expressed as a global view, the latter claim is that the entities of microphysics, as the fundamental components, determine everything to the degree it is determined at all. Writers like Weinberg are keen to emphasize that this is an *ontological* claim about the entities that make things happen. As Weinberg tells us:

(I repeat and not for the last time) I am concerned here not so much with what scientists *do*, because this inevitably reflects both human limitations and human interests, as I am with the logical order built into nature itself. (Weinberg (1992), pp. 45–6)[10]

We thus see that scientific reductionists are clear that the claims they defend are ontological in character, rather than primarily being

---

[10] Similarly, to distinguish such ontological claims from methodological ones, Weinberg tells us: "For me, reductionism is not a guideline for research programs, but an attitude toward nature itself" (Weinberg (1992), p. 52).

concerned with semantic entities like the law statements, explanations, or theories that are the focus of the Positivist framework for reduction. But what is it about compositional explanations and their compositional concepts that drive these ontological conclusions?

In response to this question, writers like Weinberg do not offer precise accounts of their underlying arguments, but they do offer suggestive remarks. Most famously, as I noted in the Introduction, the character of compositional explanations leads scientific reductionists to conclude that "Wholes are nothing but their parts." Thus the evolutionary biologist George Williams summarizes the latter point applied to biology as the claim that we can understand "complex systems entirely in what is known of their component parts and processes."[11] That is, wherever we find compositional explanations of certain higher-level entities, i.e. certain "wholes," through lower-level component entities, i.e. "parts," then scientific reductionists claim we can see that the composed entities are "nothing more than," "nothing over and above," or "nothing but," such components. For it is only the components, the scientific reductionist claims, that we should consequently take to determine anything.

These points about the entities we should take to be determinative lead scientific reductionists to matching points about laws. For example, using the term "principle" instead of law, Weinberg tells us:

Think of the space of scientific principles as being filled with arrows, pointing toward each principle and away from others by which it is explained. These arrows of explanation have already revealed a remarkable pattern: they do not form separate disconnected clumps, representing independent sciences, and they do not wander aimlessly, – rather they are all connected, and if followed backward they all seem to flow from a common starting point. This starting point is what I mean by a final theory. (Weinberg (1992), p. 6)

Scientific reductionists thus commonly claim that the laws of components we discover in simple collectives are the laws that *exhaustively* govern both components and the entities they compose in more complex collectives. Weinberg expands on this point when he also tells us:

... nervous systems ... have evolved to what they are *entirely* because of the principles of macroscopic physics and chemistry, which in turn are what

---

[11] Williams (1985), p. 1.

they are *entirely* because of the principles of the Standard Model of elementary particles. (Weinberg (2001), p. 115. Original emphasis)

We can thus see that in its common form scientific reductionism embraces deep continuities in nature across simple and complex collectives of components. Just as components are taken to be the sole entities that determine anything in a case of compositional explanation, in so far as these entities are determined, the scientific reductionist also consequently takes the laws concerning such components in simple collectives to be, in a corresponding sense, the only determinative laws in such cases involving complex collectives.[12]

Lest Weinberg's focus on a global view and a Final Theory lead to a misapprehension about the claims of all scientific reductionists, let me highlight the distinction between what I term "local" and "global" versions of this position. One can endorse scientific reductionism *without* endorsing a Final Theory. A local scientific reductionist makes a claim solely about a pair of levels of entities related by compositional explanations.[13] Thus, for example, someone might be a local reductionist about psychological and neurophysiological entities, claiming that our compositional explanations in this case establish that psychological entities really are "nothing but" neurophysiological entities, whilst rejecting similar claims about other scientific disciplines and their entities. Notice that such a local scientific reductionist may be agnostic about, or even reject the existence of, a Final Theory and the claims that the entities of physics are the only determinative entities and the laws of physics the only determinative laws.

In contrast, a global scientific reductionist such as Weinberg, or Wilson (1998), claims that we have compositional explanations for the entities of a wide range of higher sciences so that their entities can be successively shown to be "nothing but" the entities of some lower science, where these component entities are in turn shown to be "nothing but" still lower-level entities, until we come to the entities of fundamental particle physics. Thus the global form of scientific

---

[12] See Dawkins (1987), p. 15, and Wilson (1998), p. 55, for corresponding claims about the laws of physics.

[13] Such explanations can obviously involve more than a pair of levels in multi-level accounts, but in my discussion I focus on the simpler case for expository purposes.

reductionism endorses the long-standing program of unification in physics that seeks to show that the laws governing the entities we find in the simplest collectives of fundamental components exhaust the determinative laws for everything in the universe – hence supplying a Final Theory or "theory of everything.'

My primary focus will be upon *local* scientific reductionists because I am concentrating upon specific compositional explanations. But throughout my discussion of scientific reductionism I will provide accounts that still seek to cover both global and local forms of this view. This is made easier because scientific reductionists of both varieties overlaps in their commitments about a range of semantic and methodological issues in the sciences. Perhaps the most important of these claims concern the nature and status of the higher sciences studying the macro-world. To mark the different positions on this issue, let me use Daniel Dennett's term "greedy reductionism" to refer to a position that holds that the lower sciences studying components can perform all explanatory and other semantic tasks, thus including all the work of higher sciences.[14] Notice that greedy reductionism implies not only that higher sciences are *dispensable*, but also that the lower sciences studying components are semantically and explanatorily *omnipotent* – they can explain, illuminate, and perform every other theoretical task with regard to the whole of nature.

Scientific reductionists like Weinberg are now very careful and clear in repudiating greedy reductionism. As I noted earlier, Weinberg explicitly labels himself a "compromising" scientific reductionist precisely because he takes higher sciences, their laws, explanations, and predicates to be significant and necessary in a variety of ways that contrast with the claims of a greedy reductionism.[15] Nonetheless, although accepting the significance and even indispensability of higher sciences, compromising scientific reductionists like Weinberg still defend a special status for the lower sciences studying components.

For example, scientific reductionists claim that, in a certain sense, lower sciences studying components are always closer to the "frontiers" of scientific research about the fundamental questions than higher sciences. As Weinberg puts the point in the context of various

---

[14] Dennett (1996). As I detail below, greedy reductionism has both semantic and ontological facets.

[15] Weinberg (2001), p. 13. For similar sentiments, see Dawkins (1987), pp. 14–15.

levels in physics, "Elementary particle physics ... represents a frontier of our knowledge in a way that condensed matter physics does not" (Weinberg (1992), p. 59). However, the same point holds for neurobiological research over work in psychology, or for biochemistry over biology, and so on, for the science studying components is taken to be closer to the frontier of fundamental research than the science studying composed entities. Furthermore, the special status scientific reductionists claim for the sciences studying components is mirrored in their methodological claims about these disciplines. Thus, though acknowledging there are many factors in determining funding allocations, writers like Weinberg argue that lower sciences always have a claim to funding priority that higher sciences lack precisely because lower sciences are closer to the frontier of fundamental research about determinative entities and laws (Weinberg (2001), p. 23).

Overall, scientific reductionism pursues the metaphysics of science and an apparently integrated position in a set of linked commitments about ontological, nomological, semantic, and methodological issues. And this position is a familiar one from popular presentations of the great highlights of twentieth-century science and public debates over funding allocation. Running through the view is the contention that the findings of everyday reductionism, in compositional explanations and their compositional concepts, ground arguments for the more substantive claims of scientific reductionism. Although we have seen that these arguments are less than fully articulated, the driving idea appears to be that compositional explanations show, in some way, that component entities, the laws governing them, and the sciences studying them have priority in the sciences. But scientific reductionism is also claimed to be compromising in accepting the importance, and even indispensability, of higher sciences studying the macro-world.

1.1.3 Scientific Emergentism and Some of its Characteristic Claims

Versions of emergentism have returned to prominence in the sciences only in the last few decades and so there is consequently far less consensus, or clarity, about these newer positions. For my present purposes, I focus solely on the view I ultimately argue is a live option and which, for expository reasons, I am referring to as "scientific emergentism." I take Laughlin as my main focus, but as I noted in the Introduction there are a wide array of other prominent scientists who plausibly defend this position, including the chemist Ilya Prigogine, biologists such as Brian Goodwin, Franklin Harold, Richard Levins, and Richard Lewontin, the neurophysiologist Walter Freeman, and a large number of writers in the sciences of complexity such as Stuart Kauffmann, Alwyn Scott, and others, in addition to condensed matter physicists such as Philip Anderson and Laughlin.[16] To start my discussion of their views, let me again mark some preliminary points.

First, it is important to emphasize that scientific emergentists endorse everyday reductionism and the search for compositional explanation and compositional relations.[17] As Laughlin explains about his main conclusion:

One might subtitle this thesis the end of reductionism (the belief that things will necessarily be clarified when they are divided into smaller and smaller component parts), but that would not be quite accurate. All physicists are reductionists at heart, myself included. I do not wish to impugn reductionism so much as to establish its proper place in the grand scheme of things. (Laughlin (2005), p. xv)

Scientific emergentists like Laughlin, following a line defended in Anderson (1972), are thus best understood as defending *extensions* or *supplements* to everyday reductionism, rather than its replacement. And scientific emergentists thus accept the universal applicability of the compositional concepts deployed in compositional explanations and hence that nature is a comprehensive compositional hierarchy, i.e. physicalism.[18]

Second, it is equally important to note that scientific emergentism is not based around what we *cannot* or *do not* know or understand in the sciences – quite the reverse. As Laughlin explains, again focusing on the case of physics:

Ironically, the very success of reductionism has helped pave the way for its eclipse. Over time, careful quantitative study of microscopic parts has revealed at the primitive level at least, collective principles of organization

---

[16] Prigogine (1997), Goodwin (2001), Harold (2001), Levins and Lewontin (1987), Freeman (1995), (2000a), (2000b), Kauffman (1995), Scott (2007), Anderson (1972), (1995), Laughlin (2005), Laughlin and Pines (2000).

[17] Anderson (1972), p. 393.

[18] Anderson (1995), p. 2018, and Laughlin (2005), p. 6, are therefore both careful to clearly reject any uncomposed non-physical entity.

are not just a quaint side show but *everything* – the true source of physical law, including perhaps the most fundamental laws we know. (Laughlin (2005), p. 208)

If we apply our distinction between everyday reductionism and the more substantive position of scientific reductionism, then we can interpret Laughlin as rejecting scientific reductionism because of the quantitative accounts of components that we are now acquiring in many cases, building upon the work of everyday reductionism and its qualitative accounts of components. Thus scientific emergentism does not rely upon the difficulties presently surrounding consciousness, or the interpretation of quantum mechanics, or other problematic scientific cases. So I put such cases to one side in my discussion.[19] Instead, scientific emergentists focus upon mundane examples where we have *both* qualitative *and* quantitative accounts of components. Scientific emergentism claims that these examples where we have some of our most detailed accounts provide the engine for its conclusions.

We therefore find a wide range of cases cited by scientific emergentists as involving emergence where the following gives an illustrative, but by no means exhaustive, list: high-energy superconductivity (Laughlin (2005)); thermodynamical systems (Prigogine (1968)); Benard cells in chemistry (Nicolis and Prigogine (1989)); the eukaryotic cell (Boogerd et al. (2005), Harold (2001)); slime mold (Garfinkel (1987), Nicolis and Prigogine (1989)); neural populations (Freeman (2000a), (2000b)); the behavior of eusocial insects like ants and bees (Wilson and Holldobler (1988)); and flocking behaviors in vertebrates (Couzin and Krause (2003)).[20]

As my last passage from Laughlin makes clear, our putative empirical findings about the *components* in these cases, as well as emergent composed entities, putatively drive the claims of scientific emergentism. But what are the core ideas of the position? We saw above that scientific reductionism is less than clear in its arguments. Unsurprisingly, it is still more difficult to articulate the core claims of scientific emergentism. But we can again begin to illuminate this position, which has at its heart the contention that the scientific cases just noted support the

---

[19] For the reasons outlined above in note 3, there are difficulties in determining whether either compositional concepts or explanations apply to consciousness or quantum mechanical phenomena.

[20] See Scott (2007) for a more comprehensive list of cases.

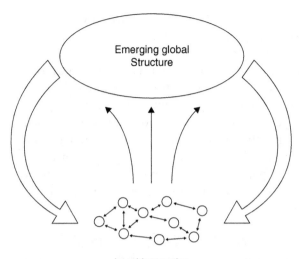

Local Interaction

Figure 1A The complexity scientist Chris Langton's illustration of the role of a composed "emergent" entity at the top and its "downward" or "self-organizing" influence on its components. (From Lewin (1992), Figure 10, p. 189).

claim that "Parts behave differently in wholes" and that "Wholes are more than the sum of their parts."

Crucially, as Figure 1A illustrates, scientific emergentism not only takes "wholes" to act at their own levels, but also defends a "downward" determinative influence that such emergent "wholes" exert upon their "parts" which is variously termed "downward causation," "part–whole constraint," "self-organization," or "circular causation," amongst other terms. Scientific emergentists thus embrace *discontinuities* in the powers of the components we find across simpler collectives and the complex collectives where components are claimed to be "downwardly" determined by "emergent" composed entities. The result is an account of emergent "wholes" *and* their "parts" that flies contrary to the scientific reductionist's contentions about the character of component entities and also the entities that are determinative in nature.

As my earlier quote from Laughlin highlighted, along with these novel claims about "parts" and "wholes" scientific emergentists also endorse similarly contrasting view of laws. Given their account of the determinative influence of "wholes" upon their "parts," we find

Laughlin, Anderson, and others claiming that there are "organizational" laws involving emergent composed entities.[21] Furthermore, scientific emergentists often argue that such organizational laws are also *fundamental* laws that hold in addition to the laws governing the simplest collectives of components that the scientific reductionist claims to exhaust the fundamental laws (Laughlin (2005)).

These ontological points feed into the scientific emergentist's methodological commitments that again oppose the scientific reductionist's contentions about the implications of everyday reductionism. For example, in his seminal paper, which is a continuing touchstone for scientific emergentism, Anderson cautions us that once we accept everyday reductionism:

[i]t seems inevitable … [to accept] what appears at first sight to be an obvious corollary of [everyday] reductionism: that if everything obeys the same fundamental laws, then the only scientists who are studying anything really fundamental are those working on those laws … This point of view … it is the main purpose of this article to oppose. (Anderson (1972, p. 393)

Here we see one of the key contentions of scientific reductionism, but why does Anderson claim this conclusion is mistaken? Once again focusing on our more detailed empirical findings about the composed and component entities in complex collectives, and the new laws we find with such entities, Anderson claims that:

[t]he behavior of large and complex aggregations of elementary particles, it turns out, is not to be understood in terms of a simple extrapolation of the properties of a few particles. Instead, at each level of complexity entirely new properties appear, and the understanding of the new behaviors requires research which I think is as fundamental in its nature as any other. (Anderson (1972, p. 393)

What we see here is a crucial rejection of the type of continuity assumed by the scientific reductionist to exist across all scales of simple and complex collectives of components. Instead, as we saw above, emergentists like Anderson are apparently committed to discontinuities in the powers of components we find in complex and simpler collectives

---

[21] Anderson (1972), p. 393, Laughlin (2005), p.xiv and p.xx.

resulting from the "downward" determination of emergent composed entities. And scientific emergentism thus takes disciplines studying nature at *many* levels to be attacking the "frontiers" of fundamental research (Laughlin (2005), pp. 5–6) and consequently denies that there is *any* special factor supporting the funding of lower sciences that is in principle lacked by higher sciences studying emergent composed entities.

Although my survey shows its claims again need further articulation, my work establishes that scientific emergentism pursues the metaphysics of science in its arguments about our empirical findings concerning the "parts" and "wholes" in complex collectives and the species of determination operating upon them. And, once more, these commitments provide a discernible backbone of ontological theses that apparently drive arguments for the wider claims of scientific emergentism about laws, methodology, or funding allocation. Scientific emergentism is therefore like scientific reductionism in focusing on everyday reductionism in pursuing the metaphysics of science, but apparently defends a strikingly different picture of nature as flowing from the empirical evidence of compositional explanations and other empirical findings in the sciences.

1.2 The Positivist Legacy in the Philosophy of Science, the Antireductionist Consensus, and New Work on Reduction

To start my survey of philosophical discussions, it is most productive to focus on the long-standing debates over reduction in the philosophy of science that were some of the most high-profile philosophical discussions of the last half century. (I therefore leave examination of philosophical work on emergence until Chapter 5, where I provide a survey of such concepts.) I start with the Positivists, in 1.2.1, whose account of scientific explanation underpinned their distinctive account of reduction as a primarily *semantic* phenomenon whose idiosyncratic features I highlight. In 1.2.2, I note the reigning antireductionist consensus in philosophy of science that is built upon the famous arguments showing Positivism's semantic account of reduction rarely applies in the sciences. And I finish, in 1.2.3, by briefly reviewing two recent strands of philosophical work on reduction in the sciences that come closer to scientific reductionism.

1.2.1 Positivism and the Nagelian Model of Reduction: Understanding the Greedy Approach and its Idiosyncrasies

After fleeing the Nazi regime, in the 1930s and 40s, the Positivists found jobs at the most influential institutions in the United States and founded the philosophy of science as we now know it. Though the Positivist program is now widely rejected, the influence of the Positivists lives on in various ways so we need to mark two of its distinctive commitments. First, Positivism rarely engaged concrete scientific cases, especially beyond physics, since the Positivists sought *rational reconstructions* of science – that is, accounts of how science *ought* to be rather than how it is. Second, the Positivists were famous for the programmatic denunciation of "metaphysics" as literal nonsense where this was largely in reaction to the Kantian and Hegelian philosophical traditions of transcendental metaphysics that dominated Austria and Germany where they trained. As we shall see, these two commitments in tandem plausibly left Positivism pursuing metaphysics for science about the nature of scientific explanation and the character of reduction.

Consider Positivism's so-called "Deductive-Nomological" account of scientific explanation (Hempel (1965)), which takes all scientific explanation to have the form of deductive arguments explaining phenomena by deriving them from the laws serving as the premises of such arguments. This account is semantic in character and plausibly a rational reconstruction since it excludes compositional, including mechanistic, explanations from counting as scientific explanations – thus ignoring vast swathes of explanations in actual science.

The situation is similar with the account of reduction that Positivism built upon its view of explanation. Abandoning metaphysics, the Positivists needed an account of reduction in the sciences not involving this proscribed area, thus forcing them to ignore any role for ontological concepts like compositional ones. Instead, the Positivists ultimately fixed upon what I am terming "semantic" or "Nagelian" reduction (Nagel (1961)), which is focused on the derivation of the law statements of the theories of higher sciences from the law statements of lower sciences in combination with identity statements.[22] Under the

---

[22] Sklar (1967), and others, each persuasively argued that rather than Nagel's favored bridge laws the model actually needed identities in order to be successful.

Positivists' Deductive-Nomological account of explanation, such derivations of higher-level laws constituted an explanation of the higher-level laws, and hence the associated theories, by the laws of the lower sciences – thus semantic reduction putatively "reduced" such higher-level laws and theories to lower-level laws and theories.

It is important to notice four idiosyncrasies of the Nagelian model of reduction that we shall see have left philosophers on both sides of their debates over reduction with a rather odd set of concerns. (i) The semantic model implicitly saddles reductionists with an account of scientific explanation at odds with scientific practice because it excludes compositional explanations. The Nagelian model of reduction thus simultaneously offers controversial accounts of both reduction *and* explanation in the sciences. (ii) Nagelian reduction is a semantic, rather than ontological, account of reduction, since it focuses on statements and theories. (iii) Semantic reduction is the epitome of Dennett's greedy reductionism, since it entails that all higher sciences are, at least in principle, *dispensable* and, on the flip side, implies that lower sciences, again at least in principle, are *explanatorily omnipotent*. And, finally, (iv) Nagelian reduction also appears to be ontologically greedy as well, since it requires we identify every predicate of a higher science with a predicate of the lowest science in fundamental physics and thus appears to entail that the whole of reality is swallowed up by the single level of entities referred to by the predicates of fundamental physics.

Given the influence of Positivism in the philosophy of science there is consequently a vast philosophical literature on the Nagelian account of reduction, its various problems, and putative modifications designed to overcome them.[23] And sophisticated descendants of the semantic approach to reduction continue to be an ongoing area of philosophical research.[24] However, before turning to the famous objections to Nagelian reduction, I want to note three wider points about the semantic model.

First, it is important to mark that the Positivist account of explanation is plausibly close to something like metaphysics for science, since

---

[23] For more detailed surveys of the twists and turns of this vast literature see Schaffner (1967), (1993), Hooker (1981), and Wimsatt and Sarkar (2006).

[24] See, for example, Hooker (1981), Churchland (1985), Schaffner (1993), and Bickle (1998) for important work in this tradition.

it jams an alien picture of explanation onto the scientific phenomena. For example, the account leaves no room for compositional, including mechanistic, explanations, fails to fit swathes of explanations in the sciences, and has been consequently widely rejected in the philosophy of science as the truth about scientific explanation. Instead, following engagement with real scientific cases, a range of scientific explanations are now recognized as legitimate which do not fit the Deductive-Nomological account, including compositional explanations.

Second, following on from these points about the underlying account of scientific explanation, and also contrasting its nature with the claims of the scientific reductionists surveyed in 1.1.2 above, one may again press the charge that the semantic model of reduction is an instance of metaphysics for science. Applying an implausible account of scientific explanation, and driven by a programmatic proscription of ontological issues or concepts, the Nagelian model of reduction, and its various features (i)–(iv), simply does not easily track the discussions of reduction we find amongst working scientists. However, I do not raise this concern about metaphysics for science as a way to undercut the Nagelian model, since the two famous objections examined in the next subsection suffice for that. My goal is simply to highlight the divergence between the scientific and philosophical discussions of reduction and its apparent source in the Positivists' "rational reconstruction" approach and programmatic commitments.

The latter points become significant because, third, there is an odd asymmetry in contemporary philosophical reactions to Positivism's account of explanation and its picture of reduction. Though the semantic model of reduction is founded upon the Deductive-Nomological account of explanation, which has been put aside by most philosophers of sciences, it is still the case that reduction is often routinely taken by philosophers of science, and philosophers more widely, to be Nagelian reduction requiring identities and its other idiosyncrasies. One must immediately ask why discussions of reduction in the sciences have not been refreshed by engagement with scientific practice in the ways that discussions of explanation have. And one can also wonder why the semantic model of reduction continues to be taken seriously to a degree that its underlying account of explanation is not. I can supply answers to these questions by turning in the next section to the philosophers of science who famously argued that we do not find Nagelian reductions in the sciences.

1.2.2 The Antireductionist Consensus and its Famous Arguments

Some of the most important debates in the philosophy of science in the last half century focused on whether or not we find semantic reductions in the sciences. These discussions generated the famous philosophical critiques of reduction which ground what I noted earlier is the reigning consensus in philosophy that reductionism is a dead position and which also support the antireductionist consensus in the philosophy of science. These criticisms were driven by an engagement with real scientific examples involving the compositional explanations overlooked by the Positivists and were presented, in differing forms, by writers such as Hilary Putnam, Jerry Fodor, Philip Kitcher, William Wimsatt, and others, in the 1970s and 1980s.[25]

Most famously, the so-called "Multiple Realization Argument" was used to attack the existence of the identities between higher- and lower-level properties needed to run the Nagelian model. Writers like Putnam, Fodor, and others claimed to show that many scientific properties are "realized" by a variety of distinct lower-level properties. However, the inter-level identities between properties required by semantic reduction imply we must have coextensivities between higher- and lower-level properties. Thus the multiple realization of higher-level properties revealed by compositional explanations was taken to preclude inter-level identities and hence to undercut the widespread existence of Nagelian reductions.

Antireductionist critics also offered a second, connected kind of objection in what I term the "Predicate Indispensability Argument," which was again based on compositional explanations from a range of sciences. This second objection challenged the implication of the Nagelian model that if reduction exists, then higher sciences are dispensable and lower sciences are omnipotent, at least in principle. Critics argued that in cases of "multiple realization" the predicates of lower sciences often only serve to express the heterogeneity of lower-level properties, thus missing the true explanations concerning the homogeneities found amongst higher-level scientific properties. Consequently, multiple realization, and other phenomena, were used

---

[25] Putnam (1967), (1975a); Fodor (1968), (1974), (1975); Wimsatt (1976), (2007); and Kitcher (1984).

to show that higher sciences are far from dispensable and hence that semantic reduction fails to apply in the sciences for a second reason.

I want to highlight two points about the aftermath of the antireductionist criticisms of Putnam, Fodor, Kitcher, and Wimsatt. One wider fact I need to emphasize is that although they used compositional explanations and their compositional notions, these antireductionist writers never explicitly offered a theoretical account of such compositional relations and so the key notions of "realization" and "multiple realization" were left unarticulated. One reason for this is that no such account was needed to run the Multiple Realization Argument: identities between properties require coextensivities between the relevant properties and the critics, like Fodor, used bare multiplicity of realization to undercut such coextensivities. However, the lack of any account of scientific composition by these pioneering writers actually engaging compositional explanations in the sciences left a vacuum that was inevitably filled. As I detail in the next subsection, and the following section, the subsequent philosophical approach to scientific notions of composition has been to shoehorn them into frameworks developed for other purposes.

More importantly, the second point I want to mark is that the antireductionist consensus, built around its two famous objections, takes reduction to be semantic reduction and thus inadvertently continues to enshrine the Positivist metaphysics for science approach to reduction and its idiosyncratic model of reduction. We can therefore see how, although the Positivist account of explanation has long been marginalized in philosophy, the connected Positivist account of reduction has not. And this issue is not a trivial one. For the antireductionist consensus is basically betting that its objections to Nagelian reduction suffice to rebut all forms of reduction in the sciences. To the degree that one can provide a plausible theoretical framework for the different approach of scientific reductionism, then this looks like an increasingly risky wager.

1.2.3 Recent Philosophical Work on Reduction in the Sciences

There has been a more recent spate of work on reduction in the sciences that is more promising with regard to its affinities with scientific reductionism. I therefore want to finish my brief survey of philosophical work on reduction in the sciences by looking at two strands in this recent research. I suggest that, in different ways, both approaches make important advances over the Nagelian account of reduction, but I also highlight how these two approaches again each manifest distinct elements of a problematic pattern found in philosophical discussions.

The first thread is found in writers such as Kenneth Schaffner (2002), (2006), and John Bickle (2003), (2008), who were both once prominent proponents of semantic accounts of reduction, but who have each independently renounced these models to more closely engage concrete scientific cases involving compositional explanations and the claims about "reduction" that working scientists make about them.[26] Thus Schaffner now emphasizes that what he terms the "sweeping" reductions of the Nagelian approach are rare in the sciences and what we get instead are only usually what he terms "creeping" reductions based around particular compositional explanations in local cases. Schaffner's and Bickle's work moves philosophical debates forward in various ways, but primarily through its engagement with concrete scientific examples, and the claims that working scientists make about them, as the relevant focus of discussion of reduction in the sciences.

However, despite such important contributions, my earlier distinction between *everyday* reductionism, which seeks compositional explanations, and a *substantive* scientific reductionism makes a problem clear. A noticeable feature of both Schaffner's and Bickle's work is the absence of any *rationale* or *argument* why the scientific cases they focus upon, and their explanations, support a substantive form of reduction. Both sides in the sciences accept that we should search for, and often locate, what I earlier dubbed "everyday reductions" in the shape of compositional explanations and the compositional relations they posit. And, as the discussion of the last section makes clear, antireductionists in the philosophy of science also endorse the search for compositional explanations and hence what I am terming everyday reductions. So pointing to such everyday reduction does not suffice as an argument or reason supporting a *substantive* reductionism that has commitments that philosophical antireductionists or scientific

---

[26] Schaffner was long one of the most important defenders of semantic models of reduction (Schaffner (1993)) and Bickle offered one of the most sophisticated formulations of such a view (Bickle (1998)).

emergentists would reject. I have also noted that the working scientists who are scientific reductionists, like Weinberg or Kandel, have yet to clearly articulate reasoning that supports their substantive reductionism about such cases. Anyone pointing to scientific cases as the grounds for a substantive reductionism, such as Schaffner and Bickle, therefore needs to outline an *argument* that shows why the relevant compositional explanations, and their compositional concepts, or other features, support something their opponents reject.

Unfortunately, neither Schaffner nor Bickle supply such an argument that, as my survey of scientific reductionists suggests, would appear to require engagement with the metaphysics of science and the nature of scientific notions of composition, since we have seen this is plausibly the basis of the position pressed by scientific reductionists like Weinberg or Kandel. It appears that Schaffner and Bickle are so suspicious of "metaphysics" that they do not pursue the work in the metaphysics of sciences that is necessary to make the underlying arguments for scientific reductionism explicit.[27] Though reaching out to the sciences, the continuing aversion of many philosophers of science to "metaphysics" thus still sunders Schaffner and Bickle's promising approach from working scientists who are plausibly up to their necks in the metaphysics of science.

The second strand of recent philosophical research on reduction has no such aversion to ontology and has been pioneered by Jaegwon Kim over the last two decades (Kim (1998), (1999)). To properly understand, as well as appreciate, Kim's contribution we need to note the presuppositions of the debates to which he contributes. The relevant work in the philosophy of science is surveyed in detail by Stephen Horst (2007) in a book aptly entitled *Beyond Reduction: Philosophy of Mind and Post-Reductionist Philosophy of Science*. Tellingly, Horst's survey follows the philosophical practice of taking reduction to be semantic and, more importantly for our present purposes, assumes scientific notions of composition are framed using standard "functionalist" machinery from the philosophy of mind. For the vacuum of work by writers like Fodor on scientific composition, such as "realization"

---

[27] For example, Bickle (2008) explicitly disavows "metaphysics" in its title, but the irony is that it appears that Bickle cannot defend his position *unless* he embraces the metaphysics of science.

relations, has primarily been filled by co-opting machinery from the philosophy of mind.

Against this background, we can properly appreciate Kim's ground-breaking contribution. Kim (1998) provides explicit arguments that reductionists must finally put aside the Nagelian model and instead focus on *ontological* forms of reduction. And Kim has gone on to offer his own detailed version of such a position in what he terms "reduction by functionalization."[28] Kim's work has thus been of seminal importance for encouraging philosophers to move away from semantic accounts, and towards ontological ones, thus drawing philosophical approaches closer to that of scientific reductionism. But it is also important to note that Kim nonetheless continues to follow the second assumption of the debates Horst ably surveys. That is, Kim follows the strategy of appropriating "functionalist" machinery from the philosophy of mind to drive his ontological model of reduction in the sciences. And subsequent work on ontological accounts of reduction offered by philosophers such as John Heil, Barry Loewer, or Alexander Rosenberg, amongst others, shares just this feature.[29] Even Rosenberg, whose work is replete with detailed engagements with concrete cases from the biological sciences, follows Kim at the key point in his arguments for reduction by deploying reasoning utilizing such "functionalist" machinery.[30]

Once again, this pioneering work on ontological notions of reduction has clearly moved the debate forward in this case by re-engaging ontological reduction, which is the species we have seen is pursued by scientific reductionists. However, despite these significant contributions, the worry about this strand of recent research is that it manifests exactly the reverse form of deficit to that of Schaffner's and Bickle's work on reduction. The work spearheaded by Kim on reduction overcomes the antipathy to ontology, or "metaphysics." But the question is whether this approach has lost touch with scientific concepts and explanations by shoehorning scientific composition into a framework developed for other purposes, thus falling into a degree of metaphysics for science. I further explore this situation, and associated worries, in the next section.

---

[28] Kim (1998) and especially (1999).
[29] See Loewer (2008) and Rosenberg (2006), amongst others.
[30] Rosenberg (2006), Chapter 6.

Overall, the recent philosophical work on reduction in the sciences nicely frames what appears to be a common type of situation in which the recent history of philosophy has left us. The tumultuous turn against "metaphysics," and especially the influence of the Positivists, meant that philosophers focused on the sciences turned away from metaphysics and continue to be leery even of the metaphysics of science. On the other side, the philosophers left pursuing metaphysics were often not focused on scientific cases or concepts. Although this historical situation is no longer so clean cut, it appears we still find reverberations of these sociological trends in our present debates. For we either find philosophers engaging scientific concepts and practice who are averse to metaphysics of science, such as Schaffner or Bickle; or we find philosophers, such as Kim, who are open to metaphysics but import frameworks from other areas to capture scientific notions. What we conspicuously lack are philosophers engaging ontological issues, concepts, or positions in the sciences, whether the nature of scientific composition or reduction, or, as we shall later see, emergence, through theoretical frameworks directly crafted in response to the nature of such scientific phenomena.

1.3 Philosophical Work on Scientific Composition

My survey of scientific reductionism and emergentism highlights their focus on the scientific notions of composition central to compositional explanation. And my work in the last section began to illuminate how the initial generation of philosophers of science focused on compositional explanation, such as Putnam, Fodor, Kitcher, and Wimsatt, left a theoretical vacuum with regard to such compositional concepts, which philosophers have filled by appropriating machinery from other areas. To situate my later work, I want to briefly outline these appropriations and begin to assess whether they are benign or whether they may also be problematic forms of metaphysics for sciences.

I therefore want to conclude the chapter by providing an opinionated survey of the two most common philosophical approaches to scientific composition, which each draw machinery from different areas of philosophy. I start, in 1.3.1, by outlining the tangled genesis of standard "functionalist" machinery from the philosophy of mind and its account of the "realization" of properties. Then, in 1.3.2, I look at the increasingly prominent manipulability-based account of the composition of processes, which uses frameworks developed for causal relations in the philosophy of science. For each of these frameworks, I am going to flag a couple of kinds of reason to be suspicious that the approach may mischaracterize scientific composition.[31]

1.3.1 The Standard "Functionalist" Machinery of Philosophy of Mind: An Approach to the Composition of Properties in the Sciences

Philosophy of mind and philosophy of psychology came into existence in their present form during the 1960s, 70s, and 80s, which saw an amazing spasm of creative work on a range of foundational questions in the philosophy of mind/psychology, the philosophy of science, and what was viewed as naturalistic metaphysics. One of the products of this activity were the versions of "functionalism," and the proprietary technical machinery associated with them, which are widely, but mistakenly, taken to have a shared metaphysical basis, since all the versions of "functionalism" talk of "functional properties," "causal roles," "realization" etc. Unfortunately, the different versions of "functionalism" were crafted to serve different purposes and have distinct proprietary understandings of "functional properties," "causal roles," "realization" etc. The result is a very confused, and ultimately damaging, situation in naturalistic philosophy. Drawing out the nuances of these problems requires more space than I can devote to it here, but I can outline the genesis of this situation and highlight reasons to be suspicious that standard "functionalist" machinery does not comfortably reconstruct scientific notions of composition.[32]

One tradition of "functionalism," including writers such as Fodor (1968), (1975) and Dennett (1969) (1978), was focused on using scientific examples of compositional explanations to provide a model for work in cognitive science or computational psychology. In contrast, another group of writers on "functionalism," such as David Lewis (1972), hailed from "analytic" areas of philosophy and focused on common-sense concepts using more traditional philosophical tools.

---

[31] I leave work using mereological frameworks, or notions of "grounding," to one side, since the concerns I raise in the next chapter are easily extended to the various accounts of scientific composition using either mereology or grounding.

[32] I provide a detailed overview of these problems in Gillett (2007b) and (2012).

Work in these traditions was seldom kept separate despite their distinct goals and methodologies.

The empirically oriented work on "functionalism" was connected to the critiques of Positivism through Fodor's research and was again focused on compositional explanation and scientific composition. For example, focusing on such concrete examples, Dennett emphasized that compositional relations in the sciences are often *many–one* relations involving *teams* of component individuals, and properties, which together compose some higher-level individual or property. Dennett explicitly accepted that the composed and component individuals/ properties are at *different levels of individuals* and *qualitatively distinct* – hence accommodating the PEP of associated explanations.[33] And a similar view of scientific composition plausibly inspired Fodor's positive account of the structure of psychology and other higher sciences, including their distinctive "realization" relations.[34]

Unfortunately, as I outlined above, pioneering writers like Fodor and Dennett, working simultaneously in the philosophy of science and the philosophy of psychology, never constructed a theoretical framework for scientific notions of composition such as the "realization" of properties. And the resulting theoretical vacuum was inevitably filled. The machinery of topic-neutral Ramseyfication developed in the analytic tradition of "functionalism" by Lewis for understanding *predicates* was unfortunately co-opted to frame *ontological* notions in the empirical traditions like that of a "realization" relation between *property instances*. Matters were made still more complicated by the still later alterations made to Lewis's machinery inspired by the features of computational explanations in psychology and other influences.[35]

These appropriations of Lewis's machinery plausibly spawned a collection of new technical concepts, like that of a "second-order

---

[33] Dennett utilized these distinctive features in articulating his influential "homuncular functionalism" as an account of an appropriate methodology for a scientific psychology. Consider Dennett's example of how we explain a capacity for face recognition as composed by a team of lower-level entities. Famously, in such cases Dennett clarifies that we can understand richly intentional entities as composed by teams of entities with more rudimentary, and hence qualitatively distinct, intentionality, which in turn can be understood as composed by teams of entities with still more rudimentary intentionality.

[34] Fodor (1968), (1974) and (1975).

[35] See Piccinini (2004) for a survey of these developments and their results.

property," now widely used by philosophers. And the new conceptions built around topic-neutral Ramseyfication, whether of a "functional property," "realization," "causal role" etc., were taken to cover *all* the versions of "functionalism" and their proprietary versions of the latter notions, though as I outline below it is dubious whether they fit the notions of the empirically oriented tradition of "functionalism."[36] Working within these appropriations of Lewis's machinery, writers such as Kim (1992b, (1998), Sydney Shoemaker (2001) and Jessica Wilson (1999) have shaped a more ontological notion of realization in what is termed the "Flat" or "Subset" view of realization. The Subset/Flat view takes realization to be a *one–one* relation between *qualitatively identical* or *qualitatively similar* properties of the *same* individual.[37] And this view of realization is now routinely used in work in the philosophy of science to articulate scientific notions of composition involving properties or the issues involving it, such as Kim's reduction-by-functionalization account.[38]

However, we can now see that there are real worries about whether the Flat/Subset account fits scientific notions of composition where such problems have plausibly been masked by the convoluted history of "functionalism." Recall that Dennett highlights how the compositional relations he finds in scientific explanations are *one–many* relations of teams of *qualitatively different* realizer properties instantiated in *distinct* individuals from the higher-level individual instantiating the realized property. In contrast, we have just seen that the Subset/Flat account takes realization to be a *one–one* relation between *qualitatively similar* properties of the *same* individual. We therefore apparently have fairly stark differences between scientific notions of composition and the standard "functionalist" machinery philosophers have recently used to articulate its nature. In the next chapter, I explore

---

[36] See Gillett (2007b) and (2012) for a more detailed defense of these claims. Under the influence of the received views in philosophy of mind, a number of writers persist in misleading talk of "the" realization relation (Polger (2007)), but there are plausibly a variety of concepts spanning ontological, linguistic, and abstract or mathematical notions. See Endicott (2006) for a survey of such notions.

[37] See Gillett (2002b) and (2003c) for an outline, and critical evaluation, of the basic features of the Subset/Flat view. The view is critically examined in more detail in Chapter 2.

[38] For examples of such work, across the last few decades, see Kim (1992a), Shapiro (2000), (2004), and Wilson (2010).

whether such problems are real by looking directly at scientific notions and their distinctive features.

1.3.2 Manipulability-Based Accounts from Philosophy of Science: An Approach to the Composition of Processes in the Sciences

Although standard "functionalist" frameworks are the dominant approach to scientific composition, it is important to discuss another increasingly prominent approach in work on compositional explanations in the philosophy of science. In a book focused on mechanistic explanation in the neurosciences, i.e. the species of compositional explanation based upon composition between processes, Carl Craver (2007) has popularized the strategy of using interventionist approaches to *causal* explanation as a way to characterize the compositional, or as Craver terms them "constitutive," relations between processes that underpin such compositional explanations.

For my purposes, the interesting feature of Craver's approach is that he claims to provide an adequate sufficient condition for the composition of processes that is built primarily around the mutual manipulability of processes. Craver is also quite explicit, and takes it as a merit of his approach, that it is an extension to compositional relations and explanations of a framework developed for causal relations and explanations. Craver tell us:

The mutual manipulability account is a plausible condition of constitutive relevance because it fits well with experimental practice and because it is an extension of the view of etiological [causal] relevance ... (Craver (2007), p. 162)

Craver outlines the heart of his account of the composition of processes in this passage (where "φ-ing" and "ψ-ing" are processes of the relevant individuals X and S):

My working account of constitutive [i.e. implementational] relevance is as follows: a component is relevant to the behavior of a mechanism as a whole when ... (i) X is a part of S; (ii) in the condition relevant to the request for explanation there is some change to X's φ-ing that changes S's ψ-ing; and (iii) in the condition relevant to the request for explanation there is some change to S's ψ-ing that changes X's φ-ing. (Craver (2007), pp. 152–3. Original emphasis)

One obvious feature of this account is that, unlike the standard "functionalist" machinery, Craver's mutual manipulability condition takes the composition of processes to be intertwined with the composition of individuals. Condition (i) demands part–whole relations between the individuals that base the composed and composing processes, i.e. between the φ- and ψ-ing individuals. It is important to emphasize that Craver deploys the sophisticated machinery of interventionism to articulate these core ideas and that Craver also adds a number of further nuances to his final account in order to address phenomena he highlights in scientific practice.[39] However, I am going to leave these nuances to one side in my discussion, since it is the adequacy of the core mutual manipulability condition that I am interested in.[40] The basic idea is that mutual manipulability suffices for the composition of processes in the sciences.

It is also worthwhile noting that Craver leverages his mutual manipulability-based sufficient condition for the composition of processes into a regulative norm for compositional explanation, primarily the mechanistic explanations explicitly deploying compositional relations between processes. Craver tells us that:

Relationships of mutual manipulability can and should replace the requirement of derivability as a regulative ideal on constitutive explanation in neuroscience. (Craver (2007), pp. 159–60. Original emphasis)

So Craver's claims about the utility of mutual manipulability in capturing composition between processes in the sciences is therefore doubly important because it is also claimed to provide a norm by which we can assess compositional explanations as well.

In this case we do not have an example of metaphysics for science, since Craver is not importing a metaphysical framework developed *outside* the sciences in order to characterize scientific phenomena. Nonetheless, Craver's strategy is structurally quite similar to that of

---

[39] For example, Craver adapts his conditions so they take account of redundancy. See Craver (2007), Chapter 4 and particularly section 8, for the details.

[40] The type of objections to Craver's account that I offer in the next chapter also apply to Craver's more sophisticated, amended account.

writers using standard "functionalist" machinery, for Craver is again seeking to shoehorn compositional relations into machinery developed for *other* phenomena, in this case causal relations. As we saw above, however, Craver is quite explicit in using a framework for causal relations in the sciences, with a couple of tweaks,[41] in application to scientific notions of composition. One can thus charitably take Craver as staking out the philosophical position that causation and composition are of a kind.

Once again, there are reasons to be suspicious that such a position leads to the mischaracterization of scientific notions of composition. Let me focus on one issue related to the Ontologically Unifying Power of compositional explanations to highlight this concern. As Craver himself notes, the relata of composition are in some sense the same, whilst causation relates entities that are wholly distinct. Mutual manipulability appears to suffice for relations, such as causation, that hold between wholly distinct entities. But one may consequently be skeptical that mutual manipulability can suffice for a relation between entities that are *not* wholly distinct. For if wholly distinct entities can be mutually manipulable, then one may immediately be worried about how the existence of mutual manipulability can safely guarantee the existence of a relation whose nature is to relate entities that, in some sense, are not distinct? In the next chapter, I assess the mutual manipulability account by exploring this issue, and others, by directly engaging scientific notions of composition and the explanations that posit them.

1.4 Building a Bridge to the Sciences and their Potential Insights

Overall, the work of this chapter confirms that there is a dislocation of scientific and philosophical discussions – and also suggests an obvious diagnosis of it. The programmatic onslaught mounted in philosophy against "metaphysics," in the middle of the twentieth century, by ordinary language philosophers, Wittgensteinians, Quineans, and

---

[41] The differences Craver highlights between compositional and causal relations include compositional relations being symmetric since they involve *mutual* manipulability, composition relating entities that are not logically independent, and composition being synchronic.

the Positivists, amongst others, appears to have left few philosophers pursuing the metaphysics of science in theoretical accounts of onto-logical concepts or issues in the sciences directly developed in response to such scientific phenomena. In contrast, scientific reductionism and emergentism are both directly focused upon the concepts of composi-tion utilized in scientific explanations – thus leaving them pursuing the metaphysics of science, albeit in a way in need of better theoretical articulation.

Given this background, I suggest the course of remedial action needed in the philosophy of science is to directly engage the notions of composition in compositional explanations in the sciences. Once we understand such concepts better, we can then also directly engage scientific pictures of reduction and emergence. In effect, this approach simply advocates refreshing discussions of scientific composition, and then reduction and emergence, in the way that we have so successfully refreshed work on scientific explanation – that is, by re-engaging the relevant scientific phenomena. Once we have supplied an account of scientific composition, and then used it in turn to construct accounts of scientific reductionism and emergentism, we will have a way to bridge the debates in the sciences and philosophy by comparatively assessing their approaches. In the next chapter, I therefore turn to the founda-tional project of exploring the character of scientific composition.

2 | A Beginning Framework for Scientific Composition

Compositional concepts in the sciences underpin some of our most ubiquitous, and successful, scientific explanations. My primary goal in this chapter is to catalog the features of the scientific notions of composition underpinning such explanations and then provide an initial theoretical framework for such concepts. Along the way, I briefly sketch a picture of compositional explanation and its basis. I also provide reasons why extant philosophical frameworks for scientific composition are inadequate, thus confirming the suspicions I raised in the last chapter. However, my main focus is upon providing a positive account of scientific composition as a foundation for the work of later chapters.

It is important for me to be clear exactly what the goal of my account is and hence how its success, or failure, should be judged, since there are now a range of superficially related projects in the philosophy of mind and analytic metaphysics. In this chapter, I am pursuing the first two steps of my methodology, focusing on articulating the features of scientific concepts of composition and then providing an adequate theoretical framework to capture their nature. The success or failure of such a theoretical account is plausibly to be judged by how well it succeeds in accommodating the features of compositional concepts and the compositional explanations that deploy them.

Overall, I suggest that composition in the sciences is best understood as involving what I term "working components" bearing relations of "joint role-filling." First, my survey suggests entities in the sciences are "working" entities because they are individuated by their relations to processes and hence productive "roles." Consequently, I suggest that scientific composition involves working components that non-causally result in the composed entity because they together result in the role of this entity. Second, I detail how a framework based around working components bearing joint role-filling relations accommodates all the main features of scientific composition. In contrast, I show two rival philosophical accounts, in Craver's manipulability framework and standard "functionalist" frameworks, each fail to accommodate key features of scientific composition and compositional explanation.[1]

2.1 Preliminary Points

The compositional notions deployed in the sciences are ontological in character, since they concern relations between entities in the world, whether individuals, properties, powers, or processes. (The reader should note that I use the term "entity" as a catch-all referring to powers, properties, individuals, processes, etc. I use the term "individual" to refer to individuals.) I therefore need a minimal ontological framework to articulate such concepts as I pursue my descriptive project. To this end, I take individuals be located in space-time and to instantiate properties and relations that contribute powers to them. I consequently take individuals to be individuated by their properties.

With regard to properties and relations, I adopt a connected account of properties in a weakened version of the causal theory of properties, since it has been shown to comfortably fit the concepts deployed in compositional explanations. I therefore use Shoemaker's (1980) account of properties/relations under which what I term a "powerful" property or relation is individuated by the powers it contributes to individuals in this world under certain conditions.[2] It is worth marking that there are other properties and relations posited in the sciences, such as spatial relations, that are not powerful and do not themselves contribute powers since they make a difference to the powers contributed by other properties. But my work is primarily concerned with powerful properties or relations.[3]

---

[1] Elsewhere I have provided more detailed treatments of compositional notions in a series of papers (Gillett (2002b), (2007c), (2013), (Unpublished)) and at book length in collaboration with Kenneth Aizawa (Aizawa and Gillett (Forthcoming)).

[2] Throughout the book I am primarily concerned with instances of properties/relations, so where I talk of a "property" or "relation" I should be taken to mean a "property instance" or "relation instance" unless I note otherwise.

[3] Although some might take these commitments to be substantive, I should note that both Humeans and anti-Humeans accept powers, though giving different explications of their natures, so my distinction is actually neutral between these two opposing frameworks.

The latter commitments mean that individuals and properties/relations are both individuated by their relation to what I term "processes" and I support this commitment in my survey of the cases. I take what I term a "productive process," or "process" for short, to consist in a power of some individual being triggered and manifested to produce some effect. When an individual's powers result in this way in a process, I shall say the individual "bases" that process. Hence, unless I note otherwise, by "process" I always mean a *productive process* and by a "power" I mean a *productive power*.

Throughout the book I am concerned with what I term "determination" or "determinative" entities, so I should mark that I take a determinative entity to be one that makes a differences to the powers of individuals. It is also important to note that all productive processes are determinative, since they result in effects that make a difference to the powers of individuals. But I should carefully mark my terminology with regard to what I term "causation" and "production." In contrast to production, I take causation to be a concept with a much broader scope, including but not exhausted by productive relations.[4] Thus, for example, Mackie's account or counterfactual analyses, amongst many others, are candidates for analyzing this wider notion I label "causation." And it appears that all productive processes will likely be causal processes depending upon one's account of causation. However, given my usage, not all causal processes are productive ones and not all causal processes are determinative. At various points, I highlight where the difference between productive and causal processes is important, but my work primarily focuses on production and the powers that underpin it.[5]

---

[4] For a related distinction see Hall (2004). However, Hall offers a tentative account of "production" distinct in nature from "powerful production." The reader should therefore treat the latter as either an alternative view of "production" to Hall's or an account of a narrower concept.

[5] There are wider reasons to distinguish these notions related to scientific practice. It has been plausibly argued that causal relations can have negative relata, but it is hard to fit such relations into a power ontology. Furthermore, although scientists make use of causal concepts, allowing for negative relata, in their ongoing methods of discovery, I contend that in their formal, published, dress versions of their explanations, scientists favor the notion of production over concepts of causation.

2.2 Two Examples of Compositional Explanations and Concepts at Work in the Sciences

My goal in this section is simply to lay out some concrete examples and begin to highlight their compositional concepts. In 2.2.1, I look at the familiar case of a diamond and its constituent carbon atoms and, in 2.2.2, I consider the inter-level explanation of a voltage-sensitive ion channel using so-called "protein sub-units." It will quickly become clear that these explanations deploy complex, multi-faceted compositional notions and I use later sections to supply accounts of their common features and broader nature.

2.2.1 Diamonds and Carbon Atoms: An Example of Intra-and Compositional Explanation

Our compositional explanation of the extreme hardness of a diamond using the bonding and alignment of carbon atoms provides a useful starting case because its details are compact and widely known. Starting with the relevant individuals, scientists obviously take the carbon atoms to be *parts* or *constituents* of the diamond. The carbon atoms are "organized" and bear spatiotemporal, productive, and powerful relations to each other, hence forming a collective.[6] The carbon atoms are also spatially contained within the diamond.[7] I should also note that the carbon atoms, and the other individuals in this example, are far from passive. We can appreciate the relevance of these productive relations to our compositional explanations if we examine Figure 2A which highlights some of the processes that underpin explanations in this case.

---

[6] I am following the usage of scientists like Laughlin, so I should carefully mark that I am thus using "collective," and also "aggregation," in the common scientific usage to mean a spatiotemporally, and powerfully or productively, related group of individuals. I therefore take all collectives and aggregations to be "organized" groups of individuals where the nature of the relations they bear makes a difference to the powers of these individuals. In contrast, there are other technical senses of "aggregation" such as that defined by Wimsatt (2007) under which an "aggregation" is a group of related individuals whose "organization," i.e. properties and relations to each other, does not alter their powers.

[7] I use the terms "part" and "constituent" interchangeably since I assume individuals are parts or constituents when we have a relation of what I term "constitution."

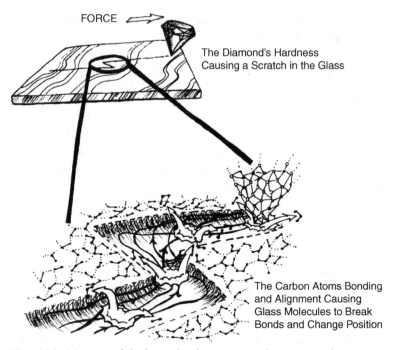

FORCE

The Diamond's Hardness
Causing a Scratch in the Glass

The Carbon Atoms Bonding
and Alignment Causing
Glass Molecules to Break
Bonds and Change Position

Figure 2A Diagram of the lower-level processes at bottom, involving
the carbon atoms, which together implement the process of the diamond
scratching a z-shape in the glass, outlined in the top of the figure.

As the top of the diagram illustrates, scientists provide *intra*-level
productive explanations of effects like the scratching of the glass using
the entities they posit at the same level. Thus they use the diamond
and its properties to explain the scratch in the glass. For when a cut
diamond is pressed with force across a pane of glass, in a Z-shaped
sequence of directions, then the diamond's hardness produces a Z-
shaped scratch in the glass. Here we have a productive process that
is based in the triggering and manifestation of one of the diamond's
powers, contributed to the diamond by its property of being very hard,
which produces a certain effect – a Z-shaped scratch on the glass –
under the relevant conditions.

As an aside, I should note that this common feature of intra-level
productive explanations suggests a flaw in a recent trend in work
on mechanistic explanation to explicitly accept *just* individuals and

processes.[8] We need to explain *why* different individuals base distinct processes. The answer given in the sciences relies upon the differing properties/relations of individuals, and the powers they contribute, which provide the bridge between individuals and the different processes they base and hence account for why distinct kinds of individuals diverge in the processes that they base. There are consequently good reasons why scientific explanations also posit properties/relations and powers alongside individuals and processes.

As the bottom part of the diagram in Figure 2A illustrates, the sciences have also provided elegant, and familiar, compositional explanations in this case, including what are now commonly term "mechanistic explanations" of the higher-level process involving the diamond. The compositional explanation of this higher process works by illuminating how it is composed by the qualitatively different productive processes posited in intra-level explanations at lower levels. For example, the sciences have shown how the many processes based by particular carbon atoms breaking the bonds between, and displacing, specific glass molecules compose, or more specifically "implement," the process of the diamond scratching the glass. For, under the relevant background and triggering conditions, the powers of the carbon atoms, to hold each other in close relative spatial relations and to break the bonds between glass molecules when under pressure, are triggered in a host of carbon atoms. The result is a large number of productive processes involving particular carbon atoms each of which produces breakages in the bonding of specific glass molecules with other molecules, as well as changes in their spatial locations. And these many lower-level processes involving carbon atoms *together* implement, and non-productively result in, the qualitatively different process involving the diamond.

I take such a mechanistic explanation to be only one species of the compositional explanation offered in the sciences, since there are plausibly other compositional explanations that work by explicitly positing compositional relations other than those between processes. For example, the properties/relations of the carbon atoms are taken to "realize" the diamond's property of being extremely hard and such realization

---

[8] For example, the explicit account of Machamer, Darden, and Craver (2000) is framed solely in terms of "entities" and "activities" or individuals and processes in my terms.

Type of Entity	Compositional Relation
Powers	Lower-level powers together *comprise* higher-level powers
Properties	Lower-level properties together *realize* higher-level properties
Individuals	Lower-level individuals together *constitute* higher-level individuals
Processes	Lower-level processes together *implement* higher-level processes

Figure 2B Table of compositional relations between different categories of entity in the sciences.

relations, alongside part–whole relations, are used to explain the diamond's hardness. Here we have a compositional explanation explicitly positing composition between properties and individuals, rather than processes, so mechanistic explanation is plausibly just one species of compositional explanation. In our examples, it is also worth noting that scientists also appear to take the powers of the carbon atoms to jointly compose, or as I term it "comprise," a power of the diamond because the powers of the former individuals base processes that together result in processes individuative of the diamond's power. Finally, the various compositional relations I have noted all only exist under certain background conditions, such as particular temperatures and pressures, amongst many others.

It therefore appears we have an array of compositional relations posited between different categories of entity in our example. These relations are summarized in the table in Figure 2B. Finding such "packages" of compositional relations is less surprising when we note that scientists only posit properties instantiated in individuals, powers had by individuals, and individuals with properties and powers – and wherever one has propertied and empowered individuals scientists accept the potential for productive processes. Scientists thus always plausibly posit "packages" of individuals, properties, powers, and processes together and hence take such entities to be what I shall term "coadunative" – that is, these categories of entity are always found together.[9] Consequently, it is no surprise that the compositional relations between

---

[9] With what strength of necessity are powers, property/relation instances, individuals, and processes (or the potential for them) always found together? My answer is that this necessity is at least nomological necessity. It remains to be established whether such coadunativity holds with some stronger necessity.

these various entities are also found together in "packages" as well and compositional relations of comprising, realization, constitution, and implementation (or the potential for it) are plausibly coadunative.[10]

One last feature of our example that is worth marking, and associated with coadunativity, is that all of the entities posited in the compositional explanations in this case are all individuated, at least in part, by the processes associated with them. Thus hardness and diamonds, and carbon atoms and their properties/relations, are each partially individuated by the processes that result from them. I term such entities "working" entities given their relations to processes, and hence "work." And I also take these entities to be individuated, at least in part, by a productive "role" that encompasses the processes associated with the entity under various conditions.

2.2.2 Voltage-Sensitive Ion Channels and Their Protein Sub-Units: Another Example of Compositional Explanation

For our second case let us consider research from the neurosciences. We know that, under appropriate background conditions, a potassium ion channel plays a key role in a neuron due to its property of being a voltage-sensitive gate contributing the backward-looking power of opening in response to a change in the charge of surrounding cells. As a result, we can appreciate how scientists offer an intra-level productive explanation of why the ion channel opens in terms of the process in which this individual is involved. For when the ion channel is under certain background conditions, such as being in the oily membrane of the cell, and there is a change in the charge of surrounding cells, then the backward-looking power of the ion channel, contributed to it by the property of being a voltage-sensitive gate, is manifested and bases a productive process that results in the ion channel opening. Thus the ion-channel's property of being a voltage-sensitive gate, and the powers it contributes, allows us to explain the opening of the ion channel in the relevant conditions.

---

[10] Note, first, that although the compositional relations always come together this does not mean all compositional explanations explicitly posit all of these relations in the same ways. Second, we always have implementation or the potential for them. And, third, I should mark that I take these coadunative relations be found together with at least nomological necessity. Whether some stronger level of necessity can be established remains to be seen.

Notice that the latter is an intra-level explanation of an effect using other entities at the same level. And we thus have an apparent commitment to individuals, properties/relations, and processes of a distinctive kind at this level. However, once again, researchers did not stop with intra-level explanations. For example, Roderick MacKinnon won the Nobel prize for his work establishing a compelling compositional explanation of how such ion channels open by illuminating the chemical and spatial properties/relations of the complex protein molecules that are "sub-units," i.e. constituents, of these channels. If we focus on why the ion channel opens, when the charge in surrounding cells changes, then the backward-looking powers, to change relative spatial position, of each of the protein sub-units are manifested together and the sub-units all swivel to adopt new spatial relations in relation to each other.[11] These lower-level processes together implement the higher-level process of the ion channel opening. The many properties and relations of the sub-units, such as their alignment and chemical properties, together realize the ion channel's property of being a voltage-sensitive gate. And the powers of the sub-units together comprise the powers of the realized property.

The latter is again a mechanistic explanation, which is a species of compositional explanation positing implementation relations between processes. And, once more, we see that the implementing processes are based by individuals, in the protein sub-units, treated as *parts* of the higher-level individual that bases the implemented process, in the ion channel. The sub-units are also again spatially contained within the ion channel. And the sub-units also again form a collective, and are "organized," since they are spatiotemporally, powerfully, and/or productively related individuals. Finally, it is important to mark that these productions of the sub-units are apparently central to their being treated as parts of the ion channel.

In our second example we thus once more find both intra-level productive explanations, and inter-level compositional explanations, where the latter are again underpinned by an array of apparently connected compositional relations. It is important to note the qualitative

---

[11] Note that there is more to the inter-level story about why the rapid ion flow results beyond the opening of the channel, since the chemical properties of the sub-units facilitate passage of certain ions and retard passage of others. However, I put those complexities to the side here and focus on the opening of the channel.

distinctness of the entities posited at the two different levels of such compositional explanations. Ion channels and protein sub-units share no common powerful properties because the productive processes at the different levels, and the properties and powers that result in them, are also qualitatively distinct. We find processes of changing relative spatial positions at the lower level, whilst we find the process of opening a voltage-sensitive gate at the higher level. Thus sub-units have powers to change their relative positions but not the power to open, whilst the ion channel has the power to open but not to changes its relative position to protein sub-units. Such qualitative differences are found in our first example as well, and we thus see the qualitative distinctness of the entities posited at different levels of compositional explanations long emphasized by writers such as Wimsatt (2007).

Connected to such qualitative differences, it is worth marking that we have good evidence for what I term productive closure across the distinct levels of individuals within a compositional hierarchy. Basically, productive relations within a compositional hierarchy are exclusively intra-level relations. That is, individuals have powers to produce effects in other individuals at the same level. For example, the sub-units have powers to produce effects in the other sub-units such as changing their relative spatial positions. However, none of the sub-units is ascribed the power to produce opening or closing in the ion channel. Instead, it is only the ion channel that is ascribed the power to open or close given its property of a voltage-sensitive gate.[12] Productive relations are thus exclusively intra-level relations within a particular compositional hierarchy.

2.3 Relations of Scientific Composition and their Features

I could easily supply many similar examples of compositional explanation from other disciplines, but my two cases are sufficient to start my examination of scientific composition. The richness and sophistication of the notions we find in compositional explanations poses serious presentational challenges, so I proceed in stages in offering an account. In this section, I lay out some of the common features of

---

[12] I want to emphasize that I am only suggesting we have productive closure within a specific, and hence local, compositional hierarchy.

compositional concepts. In the next section, I then offer a theoretical framework for such compositional relations.

Amongst their features, first, we have seen that the relata of scientific relations of composition are indeed what I earlier termed working entities. That is, the various kinds of compositional relations in the sciences all relate entities that are individuated, at least partially, by the processes with which they are associated and hence by what I am terming roles.

Second, we should note that "vertical" compositional relations are not a species of the productive determination that holds "horizontally" across time between entities. Productive determination is identical to the triggering and manifestation of powers, is temporally extended, occurs between wholly distinct entities, and usually involves the mediation of force and/or the transfer of energy. In contrast, the relations posited in our compositional explanations are not identical to the triggering and manifestation of powers (although implementation has as relata processes that are identical to such manifestations), are all synchronous, occur between entities that are in some sense the same, and do not involve the mediation of force and/or the transfer of energy.[13] I thus use the term "non-productive determination" to refer to the compositional relations involved in compositional explanations, since given my distinction between production and causation this is more accurate than the term "non-causal determination," which I deployed in earlier work.

Perhaps the most distinctive, and slippery, of the characteristics of non-productive determination is that its relata are in some sense the same. Carbon atoms are in some sense the same as diamonds, the ion channel's property instance of being a voltage-sensitive gate is in some sense the same as the properties/relations of the sub-units. Luckily, the sciences concretize this characteristic, and other features

---

[13] It is worth noting how these points apply to implementation between processes. As I marked in the text, the relata of implementation relations are processes that involve the triggering of powers, but the relation of implementation itself is not identical to a further triggering of powers. Furthermore, it appears that at each moment as a higher-level process plays out, then there are lower-level implementing processes that non-productively result in it. Thus, at all the moments over which they play out, the relata of implementation relations are again synchronously related – one never has to wait for the implementing processes to "catch up" and result in the implemented process.

of non-productive determination, in the connected feature I now want to highlight. So, third, note what I term the "mass-energy neutrality" of compositional relations. I take a relation to be mass-energy neutral when its relata have mass-energy, but the overall mass-energy of all the relata equals the combined mass-energy of the entities on just one side of the relation. As we can see in our examples, the mass-energy of a composed individual, properties/relations, and/or process is not additional to (or subtractive from) the combined mass-energy of the relevant component individuals, properties/relations, and/or processes. The mass-energy of the ion channel or diamond thus just is the mass-energy of its parts.

Mass-energy neutrality is a singular feature of scientific composition. In the absence of composition, when a new entity moves into, or comes into being within, a region occupied by certain other entities, then the mass-energy within this region is equal to the mass-energy of the initial entities *combined* with the mass-energy of the new entity. However, when a new entity is brought into being in a region of space-time through a compositional relation, then the mass-energy within the region simply equals the mass-energy of the component entities *alone*. Roughly put, the combined mass-energy of the composed entity *just is* the combined mass-energy of its components because these entities are in some sense the same. Plausibly, its mass-energy neutrality allows scientific composition to underpin the Ontologically Unifying Power of compositional explanation.

Fourth, compositional relations are also plausibly *transitive*. For example, consider the carbon atoms that are found in the diamond. Just as we have compositional explanations of the properties of the diamond using the properties of the carbon atoms, so too do we have compositional explanations of the properties of these atoms using the properties of subatomic particles. Furthermore, we often jump "levels" in scientific cases to gain explanatory power or leverage. So, for example, we can drop down to the subatomic level to gain leverage with a problem at the diamond's level. But this only works if compositional relations like the realization between properties, or the constitution between individuals, are transitive relations. Thus, in general for compositional concepts, if entities A1–An compose entities B1–Bm, and B1–Bm compose C, then A1–An also compose C.

However, fifth, compositional relations only hold in one direction. Scientists take lower-level entities to compose some higher-level entity, but not vice versa. For instance, although we can understand how the carbon atoms and their properties can together non-productively result in, and in some sense "be," the diamond, we cannot understand how the diamond and its properties could non-productively result in, or "be," the sub-units and their properties. As well as being transitive, compositional relations are thus also *asymmetric* in nature. And, sixth, connected to the latter points, compositional relations are also plausibly *irreflexive*, since we do not find scientists taking individuals to be parts of themselves, properties to realize themselves, or processes to implement themselves.

Seventh, as I earlier noted, the entities bearing compositional relations in our cases, and the many like them in the sciences, are usually *qualitatively distinct*. This is the feature that underlies the Piercing Explanatory Power of compositional explanations, since they pierce the Manifest Image by using qualitatively distinct entities to explain the entities of common sense. For example, it is very common for higher-level properties to have "novel" powers not had by any of their realizers and this commonplace feature of compositional relations underpins what we may term "Qualitative emergence" – properties of higher-level individuals not had by the lower-level individuals that are their parts.

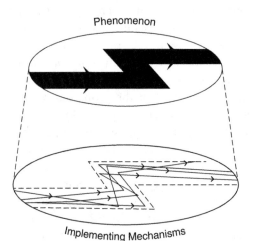

Phenomenon

Implementing Mechanisms

Figure 2C Representation of the basic nature of compositional relations in the sciences – many lower-level entities that, although qualitatively distinct from the composed higher-level entity, together non-productively result in this entity.

It may initially be puzzling how entities of one kind could compose entities of very different kinds and, consequently, it is important to note an eighth general feature of scientific composition. For, as Dennett emphasized in earlier discussions, composition is a joint affair where "teams" of many individuals, properties, powers, and processes studied by lower-level sciences *together* compose the qualitatively different individuals, properties, powers, and processes studied by higher-level sciences. As Figure 2C and my examples highlight, compositional relations in the sciences are thus usually "many–one," with many components and one composed entity. Although each component is qualitatively distinct from the composed entity, the deeper point is that numerous component entities, as a "team" or collective, jointly non-productively result in the qualitatively different entity they are taken to compose. Compositional relations are therefore not relations of identity – many entities cannot be identical to one entity.

Given the foregoing points, a ninth feature of scientific composition is that, under the conditions, components *naturally necessitate* the composed entity – that is, the components suffice for the composed entity in the relevant circumstances. This feature of compositional relations is central to their role in scientific explanations, which plausibly operates by representing these ontological relations taken to hold between entities in nature. We can thus begin to see how to understand the basic explanatory power of compositional explanations: When we have successfully illuminated certain components under the relevant conditions, then we have located lower-level entities that non-productively suffice for, and hence explain the existence of, the higher-level entity under those circumstances.

A crucial question is whether we can say more about how, and why, we have this distinctive natural necessitation, and below I offer some further ideas about its nature. But at this point I should emphasize two points. Note the contrast between the natural necessitation underlying *intra*-level productive and *inter*-level compositional explanations. Intra-level explanations plausibly work by representing productive relations between wholly distinct entities and such relations are often identical to the triggering of the powers of an individual that, under certain conditions, naturally necessitates certain effects. However, the natural necessitation of composition holds between entities that are in some sense the same, and is not identical to the triggering of powers, or the manifestation of force and/or transfer of energy that usually goes along with it. Instead, when we have a compositional relation we have non-productive determination relations between entities that naturally necessitate that we also have certain composed entities under the relevant conditions.

A tenth common feature of scientific composition is that it can give rise to what we may term "multiple composition," whether the multiple realization of properties or multiple constitution of individuals, and so on. My two examples do not neatly illustrate this feature, so consider instead the instances of the same Knoop hardness instantiated in a tooth and a filling. These two instances are realized by the distinct properties and relations of the different constituent individuals we find in the filling and tooth, i.e. one instance is realized by the

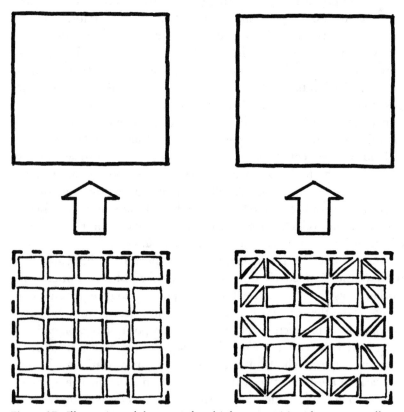

Figure 2D Illustration of the type of multiple composition that we actually find in nature – higher-level entities of the same type often result from distinct teams of lower-level entities of differing types.

bonding relations found with metal atoms, while the other instance is realized by the very different relations found with the constituents of enamel.[14] Given the noted features of compositional relations, we can easily understand why multiple composition is possible. Distinct teams of component entities, as Figure 2D seeks to illustrate, can separately be such that these distinct components each together compose "tokens" of the same type of higher-level entity. (I use "token" to refer generically to particular examples of a certain kind of process, individual, or property.) That is, one may have *multiple* combinations of lower-level entities, i.e. heterogeneous "teams" of components, which together each compose "tokens" of the *same* kind of higher-level entity. We can thus dispel any paradoxes about multiplicity of composition. For example, these broad points about the nature of scientific composition illuminate the simple reasons why predicates referring to multiply realized properties are projectible.[15]

It bears emphasis that whether, and where, we find multiple composition is a further (and empirical) question, but it is a distinctive feature of scientific composition that it allows this phenomenon. The multiplicity of composition has recently been one of the most philosophically controversial features of scientific composition.[16] However,

[14] A variety of Cobalt-Chromium and Palladium alloys used as fillings all have a Knoop hardness in the range 350–400 kg/mm². In addition, much sound, unetched human and bovine enamel (i.e. human and bovine teeth) also has a Knoop hardness in the range 350–400 kg/mm². We thus often find examples where we have individuals made of these alloys, as well as individuals made of human or bovine enamel, which each have instances of the property of having a Knoop hardness in the range 350–400 kg/mm².

[15] Kim (1992b) offers the most prominent argument that predicates commonly understood to refer to multiply realized properties are not projectible, but see also Shapiro (2000) for related claims. Though each of the teams of distinct realizers is different from the other team, nonetheless each group of realizers is such that together they result in powers individuative of the higher-level property. Consequently, a predicate referring to such a higher-level property is indeed projectible despite the instances of this property being realized by distinct lower-level properties. Fillings and teeth both crush grapes, and sunflower seeds, and so on, despite their hardness being realized by different lower-level properties and relations with powers differing from each other and the property of hardness itself. For these distinct teams of realizers still each result in the *same higher-level powers* individuative of an instance of this hardness.

[16] For objections to the existence of multiple realization see, for example, Kim (1992b), Shapiro (2000) and Bechtel and Mundale (1999), amongst others.

such opposition is often motivated by the assumption that multiple composition blocks "reduction." In the next chapter, I show that such an assumption is mistaken, since multiple composition can actually drive scientific reductionism, thus relieving one of the primary reasons for resistance to multiple composition.

Eleventh, in each of our cases of compositional explanation the composition posited only takes place under very particular *background conditions* about whose nature scientists often have either explicit or tacit working accounts. For instance, the temperature and pressure have to be within certain parameters for the carbon atoms and their properties to constitute and realize the diamond and its property of being very hard. Similarly, the properties of the protein sub-units only realize the property of being a voltage-sensitive gate under the background condition, amongst many others, of being surrounded by the oily membrane of the cell.

The role of background conditions has important implications. Even if X1–Xn are components of entity Y on some occasion, then contrary to the assumptions of many philosophers the components X1–Xn do not alone suffice for the composed entity Y.[17] In most cases, components like X1–Xn usually compose Y only under some background condition $. The components of Y thus do not form a sufficient condition for Y and we cannot assume that every entity in a sufficient condition for Y is a component, since such a condition will also include entities that are only background conditions as well as the component entities.

A twelfth common feature of compositional relations is that they always involve individuals bearing *part–whole* relations to each other of a very particular kind, since we saw that such parts base processes that implement the processes based by the individual of which they are parts. In addition, we also saw that individuals that are constituents are

---

Elsewhere I have addressed the recent philosophical objections to multiple realization in detail. (See Gillett (2003c) and Aizawa and Gillett (2009a), (2009b), and (Forthcoming).)

[17] This feature of compositional relations thus supports the conclusion that composition is not a relation of logical necessitation. With what modal strength do the components X1–Xn, under the right conditions $, result in Y? Again, I assume we can say safely that it is at least nomologically necessary that X1–Xn, under $, result in Y. Whether we are justified in accepting a stronger modality remains to be seen.

spatially contained within the individual they constitute. Furthermore, the constituent individuals are always "organized" and bear spatiotemporal, powerful, and/or productive relations to each other. And notice that the latter feature means that composition in the sciences always involves a collective of individuals in spatiotemporally, powerful, and/or productively related groups of individuals. Although it does not take a village, scientific composition does always require a collective of individuals.

In addition, and thirteenth and fourteenth amongst our features, cases of composition also always involve realizing properties and comprising powers. Thus we found that the properties and powers of the carbon atoms realize and comprise those of the diamond, and those of the protein sub-units realize and comprise those of the ion channel, under the relevant conditions.

Finally, the fifteenth common feature I want to highlight is that composition also always involves implementation, or the potential for implementation, between processes. Crucially, the higher-level process, based by the constituted individual, is implemented by the lower-level processes based by individuals that are taken by scientists to be parts of the constituted individual. Thus we saw the carbon atoms, which are parts of the diamond, breaking bonds between glass molecules and implementing the diamond's scratching of the glass. And the sub-units, which are parts of the ion channel, moving in relation to each other implemented the ion channel's opening. It is important to note that the point about the "potential" for implementation is intended to express the fact that the powers of individuals are not always manifesting themselves in processes at either the higher or lower levels. But *if* the powers at the higher level manifest themselves in a process, *then* this process will be implemented by the lower-level processes in which the powers of the parts manifest themselves.

These last four features together frame some of the coadunative aspects of the compositional relations we found in our scientific cases that in earlier work I described as the "Dimensioned" character of scientific composition (Gillett (2002b), (2007a), (2010), (2013)). As the examples highlighted, scientists posit "packages" of compositional relations, in part–whole, realization, comprising, and implementation relations, between the individuals, properties, powers, and processes at different levels. Compositional relations between various categories are thus always found together, as our last four features together imply, and we thus always have "Dimensions" in scientific composition through its always involving relations amongst various categories of entity.[18]

To summarize, I have cataloged fifteen common features of the compositional relations in our examples. These notions are all:

(i)   relations that have working entities as relata, i.e. relations that have as relata entities that are, at least in part, individuated by their relation to processes associated with them.

(ii)   non-productive determination relations that are synchronous, between entities that are in some sense the same and which are not identical either to the manifestation of powers or to the transfer of energy and/or mediation of force;

(iii)   mass-energy neutral relations, i.e. their relata have mass-energy but the overall mass-energy of the relata equals the mass-energy of the entities on one side of the relation;

(iv)   asymmetric;

(v)   transitive;

(vi)   irreflexive;

(vii)   have qualitatively different relata;

(viii)   many–one relations with "teams" of entities composing another entity;

(ix)   such that, under the conditions, components naturally necessitate, i.e. suffice for, the composed entity because the components form an interrelated team that non-productively determines that we have the composed entity;

(x)   allow cases of multiple composition, such as the multiple realization of properties, multiple constitution of individuals, and so on;

(xi)   only hold under background conditions, where the entities treated as background conditions for the compositional relation do not have powers that comprise the powers of the composed entity and do not base processes implementing the processes based by the composed entities;

(xii)   always involve teams of individuals which spatially overlap and bear constitution/parthood relations where the constituent

---

[18] What is the strength of this modality? Once again, I take it to be at least nomological necessity.

individuals bear spatiotemporal, powerful, and/or productive relations to each other and hence form collectives;

(xiii) always involve comprising powers;

(xiv) always involve realizing properties; and

(xv) always involve implementing processes or the potential for them.

This list is somewhat overwhelming and is likely not exhaustive. But it should really be no surprise that the compositional notions used in some of our most successful scientific explanations should be complex, sophisticated, and nuanced. Whether it is comprehensive or not, this list serves my purposes here, since it now allows me to provide a theoretical account of scientific composition and then to critically assess my view against rival philosophical frameworks by considering how well these various accounts do in accommodating the features of scientific composition I have now highlighted.

2.4 Composition in the Sciences: A Starting Framework

The account of scientific composition I outline is based around two broad commitments: the first is that componency in the sciences involves what I term "working components" and the second is that scientific composition is based around what I dub "joint role-filling" relations. In 2.4.1, I provide an abstract discussion of how the general framework built around working components bearing joint role-filling relations captures all the key features of scientific composition. I then build on this work, in 2.4.2, by providing a detailed account of the parthood of individuals, and also a precise theory schema for the realization of properties, focused on such joint role-filling.

In both of its central commitments, the account of scientific composition I articulate has much in common with the earlier of work of writers like Dennett and Fodor.[19] I suggest that this is no accident, since these earlier accounts were crafted in response to the relations these writers found posited in actual compositional explanations

---

[19] The contemporary account closest to my own is that of Shoemaker (2007), which is also role-based, takes compositional relations to be coadunative, and which makes processes central to the nature of entities and their compositional relations providing a "working components" account. However, there is a crucial exegetical challenge in understanding Shoemaker's position, since it is far from consistently clear whether Shoemaker takes compositional relations to be *one–one* or *many–one* relations, and hence whether he takes

in the sciences in what they termed "functional explanations". My account is also crafted in response to such explanations and so unsurprisingly has a similar shape. In fact, I contend that scientific composition involving working components, and joint role-filling, is manifest in compositional explanations and should plausibly be part of any descriptive account.

However, as we saw in the last chapter, it is a matter of philosophical dispute which kinds of relation are posited in compositional explanations. So I need to provide a defense of the centrality of working components bearing joint role-filling relations over alternative frameworks based around manipulability, sufficiency, role-playing, etc. I supply that argument in two parts. I develop my positive account in this section and then, in 2.5, I critically assess rival accounts.

2.4.1 *Working Components and their Joint Role-Filling Relations: A General Account*

Given the finding that entities in the sciences are working entities, I assume that such entities are all individuated by "roles" that, in various ways, connect back to their associated processes. Slightly different kinds of "role" individuate the different ontological categories of entity, in individuals, properties, powers, and processes, given their differences from each other, but I proceed below with a generic notion of "role" to articulate a general framework.

I contend that appreciating that entities in the sciences are working entities implies that scientific composition is best understood as involving "working components." That is, the entities that are components and composed in the sciences are all working entities and their componency is based around the "work" of component entities sufficing for the individuative "work" of the composed entity. How then can we account for why the component entities suffice for the composed entity? Putting the general idea of working components in terms of roles, it appears that working components are entities whose

such relations to span levels of individuals or not. Most plausibly, Shoemaker endorses a one–one, intra-level account close to the role-playing positions I criticize in section 2.5 below, rather than a joint role-filling account. However, if Shoemaker's framework is understood as focused on an inter-level, many–one relation, then Shoemaker's position has affinities with my own.

roles mean that these entities suffice, under the circumstances, for the role that is individuative of the composed entity.

There are presently a number of role-based accounts of working components, but my further contention is that only one of these role-based accounts provides an adequate account of scientific composition. My favored approach, plausibly following in the footsteps of Dennett and Fodor, is based around what I term "joint role-filling" in a relation where under the conditions *many* entities have roles that *jointly* fill the role of, and hence non-productively result in, the composed entity. As I outline below, the joint role-filling approach is presently a minority view, but my central contention is that only an account focused on joint role-filling accommodates all the features of scientific composition. Later, in 2.5, I critically compare such joint role-filling with the majority view focused on the role-playing relation of standard "functionalist" accounts, but let me begin by illuminating how a joint role-filling account covers key features of scientific composition.

To start, the relata of a joint role-filling relation do have mass-energy but are in some sense the same, since the entity individuated by the filled role just is, in some sense, the role-fillers. Furthermore, the "work" of the composed entity, in its individuative role, just is the combined "work" of the role-fillers. The mass-energy of the entity whose role is filled just is the combined mass-energy of the role-fillers. Consequently, joint role-filling is plausibly a mass-energy neutral relation. And joint role-filling is thus also a non-productive relation, since it is synchronous, has relata that are not wholly distinct, does not involve the transfer of energy and/or mediation of force, and it is not identical to the manifestation of powers.

Furthermore, I should note that joint role-filling relations holding between working entities plausibly only hold under certain conditions, since the powers associated with such working entities are usually only had under certain conditions. And joint role-filling is an asymmetric, irreflexive, and transitive relation. In addition, the joint role-filling account provides a nice treatment of the difference between the entities that are components and those that are merely background conditions for some compositional relation to hold. Under the joint role-filling framework, an entity is a component only if it fills the role of the composed entity, whilst an entity involved in a background condition does not fill the role of the composed entity.

We thus accommodate these important features of scientific composition by taking it to be a joint role-filling relation between working components. But it should also be obvious that we can cover another central feature of scientific composition, since we saw that composition is a many–one relation and this is central to joint role-filling which is also a many–one relation. This is important because it allows the joint role-filling framework to accommodate qualitatively distinct relata. Joint role-fillers may all be qualitatively distinct in their roles from the entity whose role they fill. That is, none of the role-fillers may play exactly the role of the composed entity. Instead, the role of the composed entity is only *jointly* filled by the role-fillers together where the roles together suffice for the qualitatively distinct role of the composed entity. In similar fashion, joint role-filling also allows the potential for multiple composition, since many "teams" of distinct entities may together fill the role of the composed entity. Consequently, three more central features of scientific composition are accommodated under the joint role-filling framework.

Next notice that joint role-filling is a non-productive relation of natural necessitation. Under certain conditions, we have interrelated entities that can serve as role-fillers and which hence suffice for, and as we saw above non-productively result in, the role that is the composed entity. Spatiotemporally arrayed teams of entities related by powerful and/or productive relations are such that their relations mean that these entities do, or would, jointly fill the productive role of the composed entity. But this role is individuative of the composed entity. Hence the existence of these lower-level entities, under the conditions at this time, non-productively determines that we also have the composed entity at this time.

The framework of working components bearing joint role-filling relations also provides a promising picture of the distinctive explanations underpinned by scientific notions of composition. If scientific composition is joint role-filling, then this distinctive non-productive form of natural necessitation can underpin an account of compositional explanation as built around the representation of such joint role-filling relations between entities. Furthermore, since joint role-filling allows qualitatively distinct relata, under this framework compositional explanations can also have the distinctive PEP that I have highlighted at various points. In addition, since joint role-filling is also a mass-energy neutral relation, with relata that are in some sense the same, the framework also accommodates the Ontologically Unifying Power of compositional explanations as well. So the joint role-filling framework also provides a plausible account of the key features of the explanations underpinned by scientific composition.

These general points suggest that taking scientific composition to involve working components bearing join role-filling relations is a promising one. I could show how still further features are covered by the framework. But it will be more productive if I now detail how two specific kinds of scientific relation of composition can be articulated within this framework since this is also the best way to show how coadunativity is accommodated under the joint role-filling account.

2.4.2 Working Parts and Working Realizers: Joint Role-Filling and the Composition of Individuals and Properties/Relations

To make the joint role-filling framework more accessible to the reader, as well as to further support its utility, I now want to provide theory schemata for the part–whole relations of individuals and then the realization of properties/relations. Let me start from a different direction by looking at individuals in our exemplar cases that are *not* taken to be parts by working scientists.

Consider a neutrino passing through our diamond and then through our ion channel. As we have seen, accounts from physics and materials science show that carbon atoms are constituents of the diamond; whilst work in cytology, biochemistry, and neurophysiology establishes that certain complex molecules are constituents of the ion channel. In stark contrast, our scientific theories tell us that the neutrino, although spatially contained within both for a certain period, is *not* a constituent of either the diamond or the ion channel.

The contrast between the carbon atoms and sub-units, and the neutrino, is nicely captured by a joint role-filling account. The atoms and sub-units each base processes that implement a process based by the relevant constituted individual – hence these individuals are each counted as parts of, respectively, the diamond and ion channel because they fill the role of the relevant composed individual. In contrast, although spatially within the diamond or ion channel at certain points, the processes based by the neutrino do not implement any of the processes based by these individuals – thus the neutrino is not counted as part of either individual since it does not bear a relation of joint role-filling to the diamond or ion channel. And we also should not be misled by the productive inertness of the neutrino. Even when an individual is productively active within some other individual this does not alone suffice for it to be a working part of that individual (Gillett (2013)). Thus a virus within a cell is productively active, but is not taken by scientists to be a part of the cell because the processes based by the virus do not implement the processes based by the cell.[20]

Appreciating both which individuals are, and are not, taken to be parts in the sciences suggests, putting things roughly and leaving aside other necessary conditions, that individuals are parts or constituents of some other individual s^* when they together fill the role of s^*. We can thus see that constituents in the sciences are plausibly what I am terming "working parts" in individuals, which are members of "teams" spatially contained within some other individual s^* that do, or would, base processes that through their joint role-filling together non-productively do, or would, result in the processes based by s^*.

Once we appreciate the character of working parts, and their joint role-filling relations, then we can also better understand a number of other features of scientific constituents that I noted earlier. First, we can begin to see why the members of a "team" of working parts are always "organized" and bear spatiotemporal, powerful, and/or productive relations to each other. Such relations are central to composition in the sciences, since these relations underlie the integration of the individuals into a "team" and allow these individuals to together accomplish the "work" of, and fill the role of, the constituted individual. Second, we thus confirm that composition always requires a *collective* of individuals in a spatiotemporally, powerfully, and/or productively related, and hence "organized," group of individuals. And, third, the notion of joint role-filling frames a tight connection between the constitution relations of individuals and the implementation of processes, hence accommodating the coadunativity of the compositional relations between individuals and processes.

---

[20] We thus see the problems with Mellor's (2008) notion of a "working part" as simply a productively active individual contained within some other individual. See Gillett (2013) for more discussion.

However, if we reflect a little further on my constant caveat that something is a part if its processes *would* implement the processes of the relevant whole, then we see that constituency relations between individuals are also tightly interconnected with more than just implementation. It is now a familiar point that individuals have properties and powers even when they are not manifesting. And it is important to note that, apparently reflecting this point, carbon atoms are taken to be parts of diamonds, or sub-units to be constituents of ion channels, even when these individuals are *not* basing any processes because their relevant powers are not being manifested. Consequently, a certain carbon atom might now be taken to be a part of a diamond even though the atom is not now basing any processes that implement a process of the diamond. Individuals are thus often parts of other individuals because they have the *potential* to base processes that would implement processes of the higher-level individual and hence fill a certain role.

These observations might appear to generate a puzzle. If we merely have the potential for implementation relations, and hence various role-filling relations, then what is it that *now* underwrites the part–whole relations scientists routinely ascribe? For example, what now underwrites part–whole relations between carbon atoms and diamonds if the implementation relations do not presently exist? In response, we can again appreciate why scientific explanations posit properties and relations, and not just individuals and processes, for relations between their properties/relations plausibly underwrite such potential implementation relations and hence also part–whole relations.

As we saw in the examples, the properties and relations of the carbon atoms, or protein sub-units, contribute certain powers to these individuals that, if they are manifested, base processes that implement the processes that would be based by the manifestations of the powers contributed to the constituted individual by its properties and relations. We consequently take the properties and relations of the atoms or sub-units to *realize* the properties and relations of the diamond or ion channel, and the relevant powers of the atoms or sub-units to *comprise* the powers of the diamond or ion channel. In the next section, I provide a more detailed account of such compositional relations of properties/relations. But the point I want to foreground here is that it is these realization relations, between the properties/relations of the individuals, that underwrite both their potential for implementation relations and hence their part–whole relations. Consequently, we can further confirm the coadunativity of compositional relations in the sciences.

At this point, I can offer a theory schema for scientific parthood/constituency to guide my later work. The account accommodates the coadunativity just noted, but the coadunative character of compositional relations makes presenting the account challenging given its complexity. So I am going to offer a conceptually non-reductive account avoiding explicit reference to any specific compositional relation and simply using relations of non-productive determination through joint role-filling.[21] To this end, consider this schemata:

(Constitution/Parthood – JRF) Individuals $s1$–sn constitute, and are parts or constituents of, an individual s^*, under conditions $, *if and only if* (a) $s1$–sn are spatially contained within s^*, (b) $s1$–sn bear spatiotemporal, powerful, and/or productive relations to each other, (c) the powers of $s1$–sn through their joint productive role-filling together non-productively result in all of the powers of s^*, but not vice versa, (d) the properties/relations of $s1$–sn through their joint productive role-filling together non-productively result in all of the powerful properties of s^*, but not vice versa, and (e) the processes that are or would be based by $s1$–sn, under $, through their joint role-filling non-productively result in all the processes that are or would be based by s^*, under $, but not vice versa.

The account is a little daunting with its many conditions, but it simply frames the features of constitution that I have now illuminated in more precise terms. Conditions (a) and (b) frame the two important features of constituents we have noted in our examples, whilst (c)–(e) articulate the coadunativity of relations of joint role-filling between powers, properties, and processes that I have shown to be required for constituency.

Being built around joint role-filling relations, the schema accommodates the various features of scientific composition (including having relata that are not wholly distinct, being a mass-energy neutral

---

[21] The account is intended to highlight the nature of the joint role-filling picture in a conceptually non-reductive form. In earlier work, I presented my account of constitution in a different format where I offered inter-defined accounts of "realization," "parthood," etc. See Gillett (2007a) (2013). Shoemaker (2007) has a related account of constituency framed in terms of realization.

relation, being a relation of natural necessitation, etc.) in the ways outlined in the last subsection. My theory schema for scientific constituency also performs well with the positive ascriptions of constituency made by scientists in the cases of the diamond and the ion channel, and similar points plausibly hold with a range of cases from a variety of disciplines. And we have also begun to see that the schema also fits the withholding of ascriptions of parthood in cases like the neutrino or virus. The schema is thus plausibly a promising account of parthood/constitution in the sciences.

A similar presentational approach also allows me to present a schema for the realization of properties/relations in the sciences that draws together my earlier observations about this relation.[22] Again seeking to provide a conceptually non-reductive account avoiding explicit reference to other compositional relations, and instead relying on relations of joint role-filling, I endorse this theory schema for realization:

(Realization – JRF) Property instances F1–Fn, in individuals $s1$–sm realize a property instance G, in individual s^* under background conditions \$, *if and only if*, under \$, (a) $s1$–sm are members of, or are identical to, a group of individuals $s1$–sn spatially contained within s^*, (b) $s1$–sm bear spatiotemporal, productive, and/or powerful relation to one another, (c) $s1$–sn through their joint productive role-filling together non-productively result in s^* under \$, but not vice versa, (d) the powers contributed by F1–Fn to $s1$–sm together through their joint productive role-filling non-productively result in the powers individuative of G, in s^* under \$, but not vice versa, and (e) the processes based by F1–Fn under \$ are or would jointly non-productively result in all the processes that are or would be based by G under \$ but not vice versa.

Here condition (a) demands the existence of individuals spatially contained within the individual instantiating the realized property instance and (b) requires that these individuals form a collective by being "organized" and interrelated. Once more (c)–(e), then require that we have compositional relations between the relevant individuals, powers, and processes. Utilizing joint role-filling as it does, the schema

---

[22] Once again, in earlier work I offered my account of realization in a different format where its definition was inter-defined with accounts of parthood, implementation etc. See Gillett (2002b), (2007a), (2010).

again accommodates the features of scientific composition (including having relata that are not wholly distinct, being a mass-energy neutral relation, being a relation of natural necessitation, etc.) in the ways sketched in the last subsection.

It is worth marking why in the schema the individuals in conditions (a), (b), (d), and (e) are allowed to be a *subset* of the individuals in (c). The reason is that it is often the case that only the properties/relations of some of its parts are involved in the realization of one property of some individual, whilst other parts have properties realizing other properties of this individual. Thus different constituents of a kidney realize its properties of rigidity and cleansing blood, respectively. However, all of these parts together constitute the kidney. The conditions in my schema are consequently constructed to allow for this common situation.

The schema for realization nicely brings out why realizers are *working realizers*, since these entities are again members of teams of entities whose roles together fill the role individuative of the composed entity – that is, the associated productions of the realizers do or would implement the productions of the composed entity. And it is also worth noting that the theory schema again appears to nicely capture the positive ascriptions, and withholdings, of realization relations by working scientists in our examples.

2.4.3 A Promising Initial Framework and the Need for Future Work

Understanding scientific composition as involving working components bearing relations of joint role-filling provides a broad account capturing all of the key features of scientific composition. The coadunative character of scientific composition leads to some complexity in providing a theoretical framework for particular compositional relations, but I have now shown that this broader framework provides fruitful theory schemata for the particular compositional relations between individuals and properties/relations that successfully capture the features of such relations, including their coadunative aspects. My work thus plausibly provides a promising theoretical account of scientific composition.

However, I should also immediately note that my framework here only provides a starting account and a lot more work needs to be done to provide a complete framework for scientific composition. For example, a detailed account of the implementation of processes needs to be provided. And we also need to provide a connected account of the compositional levels founded on the kinds of compositional relations I have been examining. In addition, more work needs to be done on articulating the nature of the explanations built around such compositional notions.

Elsewhere I have pursued some of these tasks, but here I have given good general reasons to think that the nature of scientific composition is accurately captured by understanding it as involving working components bearing joint role-filling relations, since this framework accommodates the various features of compositional notions in the sciences. In order to establish that my framework is the *best* account of scientific composition, I now want to return to rival philosophical treatments of scientific composition.

2.5 Rival Views of Scientific Composition and their Problems: Counting the Costs of Metaphysics for Science

Theory appraisal, whether in science or philosophy, is always comparative, so I now want to briefly examine rival accounts of scientific composition. It bears emphasis once again that I am assessing these philosophical accounts as theories about the nature of scientific notions of composition and, in fairness, I should again note that these views were initially developed for other phenomena. When assessing whether these frameworks succeed, or fail, as accounts of scientific composition the appropriate measure is how well they do in accommodating the features of scientific composition, but I show these philosophical accounts fail to accommodate central features of scientific composition.

2.5.1 Why Manipulability (and Counter-Factual Dependence or Supervenience or Sufficiency) Accounts Fail: Vitalism, Manipulability, and Mass-Energy Neutrality

Craver's manipulability account can be interpreted as seeking to provide a sufficient condition for the compositional relations between processes. To start the assessment of this account I am going to focus on a type of hypothesis from historical debates in the sciences that is a paradigm of a position rejecting composition and denying the existence of compositional explanation, including mechanistic explanation. In debates in nineteenth-century biology, there had been longstanding problems in providing compositional accounts of biological entities. Consequently, one of the prominent positions of this period was that of so-called "Vitalists" of different kinds who offered hypotheses under which various biological entities properties or processes were uncomposed by cellular and molecular entities, but where such uncomposed biological entities were still nomologically connected to cellular and molecular entities.

There were a wide variety of differing Vitalist hypotheses, but for my purpose of critically engaging Craver's manipulability account let me stipulate the following as our Vitalist hypothesis. This Vitalist account takes organisms or organs to be biological individuals that are composed by cellular or molecular individuals that are spatially contained within these biological individuals. To this end, we may assume that these biological individuals have some non-biological properties, for example their rigidity or hardness, realized by the properties/relations of cellular or molecular individuals. And we may further assume that processes based by the biological individual's realized non-biological properties are implemented by processes based by the properties of the cells or molecules serving as realizers. However, our Vitalist hypothesis also assumes that there are some *purely biological* properties, like digestion or metabolizing, of organs or organisms that are unrealized and hence uncomposed. Call these uncomposed biological properties "B-properties." And let us take the B-properties of organs or organisms to result in what we may term B-processes of the organs of the organisms. The Vitalist also takes B-processes to be unimplemented, and hence uncomposed, by processes based by the properties and relations of cells or molecules – although the latter individuals and processes are spatially contained within the former individuals and processes. Lastly, this Vitalist hypothesis assumes that there are brute laws of nature that when there are changes in the B-properties and B-processes of organs or organisms, then there are changes in the properties and processes of the cellular or molecular individuals contained within the relevant organ or organism, and vice versa, even though the B-properties and B-processes are not composed by these cellular and molecular properties and processes.

Although rather elaborate, this kind of Vitalist hypothesis has the virtue of highlighting important concerns about Craver's or other manipulability-based account of scientific composition. Notice that under the Vitalist hypothesis cells and molecules plausibly are *parts* of the organ and organism under either a mereological account given their spatial containment, or under some weakened versions of a working parts account, because some properties and processes of the organ or organism are composed by properties and processes of the cells and molecules. So Craver's clause (i) demanding part–whole relations between the relevant individuals involved in the processes is satisfied under various understandings of parthood.[23] And so too is Craver's demand for mutual manipulability. Given the brute laws of nature in play, changes in the properties and processes of the spatially contained cells or molecules result in changes in the B-properties and B-processes of the organ or organism, and vice versa.

Given the latter points, Craver's manipulability-based sufficient condition for the composition of processes is satisfied in the scenario outlined by the Vitalist hypothesis and counts the B-processes as composed, and implemented, by the processes of molecules or cells. The obvious problem is that the Vitalist hypothesis frames a paradigm example where we lack any such scientific composition of processes, so Craver's manipulability criterion is a mistaken account of the composition of processes in the sciences.

In addition, Craver's manipulability-based "norm" for the mechanistic species of compositional explanation is similarly flawed. Under the Vitalist hypothesis, Craver's norm implies we have a good mechanistic explanation of the B-processes based around the processes of cells or molecules, since these processes are mutually manipulable and

---

[23] What if a defender of the manipulability account insists that condition (i) be read using the stronger joint role-filling account of parthood I offered earlier? In this case, given the coadunativity of scientific composition, condition (i) actually demands we have joint role-filling for properties, powers, and processes as well as individuals. Consequently, condition (i) brings along a joint role-filling account of implementation, hence leaving us with a thoroughly ontological account of implementation, which Craver sought to avoid with the manipulability framework. Basically, Craver's position would then be claiming that composition and manipulability suffice for composition – which is certainly true but trivial. I therefore use the weaker readings of parthood outlined in the text in order to assess the ontology-lite manipulability account Craver apparently intends.

involve individuals bearing part–whole relations construed in certain ways. However, the Vitalist hypothesis is a paradigm where we *lack* compositional explanations including mechanistic explanations – a claim the Vitalists and their opponents each explicitly endorsed. So Craver's manipulability-based norm also fails to provide an adequate prescriptive account of mechanistic explanation.

Simply offering counterexamples to an account is rarely fully satisfying or convincing, but my earlier work illuminates the deeper flaws of the manipulability account that lead to these mistaken attributions of compositional relations. As I highlighted earlier, scientific composition involves relata that are in some sense the same where such relations are mass-energy neutral. But a manipulability-based condition can be satisfied by entities that are wholly distinct and fails to guarantee that we have relations that are mass-energy neutral. Thus Craver's manipulability-based condition implies there is a compositional relation between the B-processes and the processes of the cells or molecules. But in this case the B-property is an uncomposed property with its own fundamental biological energy that underpins the uncomposed B-process, so we do not have a mass-energy neutral relation. And the latter fact was why so many Vitalists explicitly endorsed fundamental, and hence uncomposed, vital energies. The latter point means that Craver's account fails to provide an adequate account of scientific composition. Given the connection of these features to the Ontologically Unifying Power of compositional explanation, Craver's manipulability framework also fails to accommodate this central feature of such explanations.

Establishing we have manipulability relations between certain entities simply does not establish that these entities are in some sense the same. And Vitalist hypotheses of various configurations also plausibly establish, for similar reasons, the inadequacy of accounts of compositional relations as identical to relations of counterfactual dependence, sufficiency, or supervenience. Each of these accounts can also be satisfied by relations holding between entities that are wholly distinct where the relations between these entities are not mass-energy neutral. Consequently, these kinds of accounts thus also fail to accommodate key features of scientific composition and hence fail as accounts of its nature. Furthermore, we can see that these accounts once more plausibly fail to accommodate the Ontologically Unifying Power of compositional explanations.

One lesson from the failure of these accounts is that although manipulability, counterfactual dependence, sufficiency, or supervenience relations may underpin *methods* for discovering, or illuminating, the existence of specific relations of scientific composition, it is a mistake to take compositional relations in the sciences to *be* relations of manipulability, counterfactual dependence, sufficiency, or supervenience. Scientific composition is an ontologically richer, and more complex, relation than mere manipulability, counterfactual dependence, sufficiency, or supervenience.[24] A second lesson is consequently that appropriating frameworks for causal relations and trying to shoehorn scientific composition into such machinery fails to accommodate the singular features of compositional relations in the sciences.

2.5.2 Why Standard "Functionalist" Accounts Fail: Role–Playing, Qualitative Distinctness, and a Lack of PEP

In contrast to manipulability accounts, "functionalist" treatments from the philosophy of mind have the great virtue of being role-based, hence accommodating the "working" nature of scientific entities. As a consequence, standard "functionalist" accounts of scientific composition do successfully accommodate the troublesome features of scientific composition that trip up manipulability accounts.

For example, the Flat/Subset view of the realization of properties is based upon what I termed "role-playing" in a one–one relation between property instances that are qualitatively the same or similar in either matching or overlapping in their contribution of powers under some condition. Notice, first, that the realizer and realized properties instances under the Flat/Subset view are plausibly entities that are in some sense the same, since the powers of the one just are powers of the other.[25] Second, I should emphasize that role-playing is also plausibly a mass-energy neutral relation because the processes that result from the powers of the realized property instance just are, or are amongst, the processes that result from the powers of the realizer

---

[24] Notice that the joint role-filling framework does not take composition to be a mere sufficiency relation, rather it takes composition to be a joint role-filling relation, which is a sufficiency relation but has more characteristics than a bare sufficiency relation.

[25] See Pereboom (2011) for a detailed defense of this point.

property instance – hence the "work" of the realized instance just is the "work" of the realizer.

Given the failures we have seen in other approaches with regard to these two key features of scientific composition, in having relata that are in some sense the same and being a mass-energy relation, one can consequently better appreciate the recent popularity of standard "functionalist" accounts from the philosophy of mind in application to various topics in the philosophy of science. Unfortunately, once we delve more carefully into the success of the Flat/Subset view with *other* central features of scientific composition we find problems with a different set of connected features.

To begin, notice that role-playing is a one–one relation and hence fails to accommodate the many–one character of scientific composition. And this is linked to a still more troubling problem, since as a one–one relation role-playing consequently cannot have qualitatively distinct relata. Role-playing works through powers and roles that match exactly or are overlapping – so by its very nature role-playing necessarily has relata that are qualitatively the same or similar. Although role-playing does well with mass-energy neutrality, it therefore fails to accommodate other key features of scientific composition. And it is important to contrast role-playing and joint role-filling in this regard. For the many–one character of joint role-filling allows it to have qualitatively distinct relata because although none of the realizers plays the exact role of the realized it is still the case that the realizers can jointly fill that role.

Sometimes proponents of the Subset/Flat view seek to save their frameworks by challenging my descriptive account of cases of compositional explanation as involving many–one relations or even qualitatively distinct relata. I contend that such alternative interpretations are descriptively defective and lead to all manner of problems.[26] I can most swiftly highlight such difficulties by accepting for argument's sake that scientific composition is neither a many–one relation nor one with qualitatively distinct relata. The problem is that we are then left committed to a PEP-less characterization of compositional explanation. If we take such explanations to be based upon realization relations that

---

[26] There are potential lines of response about my interpretation of the cases based around alternative characterizations using structural properties or states of affairs to interpret cases as involving only one–one relations between

do not relate qualitatively distinct entities, then such explanations lack Piercing Explanatory Power. But PEP is one of the most central, and distinctive, features of compositional explanations, so we see that the proposed alternative descriptive interpretation is mistaken. And we also consequently see another inadequacy of the Flat/Subset view, and other standard "functionalist" accounts, that fail to accommodate the PEP of compositional explanation in the sciences.

Although standard "functionalist" approaches in the philosophy of mind, through their role-based approach, have hit on key insights about scientific composition, the latter points show that the *kind* of role-based relation is important. As a result of the twists and turns of debates over the mind–body problem in philosophy that generated them, standard "functionalist" accounts have ended up focused on role-playing, which fails to accommodate central features of scientific composition and compositional explanation. In contrast, I have shown that joint role-filling, which was plausibly the basis of the empirically inspired strand of "functionalism" in writers such as Fodor and Dennett, comfortably covers all the features of scientific composition.

2.6 A Beginning Framework on a Foundational Issue

My critical work in the last section confirms the suspicions I raised in the previous chapter about accounts of scientific composition that seek to shoehorn scientific composition into frameworks developed for other phenomena. Such accounts routinely fail to pay careful heed to the nature of scientific notions of composition, hence ignoring the first step of the methodology I have followed. But there is a price to pay for ignoring such descriptive work, since we end up pursuing metaphysics *for* science, which we have now seen mischaracterizes scientific composition and misses key features of compositional explanation in either their Piercing Explanatory Power or Ontologically Unifying Power.

In contrast, I have also shown there are substantive benefits that accrue from my alternative approach, which engages the metaphysics of science through the approach of describing scientific notions

property instances of the same individual. I have argued elsewhere that such interpretations of our examples using structural properties or states of affairs are flawed for a variety of reasons (Gillett (2010)).

and then crafting a framework to capture the features illuminated. For I have shown that my joint role-filling account crafted in response to a careful descriptive account of actual notions of composition in the sciences comfortably accommodates all the features of such concepts.

The work of this chapter has been primarily descriptive in providing an overview of the features of scientific composition and then a theoretical account to capture its nature. However, if one endorses scientific realism, then this account also has strong credentials for being an account of composition in nature itself. Basically, the scientific realist takes the success of explanations to be best accounted for by their being true. And our compositional explanations are amazingly successful. Consequently, those inclined to scientific realism have a prima facie plausible justification for taking compositional explanations to use concepts reflecting actual relations in nature. In wider debates in metaphysics over the nature of composition, parthood, or grounding, the framework of working components bearing joint role-filling relations thus offers an important resource and/or competitor.[27]

More pertinent to my concerns in this book, however, is that just this kind of scientific realist justification also appears to implicitly underlie the widespread acceptance of the Scientifically Manifest Image framed in Figure 2E. Through my descriptive work, I have confirmed that scientists routinely posit determinative entities *within* levels, in giving intra-level productive explanations, to explain effects at the same level. In addition, I have now shown at length that researchers also go on to offer compositional explanations that explain the entities at one level by positing compositional relations to the qualitatively different entities at lower levels. Applying the scientific realist approach, and combining the commitments manifest in accepting both intra-level and compositional explanations, we are left endorsing a nature with *multiple* levels of compositionally related, but nonetheless determinative, entities. This plausibly explains why so many working scientists

---

[27] My arguments in this chapter are neutral between scientific realism and anti-realism, since I simply provide descriptive accounts of certain concepts rather than about their truthmakers. However, even when I turn to the claims of both scientific reductionism and emergentism about the truthmakers of compositional explanations, then I should note there are examples of such explanations where we have observable composed and component entities. Any anti-realist troubled about cases of compositional explanation involving unobservables that I discuss should therefore simply focus on such observable cases and all of the important arguments will proceed.

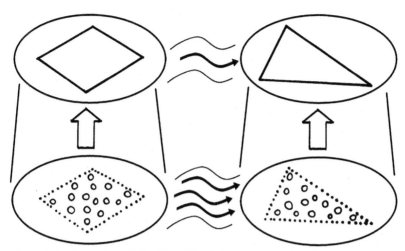

Figure 2E Illustration of the Scientifically Manifest Image of nature involving levels of determinative component and composed entities.

endorse the Scientifically Manifest Image and why so many naturalistic philosophers endorse structurally similar versions of non-reductive physicalism.

The latter points immediately begin to highlight the potential fruits of our better account of scientific composition, since this route to the Scientifically Manifest Image is the *target* of scientific reductionists, who argue it is based on an overly quick interpretation of the import of compositional explanations and provides a mistaken picture of the structure of nature. My work providing a beginning theoretical framework for compositional concepts in the sciences thus promises a fruitful platform from which to illuminate scientific reductionism, so I now turn to that task.

PART II

The Roots of Reduction

3 | How to be a Scientific Reductionist: Defending "Nothing But" Claims and an Ultimate Scientific Image

Scientific reductionism, as Weinberg makes clear, is directly concerned with the ontological structure of nature.[1] And, unlike the many philosophical reductionists fixated on identities, the scientific reductionist accepts the role of compositional concepts in explanations in the sciences and the broad analysis of them offered in Part I. The driving idea of the scientific reductionist is that such compositional concepts ground arguments that show that "Wholes are nothing but their parts" and establish what I term "compositional reductions" in our ontological commitments.[2] But the scientific reductionist still accepts a macro-world and that higher sciences are indispensable for expressing the truths about it. The resulting position is thus very different, as Weinberg repeatedly stresses, from the semantic reduction that philosophers of science have focused upon for a generation, and a fresh approach is therefore needed to understand scientific reductionism.

I begin by using my framework for composition to reconstruct the reasoning underlying the scientific reductionist's claims in a kind of ontological parsimony reasoning I term the "Argument from Composition."[3] It is often claimed that scientific reductionism implausibly entails there are only "atoms in the void," but I show that such a crude construal does a deep injustice. My treatment of compositional

---

[1] For example, Weinberg contends that "reductionism, as I've described it in terms of the convergence of the arrows of explanation, is not a fact about scientific programs, but a fact about nature" (Weinberg (2001), p. 18).

[2] Note that by "compositional reduction" I am not therefore referring to the search for compositional explanation and compositional relations, i.e. everyday reductionism. The latter is what Wimsatt and Sarkar (2006) refer to as "compositional reduction" and their usage is therefore different from my use of the term.

[3] Throughout our discussion, I shall continue to talk of "component" entities under scientific reductionism. By this I mean the entities at the level of entities commonly understood to compose other entities as outlined in section 3.1.2 below.

explanations shows that the reductionist endorses what I term a "Collectivist Ontology" embracing *both* isolated components *and* collective phenomena. I thus show that what I term the "Fundamentalist" account of the scientific reductionist does *not* deny that there is a macro-world, but only offers an alternative account of macro-phenomena to the picture supplied by the Scientifically Manifest Image. Using her Collectivist Ontology, I outline why the scientific reductionist takes compositional explanations, and other statement of higher sciences, to refer to collective phenomena and hence to be true.

Rather than rejecting the macro-world, or endorsing the dispensability of higher sciences, as the semantic reductionist does, the chapter shows that the audacious approach of the scientific reductionist is instead to argue that her Fundamentalist picture provides the best account of the structure of nature and the sciences at both lower and higher levels. And, paying attention to the nature of compositional explanations and its concepts, I show that scientific reductionism plausibly offers a live view of cases of compositional explanation in the sciences, contrary to the received philosophical wisdom.

3.1 Scientific Composition and "Nothing But-tery": Appreciating Ontological Parsimony Arguments Utilizing Compositional Concepts

The basic idea behind scientific reductionism is one that has struck people for centuries even using the less sophisticated making-up concepts of common sense. Famously, for example, parsimony arguments focused upon everyday notions of composition were one of the elements that drove the ontological reductionism of various Buddhist positions.[4] After the Scientific Revolution, the increasingly sophisticated, and pervasive, concepts of composition used in explanations in the sciences evolved. And the working components, and mass-energy neutral relations, utilized in the sciences only deepen the intuitive pull toward ontic reductionism. The goal of the scientific reductionist such as Weinberg is plausibly focused on making us see that, once we reflect more carefully, compositional concepts in the sciences ground unforced, and well justified, ontological parsimony arguments that

---

[4] See Siderits (2007) for an excellent introduction to the main strands of ancient Buddhist reductionism.

establish that we should only accept the existence of components and related entities – thus forcing us to abandon the Scientifically Manifest Image.

To appreciate the reductionist's reasoning consider our compositional explanation in the case of the diamond and the carbon atoms. I start by considering what *individuals* the reductionist argues we should accept and then consider their claims about properties and relations in this case. Taking this to be a local case involving a compositional explanation relating only a pair of levels, and hence only considering the claims made at the level of materials science and chemistry, the competing hypotheses about individuals are easy to state.[5]

On one side, we have the hypothesis of the Scientifically Manifest Image, which claims that we have many carbon atoms, say "n" in number, but one further individual in a diamond which is constituted by these atoms. When we consider the individuals at these two levels, then the hypothesis of the Scientifically Manifest Image is that we have n+1 individuals. On the other side, the scientific reductionist claims that, at least after more careful reflection, we must reject this picture of

---

[5] Potentially, one can focus on compositional explanations involving more levels, or the implications of many such explanations at once, but I start with the simplest case that illuminates both local and global forms of scientific reductionism. And note that in such cases where we compositionally explain a higher-level entity Y using components X1–Xn, then the two plausible competing hypotheses are always:
(i) Only component entities X1–Xn exist.

Or:

(ii) Component entities X1–Xn and a composed entity Y exist.
The claim that (i) and (ii) are usually, at least ceteris paribus, the only plausible hypotheses about such a case may prompt questions about a third hypothesis. This is the claim there is only a composed entity in this thesis:
(iii) Only the composed entity Y exists.
There are good reasons to put (iii) to one side as implausible in the scientific cases at hand. First, we usually have other lower-level *intra*-level explanations positing some, or all, of X1–Xn, in cases where Y plausibly does not exist, so we have evidence that any of X1–Xn can exist without Y existing. Second, given the structure of a compositional explanation, we cannot account for the powers of the components, or the lower-level effects that their manifestation produces, solely using the composed entity Y and its powers. In the local cases we are considering, we thus plausibly have to accept that the components X1–Xn exist. The question is then whether we also accept a composed entity like Y as well, as (ii) claims, or solely accept the components exist, as (i) asserts. (Later I critically discuss global positions that might be thought to hold analogs of claims like (iii) such as Schaffer (2010).)

nature and conclude that we have "nothing but the components" and thus only have n individuals in the carbon atoms.

The reasoning that the reductionist uses to defend her hypothesis, and reject that of the Scientifically Manifest Image, is now easier to understand. As we saw in the last chapter, a good compositional explanation allows one to account for the properties and powers of the constituted individual, as well as the processes and effects that result from the individual's properties and powers, using the properties and powers of its constituents. Thus we compositionally explain the powers and properties of the diamond using the properties and powers of its constituent carbon atoms, but not vice versa. Given the nature of such compositional relations, the reductionist concludes that our compositional explanations mean that we can account for all the powers, properties, and events at both higher and lower levels using the components alone. But given this sub-conclusion, argues the reductionist, we can apply the so-called "Parsimony Principle" in this case – that is, the principle that when we have two equally explanatory hypotheses about some phenomenon, then we should accept the hypothesis committed to fewer entities.[6] We regularly use the Parsimony Principle in the sciences and have good reasons to think we are justified in using the principle in applications to scientific hypotheses.[7] So the reductionist concludes that we should only accept the simpler hypothesis that there are n individuals in the constituent carbon atoms. Putting this more colorfully, and as we shall shortly see a little roughly, the scientific reductionist concludes that we should accept that there is really *nothing but* the carbon atoms.

The scientific reductionist also presses similar claims about other categories of entity, so let me provide a still more precise version of such parsimony reasoning applied to property/relation instances and their realization relations. To this end, consider a general case where once more we have a pair of levels and we have provided a successful

---

[6] The principle has various names including the "Simplicity Principle" and "Occam's Razor," amongst others, in the Western tradition and the "Principle of Lightness" in Buddhist traditions.

[7] There is a fairly large literature on the scientific uses and justification of the Parsimony Principle. In fact, there are now a number of independent justifications of the use of the Parsimony Principle in the sciences, which vary in a many dimensions. A useful recent survey is found in Huemer (2009).

compositional explanation of a higher-level property instance H of an individual s^* using instances of lower-level properties/relations P1–Pn had by the constituents of s^* which are taken to realize H. What properties/relations should we accept once we reflect upon this explanation and its implications?

In answering this question, the scientific reductionist argues as follows. Our compositional explanation allows us to account for the powers and resulting effects of H in terms of the joint effects of the powers contributed by P1–Pn to lower-level individuals. Furthermore, the contributions of the powers of P1–Pn account for all the lower-level effects of the relevant lower-level individuals. Therefore, the reductionist's sub-conclusion is that we can account for *all* of the effects and powers of the relevant higher- and lower-level individuals simply using the contributions of powers by the realizer property/relation instances P1–Pn. But, continues the reductionist, there are two competing hypotheses about the properties we should endorse in this case in light of our compositional explanation. One hypothesis is that we should accept that both H and P1–Pn contribute powers to individuals as the Scientifically Manifest Image implies. The other is the reductionist's favored hypothesis that we should only accept that the component properties P1–Pn contribute powers to individuals. However, continues the reductionist, given the sub-conclusion that we can account for *all* of the powers and effects of the relevant higher- and lower-level individuals simply using the contributions of powers by the realizer property instances P1–Pn, these hypotheses are equally explanatorily adequate. Applying the Parsimony Principle the reductionist therefore concludes that we should *only* accept that the realizer property instances P1–Pn contribute powers to individuals. But it is also true that we should only accept the existence of properties that make a difference to the causal powers of individuals. Therefore we should only accept that the realizer property instances P1–Pn exist.

Though I have only given examples of such reasoning applied to individuals and properties/relations, parsimony arguments can also be applied to powers and processes – and plausibly need to be pursued by the scientific reductionist given the coadunativity of the different categories of entity and their compositional relations. I dub such reasoning the "Argument from Composition," regardless of the category of entity it is applied to. From this point onwards, I also use the term "compositional reduction" to refer only to the brand of reduction that

utilizes the Argument from Composition, or the more complex parsimony argument outlined in later chapters, unless I note otherwise.

As I show in the remainder of this chapter, and in the following one,
such ontological parsimony arguments allow one to support many of
the wider claims made by scientific reductionists such as Weinberg,
Dawkins, E. O. Wilson, and others. I therefore contend that such reasoning provides the best explicit account of the arguments driving the
positions of such writers that I will call, following Cartwright (1999),
"Fundamentalist" views. Most importantly, ontological parsimony
arguments have the shape and conclusion that Weinberg, and others,
articulate for scientific reductionism, since these arguments are driven
by the concepts deployed in compositional explanations and force us
to look downward to components just as the arrow of such explanations leads us to do. Nonetheless, just as Weinberg stresses, the conclusion of such reasoning is not about explanations or theories, but
the ontological structure of nature. And such reasoning allows us to
reduce our commitments about the number, and kinds, of entities in
nature to drive the reductionist's famous claim that "Wholes are nothing but their parts."

It is also useful to mark a division between what I shall term "strict"
and "permissive" scientific reductionists. The strict reductionist accepts
the Argument from Composition in full by accepting the premise of
this reasoning, which Armstrong terms the "Eleatic Principle":

(Eleactic Principle) The only entities that exist make a difference to the powers of individuals.[8]

The Eleactic Principle frames the idea that only determinative entities
exist, since it is unclear what kind of wraithlike existence could be
had by an entity that made no difference to powers. However, what
I term "permissive" reductionists baulk at the Eleatic Principle and the
conclusion that we should reject the existence of composed entities.
Permissive reductionists instead only conclude that we should accept
that components are the only determinative entities in a case of compositional explanation – thus continuing to accept the existence of
composed entities that are not determinative.

---

[8] Armstrong (1997).

Whether permissive scientific reductionism is a position one ought to accept is not an issue I shall pursue further here. It is important to note that strict and permissive reductionists both still accept that component entities are the *only determinative entities*. And this common conclusion suffices for the key commitments of scientific reductionism about wider ontological, semantic, and methodological issues that I explore. In my discussion, for simplicity of exposition, I focus primarily upon the strict reductionist, but I shall try to be careful to allow for the permissive view as well where this is possible.

Overall, the Argument from Composition underpins this claim about compositional reduction:

(Compositional Reduction) If we know entity Y is composed by entities X1–Xn, under \$, then we ought to reduce our commitment to both Y and X1–Xn to a commitment solely to X1–Xn.

The thesis frames how the scientific reductionist contends that merely having a compositional explanation, as both sides in the sciences assume we often do, suffices to establish Fundamentalism. As this thesis also highlights, scientific reductionism, its parsimony reasoning, and the resulting compositional reductions all differ from earlier pictures of "reduction," so let me briefly contrast the features of scientific reductionism and these earlier accounts.

Rather than being focused upon any *semantic* conclusion or sub-conclusion, such as the adequacy or dispensability of explanations, the projectibility of predicates, or the like, the Argument from Composition is thoroughly ontological in nature. In contrast to the reasoning used by semantic approaches to reduction, the Argument from Composition, and similar arguments, is focused on the *ontological* relations posited between *entities*. This reasoning simply seeks to reduce our initial commitments to both parts and wholes, or to realizer and realized property instances, or other categories of composed and component entities, down to an ontology encompassing only parts, realizers, or the relevant component entities. The core reasoning in defense of compositional reduction is thus not committed to an assault on the need for higher predicates, explanations, and theories, and hence need not endorse the greedy reductionism that claims lower sciences are semantically omnipotent. As I outline in the remainder of this chapter, and in the next chapter, this feature looms very large when we examine the wider commitments, and plausibility, of scientific reductionism.

I should also emphasize that the Argument from Composition is unlike many of the recent arguments against mental causation that seek to express similar underlying ideas, although a few philosophers have utilized such arguments on various topics.[9] The Argument from Composition is not focused, at least in the first iteration, on *causation*, or causal exclusion principles, or causal explanation, or concerns about causal overdetermination, but instead argues to a conclusion about the putative implications of the *compositional* relations between the individuals, properties, and other categories of entity posited by compositional explanations.

These points provide a useful backdrop for examining how the Argument from Composition relates to Jaegwon Kim's various formulations of the arguments driving ontological reduction.[10] The important point to note is that one can charitably interpret Kim's Exclusion Arguments as pressing the same underlying parsimony concerns that are articulated by the Argument from Composition. However, I have already noted a couple of divergences, since the Argument from Composition uses scientific concepts of composition, whilst Kim's arguments deploy "functionalist" machinery. And the Argument from Composition comes out and directly addresses the implications of scientific composition, while Kim's arguments instead focus upon explanation, causation, or concepts of supervenience to mediate the underlying concerns (see the papers in Kim (1993a)). And the latter point is significant when one turns to assessing these different arguments, for the Parsimony Principle has long played a successful role in scientific practice and its use can be stoutly defended. However, as the large literature on Kim's arguments makes clear, his alternative theses

---

[9] For example, John Heil (1999) uses the Argument from Composition in arguing against realized properties, albeit assuming the Subset/Flat account of realization.

[10] I want to emphasize that there are a number of strands of argumentation in Kim's corpus, for we shall see later that Kim endorses arguments closer in nature to the Argument from Composition and Completeness. But, for the moment, let me therefore simply mark some affinities and divergences between the Argument from Composition and some of Kim's other arguments.

are far more contentious, less clearly grounded in successful scientific practice, and are open to a number of plausible responses.[11]

I should finally note that these differences mean that Kim's centerpiece account of reduction in the sciences also diverges from compositional reduction in a number of important respects. First, Kim uses standard "functionalist" apparatus in framing his account of "reduction-by-functionalization." Second, Kim's commitment to such "functionalist" machinery plausibly explains why he continues to follow the Nagelian in using an argument based around *identities* to argue for ontological reduction – although only identities between property instances (Kim (1998), (1999)). But, as we have now seen in detail, when one looks at our compositional explanations, then it is very plausible that their compositional concepts are not identities whether between the properties or property instances posited at different scientific levels.[12] However, such points leave the Argument from Composition, and compositional reduction, completely untouched given their reliance solely upon compositional concepts and their avoidance of any use of such inter-level identities.

3.2 Collectives, Aggregation, and the Macro-World of Scientific Reductionism: An Ultimate Scientific Image and its Sophisticated Account of Nature

A common objection to scientific reductionism is that it is left with "atoms in the void," or a "dust cloud" of isolated unrelated components, which is less than even a desert landscape. But given our empirical findings supporting the existence of a highly complex macro-world, concludes the objection, we can therefore reject scientific reductionism.

[11] For example, responses based on illuminating benign forms of causal overdetermination (Bennett (2003)), or using interventionist frameworks for testing causal claims (Shapiro (2010)), have little purchase with the scientific reductionist's parsimony arguments but pose real challenges to Kim's arguments.

[12] One nuance that we need to carefully mark is that Kim (1998) and (1999) uses structural properties, in what he terms "micro-based" properties, which I have put to one side since there are reasons to resist such entities and they fail to ameliorate the problems of standard "functionalist" frameworks in providing adequate interpretations of cases of compositional explanation. See Gillett (2010).

However, the scientific reductionist explicitly denies that she is led to any such arid view of nature. For example, Weinberg tells us that:

> As we look at nature at levels of greater and greater complexity, we see phenomena emerging that have no counterparts at simpler levels, least of all at the level of elementary particles … The idea of emergence was well captured by the physicist Philip Anderson in the title of a 1972 article: "More Is Different." The emergence of new phenomena at high levels of complexity is most obvious in biology and the behavioral sciences, but it is important to recognize that such emergence does not represent something special about life or human affairs; it also happens within physics itself. (Weinberg (1992), p. 39)

Scientific reductionism thus putatively accepts the kind of macro-world that the sciences have illuminated and even goes so far as to endorse "emergence" in the claim that the macro-world, i.e. "more," is different from the micro-world of components in isolation or simpler collectives.

However, the challenge a scientific reductionist faces is to explain *how* to make sense of these claims alongside her acceptance of compositional reductions? One benefit of my exploration of the compositional explanations utilized by the reductionist is that it helps us understand her account of the macro-world and eventually the higher sciences that study it.[13] Central to the reductionist's answer, I show in 3.2.1, are collectives and the aggregation that produces them. For the reductionist accepts isolated, unrelated component-level entities *and also* the collectives of these entities that we saw in the last chapter always underpin compositional explanations. In addition, I outline the scientific reductionist's accompanying account of aggregation in what I term the "Simple" view of aggregation. Building on these points, in 3.3.2, I illuminate the basic picture of the macro-world endorsed by the reductionist in any case of compositional explanation. And I briefly note how scientific reductionism can accept complexity

---

[13] A number of earlier philosophers sympathetic to scientific reductionism have recognized the need to better articulate what this position should say about macro-phenomena. For example, one can charitably interpret Dennett's (1991) famous suggestion that the reductionist needs to accept that there are "real patterns" in nature at the macro-level, as well as component entities, to be directed toward this important issue.

in the macro-world in "multiple composition," feedback loops, and non-linear dynamics. Overall, I thus show how, and why, scientific reductionism embraces a rich and complex macro-world – just not the macro-world the Scientifically Manifest Image portrays.

3.2.1 Scientific Reductionism, its Collectivist Ontology, and its Picture of Aggregation

We must start by recognizing a simple, but foundational, point about the entities we find in compositional explanations that is endorsed by the scientific reductionist. Namely, as Dawkins ((1982), pp. 113–14) emphasizes, in such cases component individuals have formed powerful relations and/or productive relations to each other and hence result in collectives – that is, "organized" and *related* arrays of components, rather than isolated components.[14] Anyone focused on compositional explanation must accept such collectives, since we saw in the last chapter that collectives are always involved in such explanations at the lower level. Rather than just accepting "atoms in the void" in isolated, unrelated component individuals, the reductionist's appreciation of compositional explanations, and other evidence, thus means they endorse the existence of *both* component individuals *and also* collectives of such individuals. Let me highlight this crucial commitment of the scientific reductionist by terming it her "Collectivist Ontology."

I should emphasize that the scientific reductionist does not take a collective to be a *further* individual, for the reductionist collectives are nothing more than teams of spatiotemporally, powerfully, and/or productively related component individuals. Furthermore, the reductionist does *not* take a collective to be a *determinative* entity. The component individuals found in collectives are determinative, but applications of the scientific reductionist's parsimony arguments putatively show that the collective itself is not determinative. Given the latter points, the reductionist does not take a collective *itself* to have powerful properties/relations and hence powers, although the component individuals

---

[14] Remember we are thus using "aggregation" and "aggregate" in the standard scientific way, rather than with the technical meaning of philosophers such as Wimsatt (2007).

within the collective have powerful properties/relations and hence powers.[15]

Nonetheless, the scientific reductionist accepts that collectives exist, since they accept that the sciences have shown that component-level individuals often form powerful relations to each other and such collectives are deployed in compositional explanations. Consequently, we should note that when it is first formed any collective is a *novel*, mind-independent phenomenon in nature. A collective, whether of bonded carbon atoms, interrelated protein sub-units, or some vast collective of microphysical particles, is a real, mind-independent phenomenon that is different from simpler collectives – and utterly different from any "dust cloud" of unrelated individuals or "atoms in the void." We can consequently begin to see how, and why, scientific reductionists such as Weinberg should accept that "More is Different" under their Fundamentalist account for mundane empirical reasons.

First, at the component level, I should note that component individuals in collectives are different in their powers than unrelated component individuals for the simple reason that relations often contribute powers to these individuals. Often forming relations to other components thus reduces the degrees of freedom of this individual. For example, in our exemplar case the properties and relations of the carbon atoms in the relevant collective contribute powers to these individuals that are different from the powers had by isolated or unbonded carbon atoms. And similar points hold for components in all manner of collectives. So for the components in collectives the scientific reductionist accepts that "More is Different."

Second, at the collective level, I should also emphasize that we have novelty. As we saw in the last chapter, the relations between individuals in a collective integrate these individuals in ways that mean that they jointly behave in a novel fashion. For instance, the carbon atoms in a covalently bonded collective, with certain spatiotemporal relations, jointly behave in ways that the unrelated atoms in carbon dust do not behave given their unbonded state. The collective carves a path through nature, and literally carves a path through glass under pressure, which simpler collectives or isolated individuals never do.

---

[15] Notice that the reductionist may take collectives to be *causal* even though they are not determinative. Remember that the concept of causation is wider than that of production and notice that collectives may support counterfactual relationships to each other.

Consequently, at the collective level we have novel joint behaviors that also again mean that even under the Fundamentalist picture, as Weinberg rightly emphasizes, "More is Different."

The latter points bring out both the importance of the Collectivist ontology in understanding the scientific reductionist's Fundamentalist account and also the fact that the reductionist will plausibly be committed to *other* collective phenomena beyond collectives of individuals. The most important of these phenomena are the propensities of various collectives of individuals to jointly behave together in certain ways. Let me therefore define a new term and say that in such cases a collective has a "collective propensity" to jointly behave in certain ways.

Once again, I should emphasize that by ascribing a collective propensity to a collective the scientific reductionist is not positing a property. The scientific reductionist is simply saying that the component properties and relations, of the interrelated component level individuals in this collective, are such that these individuals behave jointly in a certain way, such as the bonded carbon atoms jointly carving a groove through glass molecules under pressure. Collective propensities are therefore not determinative and do not contribute powers – only the properties/relations of the component individuals contribute powers. But collective propensities do exist and are objective, mind-independent phenomena even if they are "nothing more than" component entities together. Given the nature of collective propensities, we can take them to be individuated by the macro-effects that result from the joint behaviors of the collective. Thus two collective propensities are of the same type when they are associated with exactly the same macro-effects under the same conditions.

The scientific reductionist thus accepts collective propensities, as well as collectives of individuals, in their Collectivist Ontology, and we could plausibly articulate similar collective phenomena involving the teams of component processes associated with collectives. But collectives and their collective propensities will suffice for my illustrative purposes here. However, to properly appreciate the Fundamentalist picture of standard scientific reductionists such as Weinberg we finally need to articulate the picture of aggregation they endorse.

This account of aggregation is arguably one of the scientific reductionist's epochal assumptions – that is, one of her deep, unarticulated ontological commitments. As should now be obvious, the nature of collectives and aggregation is a central focus in cases of compositional explanation and hence in our new scientific debates over reduction and emergence. One key issue concerns the powers contributed to component individuals when they aggregate, i.e. when they form powerful and other relations to each other. For example, what determines the contributions of powers by their new properties as component individuals aggregate in this fashion? Articulating the scientific reductionist's stance on this issue is especially important, since it highlights her overall position and the limits to the ways in which "More is Different" under the standard Fundamentalist picture.

Given her commitments, the standard scientific reductionist implicitly assumes, first, that the powers and properties/relations of aggregated individuals are *determined only by other entities at the relevant component level or still lower levels*; and, second, that the determination of these new properties/relations and powers of aggregated individuals is *continuous* across both simpler and more complex collectives. If we put these points in terms of the laws or "principles of composition" that cover aggregation, then the scientific reductionist takes such laws or principles of composition to refer only to component entities (at the same or still lower levels) and to be continuous in nature across all collectives of individuals however simple or complex. We thus have a picture of aggregation with elegantly simple commitments, so I term this the "Simple" view of aggregation.

We can see the commitment of the scientific reductionist to the Simple view of aggregation by noting that it is a precondition for the truth of a Final Theory of the kind explicitly vaunted by writers such as Weinberg (1992) and Wilson (1998). If one assumes that the laws governing the fundamental component entities in the simplest collectives are the complete set of fundamental laws, then one assumes two things. First, when component entities aggregate they must continue to have their contributions of powers determined by other entities in the same ways they are determined in the simplest collectives. Otherwise the laws in the simplest collectives will not exhaust the fundamental laws. And, second, only component entities can ever determine the powers of component entities as they aggregate. Otherwise some entity other than a component will be determinative and there will consequently be some fundamental law referring to this entity, thus implying that the fundamental laws are not exhausted by those illuminated by looking at the simplest collectives, which obviously only refer to the component entities we find in such systems. We can thus see that the truth of the Simple view is a precondition for the existence of a Final Theory.

Unsurprisingly, similar points hold for the related thesis which philosophers term the "Completeness of Physics." To bring out the ways in which it is connected to the key aspirations of scientific reductionism I frame the thesis as follows:

(Completeness of Physics) Microphysical entities are determined, in so far as they are determined, solely by other microphysical entities according to the laws covering their behavior in isolation or the simplest collectives.[16]

Given the nature of the Completeness of Physics, the truth of the Simple view of aggregation is a precondition for its truth for the same reasons that it provides a precondition for the existence of a Final Theory. The Simple view thus articulates the scientific reductionist's general view of the manner in which components form collectives at every scale and this thesis is central to the reductionist's Fundamentalist picture of nature.

Although scientific reductionists such as Weinberg happily agree with scientific emergentists like Anderson that "More Is Different," appreciating the Simple view of aggregation clarifies the limits of this contention. As individuals aggregate, then new collectives and collective propensities come into existence under the scientific reductionist's Fundamentalist picture. Furthermore, as individuals aggregate these component individuals have new powers they never had in isolation or even in simpler collectives, so "More is Different" under Fundamentalism. But the scientific reductionist contends that neither collectives nor collective propensities are determinative – only the component individuals, and their properties/relations, are. Furthermore, the scientific reductionist contends that the contributions of novel powers of component individuals in collectives always exhaustively follow the same laws or principles of composition holding in the simplest collectives and that these contributions of novel powers are only ever determined by other component entities. Consequently, we can begin to

---

[16] I have framed the Completeness of Physics to refer to the laws holding in the simplest collectives, rather than the usual reference to the "laws of physics," to capture the intended content of the thesis that mirrors the scientific reductionist claims.

see how the scientific reductionist can accept that "More is Different" whilst still only taking component entities to be determinative.

3.2.2 The Macro-World of Fundamentalism, its "Composition" and "Levels," Multiple Composition, and Complexity

To better illustrate Fundamentalism's account of the macro-world let me consider an example involving a compositional explanation. Looking at an empirical case would be too lengthy and in almost every instance, as we shall see in later chapters, it would beg important interpretive questions about such cases in dispute with the reductionist's opponents in the sciences.[17] So I am going to use a scientific thought experiment to illustrate this view – that is, an example using well-confirmed concepts, in scientific notions of composition, and not making any background assumptions at odds with our empirical evidence. In this way, I can avoid begging any of the questions I show in later chapters are central to ongoing disputes.

To this end, I am going to look at fancifully named individuals in "blobs" and "globules" to remind us the example is merely illustrative. In the example let me assume that blobs are individuals that can exist in isolation or in simple collectives, where they form bonds to other blobs in a certain way and also have the property, call it P1, of exerting a weak attractive force on other blobs. The nature of collectives of blobs, following the Simple view, is outlined in Figure 3A. In addition, let us assume that when aggregated in the right way in simple collectives, then in their compositional explanations scientists take blobs to constitute the individuals I am terming "globules." Let us further assume that the simple globules that result have a range of realized properties, including the realized property, call it "F," of exerting a weak attractive force on globules in all directions, for this is simply the attractive force that results from the powers together of all the instances of property P1 found in the blobs that constitute such a simple globule. Notice that in this situation, as we aggregate more and more blobs, i.e. as blobs bond

---

[17] If the ion channel case is too complex, the reader might ask, why not look at a case like the diamond discussed in Chapter 2? The answer is that even the nature of such mundane cases as crystals is hotly contested (Laughlin (2005)) in ways that Part III illuminates.

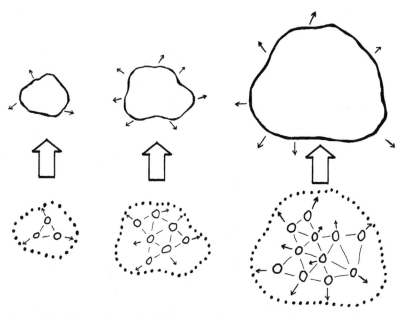

Figure 3A Illustration, with blobs and globules as our example, of the Scientifically Manifest Image of how aggregation and compositional relations proceed under the Simple view. (Small diagonal arrows are forces, dotted lines indicate collective phenomena and vertical enclosed arrows are compositional relations)

to those already bonded to each other, then given her Simple view of aggregation the reductionist assumes that the blobs continue to follow the same principle of composition. That is, as Figure 3A highlights, the newly bonded, as well as existing, blobs contribute powers determined by the same principle of composition applying to simpler collectives and which only refers to other components.

However, the scientific reductionist argues that on reflection we should *not* ultimately accept the existence of composed individuals like globules or their realized properties like F. Applying her parsimony reasoning to the series of collectives in Figure 3B, the scientific reductionist concludes that Figure 3B ultimately better reflects what exists. We do have a higher level, in a macro-world, but it only involves *collective phenomena* rather than *composed entities* as commonly understood in the Scientifically Manifest Image. And the only determinative entities are found only at the lower level in component entities.

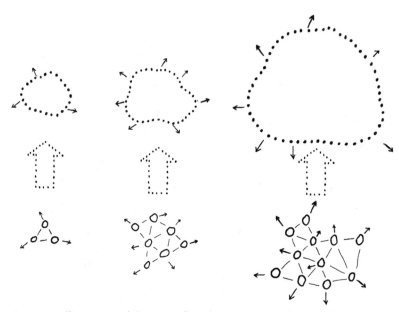

Figure 3B Illustration of the scientific reductionist's conclusions about the actual implications of aggregation under the Simple view, using blobs and globules as our example. At bottom we have related components, in blobs, and at the top we have collectives of entities and their collective behaviors. (Small diagonal arrows are forces and dotted lines indicate collective phenomena)

In contrast to the Scientifically Manifest Image, as Figure 3C below illustrates, under Fundamentalism we have a different picture of nature, but one where we do still have "levels" in a case of compositional explanation where at the lower level we find component individuals, properties/relations, and processes, whilst at the higher level we have collective phenomena in collectives and their collective propensities. And, as aggregation continues, we have greater and greater scales of collectives. Under the Fundamentalist picture, when ultimately understood the terms "compose," or "constitute," or "realize," refer to a relation between components and the collective phenomena they form. And when ultimately understood the Fundamentalist takes a "level" to refer to a particular scale of collective.

Furthermore, Fundamentalism can make sense of the distinctive character of, and differences between, such levels in terms of the differing scales of aggregation, and hence collectives, found at these levels. One of

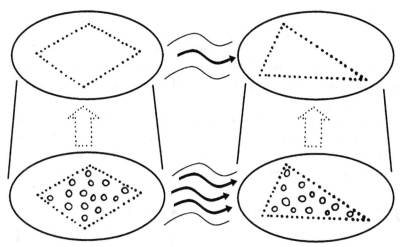

Figure 3C The scientific reductionist's ultimate image of the overall structure of nature. Note that we only have component entities at the "lower" level and that we have collective phenomena at the "higher" level. (Dotted lines indicate collective phenomena, the wavy horizontal line at the top is only a causal relation and at the bottom indicates causal and productive relations)

the defining characteristics, highlighted in chapter 2, is that each "level" has distinctive kinds of individual properties and processes. And scales of collective have characteristic types of collective with differing collective propensities where under some accounts of causation we may take causal relations to hold between such distinctive collectives in virtue of their specific collective propensities, which differ between the scales of collective and hence levels so understood. (Remembering, of course, that there are no productive processes between these collective phenomena since they lack powers.) So under Fundamentalism we have levels with some of the central features of such macro-phenomena.

It is also briefly worth marking some of the various ways in which scientific reductionism can also embrace complexity in the macro-world under this picture. To begin, consider in what form the scientific reductionist can endorse "multiple composition." As I noted earlier, we have good evidence in various areas that we have multiple composition. The scientific reductionist will characterize such multiple composition as macro-homogeneity amongst collective propensities of collectives across micro-heterogeneity amongst the component entities, both individuals and their properties/relations, forming these collectives.

To better understand the Fundamentalist's basic account of multiple composition, consider again the tooth and the filling. In such cases, the scientific reductionist only accepts that we have two collectives of component individuals of heterogeneous kinds bearing heterogeneous properties and relations. With the "tooth" we have different component individuals with different relations and properties than we have with the "filling." As Figure 3D outlines, although the two collectives are built from heterogeneous component individuals, with different component properties and relations, these two teams of diverse component individuals and their properties/relations are each such that

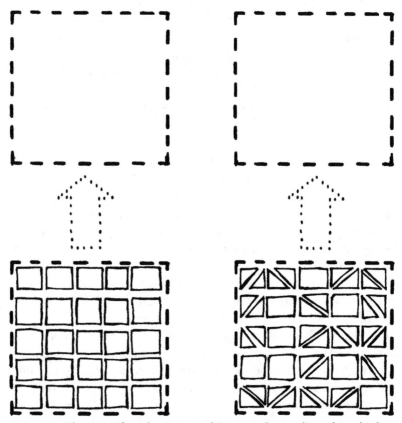

Figure 3D The scientific reductionist's ultimate understanding of "multiple composition" as heterogeneous collectives of component entities that each result in the same collective phenomenon. (Dotted lines indicate collective phenomena)

they jointly result in the same macro-phenomenon in the crushing of grapes, broccoli, etc. The scientific reductionist thus accepts that these distinct collectives share a common collective propensity, for both collectives jointly result in grapes or broccoli being crushed, etc.

For the scientific reductionist the foundation of the relevant macro-homogeneity, and hence multiple composition, is again not the one the Scientifically Manifest Image presses upon us. The reductionist's understanding of such macro-homogeneities, and multiple composition, obviously involves no composed property instance or individual, or any determinative property beyond the component properties and relations. But such macro-homogeneities are still real, mind-independent, commonalities in nature in a common collective propensity found across the relevant collectives. In this way, the scientific reductionist can therefore endorse multiple realization, and the underlying macro-homogeneities across micro-heterogeneities, understood using her Collectivist Ontology.

At this point it is also worthwhile noting some other ways in which Fundamentalism can accept a complex macro-world. To this end, let me carefully mark how, and why, scientific reductionism is different from an influential view in earlier debates that takes aggregation to be *linear*. C. D. Broad (1925) outlined a position he termed "Mechanism," in opposition to varieties of what he termed "emergentism," where Broad's "Mechanism" has affinities to scientific reductionism in taking components to be the only determinative entities, but where this earlier position is implicitly committed to the *linear* character of principles of composition and hence aggregation. But I should note that even in the early twentieth century when Broad was writing scientists had already accepted non-linear accounts of the nature of components across their aggregation. For example, gravitation was taken to be non-linear in character and solutions to many body problems were also found to be non-linear. Given such empirical findings, scientists had thus plausibly abandoned linear views of aggregation even at the time Broad was working. However, such linear accounts of what "reductionism" says about aggregation are still common, so we need to be careful to mark how they relate to scientific reductionism.

To start, I should emphasize that the Simple view of aggregation is not committed to its applications to real collectives being simple – far from it. In real collectives we often have many, many components of many, many kinds that bear all manner of powerful and productive relations to each other. In real examples we will also often have many differing principles of composition applying to the same collective just as a number of laws may apply to the same system, hence piling on further complications. We also know that when we have even small numbers of interacting, or otherwise related, components, then we quickly have situations only describable in non-linear ways. As a result, non-linear frameworks are plausibly necessary for us to understand such collectives and their aggregation. And let me mark that collectives whose components are only describable by non-linear principles of composition *still* fall under the Simple view of aggregation so long as the properties/relations of these components, and the powers they contribute, are only determined by other components in a continuous fashion across all collectives. Under one intuitive, albeit rough, understanding of "continuity" many non-linear frameworks are continuous because under them entities still are determined by the same determinative function across all conditions. So the Fundamentalism of scientific reductionism does not follow Broad's "Mechanism" in its empirically implausible claims that aggregation is always describable in linear frameworks.

There are many other features of a complex macro-world that Fundamentalism can, and does, accommodate. For example, Fundamentalism always acknowledges the role of background conditions given their centrality in the sciences. And let me lastly note that scientific reductionism also plausibly endorses feedback loops for similar reasons. Once again, we have good empirical evidence for such feedback loops in many kinds of component and so scientific reductionists accept such loops at the level of components. The role of background conditions, and feedback loops amongst components, will consequently be reflected in the macro-world of collectives endorsed by Fundamentalism – thus adding still more layers of complexity to the scientific reductionist's picture of the macro-world.

3.3 The Higher Sciences under Scientific Reductionism: Collective Phenomena and the Predicates of Higher Sciences

Appreciating her account of the macro-world, I can now sketch the backbone of the scientific reductionist's understanding of higher sciences and their predicates. This is an important issue to clarify, both since higher sciences are so significant and also given that compositional explanations use the predicates of higher sciences. As we have seen, the scientific reductionist *endorses* compositional explanations and uses them to craft her view of nature, but the reductionist also rejects the common construal of compositional explanations associated with the Scientifically Manifest Image. So it is important to clarify the scientific reductionist view of higher sciences and compositional explanations in particular.

Given her view of the macro-world, we can better appreciate what the scientific reductionist ought to say about the predicates and sentences of higher sciences. The basic point is that the reductionist will plausibly endorse *two* ways of understanding the sentences of higher sciences – one associated with the Scientifically Manifest Image, the other deriving from the reductionist's own account of the macro-world. There are various competing frameworks in the philosophy of language I could use to frame these two stances, so I put the relevant distinction quite simply to be neutral amongst these positions.

On one side, there is what I term the "common understanding" of the subjects and predicates of sentences of the higher sciences that reflects the usages of scientists in everyday practices of discovery and the Scientifically Manifest Image. In their research, scientists routinely, and quite sincerely, take the predicates of higher science sentences like "Ion channels are voltage-sensitive gates" to be about an instance of a realized, higher scientific property instance such as being a voltage-sensitive gate. Working scientists thus take such a property instance to be a unitary, composed entity that contributes higher-level powers that manifest in processes such as an ion channel opening. Similarly, under the common understanding, the subjects of such higher science sentences are taken to be unitary, composed higher-level individuals, like ion channels, which are taken to have such realized properties and to base associated higher-level processes. However, the scientific reductionist denies the existence of composed higher-level processes, properties, and individuals, as commonly understood, and hence rejects the Scientifically Manifest Image as mistaken – thus taking higher science sentences, such as compositional explanations, to be *false* or *meaningless* when commonly understood.

My work in the last section shows that the scientific reductionist should not simply defend this negative point, but should also offer an alternative positive account of higher science sentences as describing the macro-world as the scientific reductionist understands it. This involves what I term the "ultimate understanding" of predicates of sentences in higher sciences built around the Collectivist ontology of the scientific reductionist's Fundamentalist position. Under this ultimate understanding, the predicate of a sentence of a higher science, for example "Ion channels are voltage-sensitive gates," is about a collective propensity. And, under the ultimate understanding, the subject of such a higher science sentence is about the collective of component individuals having this collective propensity. Rather than higher science sentences being about composed individuals, properties/relations, and productive processes, under the ultimate understanding such sentences are therefore about collectives, collective propensities, and often the causal relations between them.

Applying the distinction between the common and ultimate understandings of higher science sentences, we can see that the scientific reductionist will consequently often take such sentences to express truths. For when *ultimately understood*, sentences of higher sciences articulate truths about the collective phenomena that the scientific reductionist argues that the sciences have shown to exist and which comprise the macro-world. We can thus begin to appreciate scientific reductionism's view of higher sciences as disciplines needed to study collective phenomena - expressing truths about collective phenomena, articulating the causal relations involving them, and framing the true explanations and laws holding of these collective phenomena.

These points also apply to compositional explanations. A common complaint often raised about scientific reductionism is that it uses compositional explanations to drive its picture of the natural world but simultaneously undercuts these explanations because it shows there are no composed entities or compositional relations. Hence, concludes the complaint, scientific reductionism is self-defeating or at least inconsistent. But we can now see where this complaint goes awry. It is indeed the case that the scientific reductionist argues that we ought not to accept compositional explanations as *commonly understood*, since we ought not to accept that composed entities either exist or are determinative. However, the scientific reductionist still takes compositional explanations when *ultimately understood* to articulate truths about nature and the components, collectives, and other entities that it contains. Hence we can see why the scientific reductionist endorses compositional explanations, and uses them in her arguments, since she accepts that when ultimately understood such explanations are true, although one of her main goals is to show that when commonly understood these explanations are false or meaningless.

One might also be puzzled about how scientists using the common understanding of higher science predicates, which leaves these researchers expressing or thinking false or meaningless claims, could be such a successful practice. But my work now makes this far from puzzling. A compositional explanation, even if commonly understood, still has predicates that track collectives and collective propensities when these predicates are ultimately understood. Although scientists mistakenly take there to be composed individuals, and composed properties/relations, using the common understanding of higher science predicates still allows working scientists to successfully construct explanations that express truths about collective phenomena when the predicates of these same explanations are ultimately understood.

We can thus see what the scientific reductionist assumes about scientific explanations in the higher sciences and the ontology that underpins their success. And we also begin to get an inkling of how, and why, a scientific reductionist takes higher sciences to be significant. In the next chapter, I explore the scientific reductionist's view of higher sciences in more detail and show that scientific reductionism should even accept that sentences of higher sciences, when ultimately understood, are *in principle indispensable* for expressing many truths about nature. The scientific reductionist's novel account of the macro-world thus opens up the possibility of an account of the higher sciences as significant, and even indispensable, disciplines that better fits what we know about these sciences – in sharp contrast to the Positivist model of reduction that implies lower sciences are omnipotent and higher sciences are dispensable.

At this point, I can now explain how, and why, I contend scientific reductionism passes an important hurdle in what I term the "Meta-Explanatory Adequacy Test" or "MEAT." MEAT is a criterion of adequacy and states that a view about the structure of a compositional explanation, and/or nature, is acceptable *only if* this view can successfully show that its ontology is compatible with, and accounts for, the success of the relevant scientific explanations. One might think that passing MEAT is a simple matter, but it is actually a challenging requirement because it involves both crafting an ontology for the relevant cases of explanations and then showing that this ontology would explain the success of our actual explanatory practices in the sciences.

With regard to the latter burdens, scientific reductionism has a built-in advantage that allows it to make passing MEAT look far easier than it really is. The advantage of scientific reductionism is that it starts with scientific explanations, in particular compositional explanations, and bases its ontology on the entities posited in such explanations at the lower levels. Although scientific reductionism argues against the existence of the entities posited at the higher level in compositional explanations as they are commonly understood, i.e. composed entities, scientific reductionism retains the entities posited at the lower level. Using these components, and crucially the collective phenomena that they form, the scientific reductionist can deploy her Collectivist Ontology in the ways I have now begun to sketch to account for the success of compositional explanations and the truths they often express – hence passing MEAT with flying colors.

Overall, scientific reductionism thus provides a far more plausible picture of nature and the sciences than the Positivists provided under their semantic model. Crucially, the scientific reductionist does not deny the existence of a macro-world or the need for higher sciences to study it. Instead, scientific reductionism simply seeks to show that its account of these phenomena is better than that provided under the Scientifically Manifest Image. And I have begun to show that the Fundamentalist framework of scientific reductionism provides a rich and plausible account of the macro-world as consisting of collective phenomena. And, as we have now also seen, scientific reductionism consequently passes MEAT by providing an account of the success of compositional explanations, and other explanations in the higher sciences, ultimately understood as focused on such collective phenomena.

3.4 Objections to Scientific Reductionism: An Initial Assessment

With a basic framework for scientific reductionism in place, we can now examine some common objections to this view. I consider a range of arguments that have been made to "reductive" positions of various kinds and assess whether they work against scientific reductionism. I suggest that appreciating the novel views, and richer resources, of scientific reductionism reveals that it has plausible responses to all of these concerns. Most importantly, we need to clearly distinguish the scientific reductionist's ontological reduction from the semantic reductionism philosophers have often focused upon, and also factor in the scientific reductionist's rich account of the macro-world using her Collectivist Ontology and her connected account of the higher sciences.

3.4.1 The Mere Metaphysics Objection

One response to scientific reductionism mirrors that offered to philosophers defending ontologically reductive conclusions about the higher sciences. For example, when Jaegwon Kim's arguments attacking the causal efficacy of mental properties are extended to properties of the higher sciences, then it is often argued in response that we may ignore such *mere metaphysics* as irrelevant, since the reasoning clashes with successful scientific practices.[18] In the same way, it can be objected that successful scientific practices posit composed entities, so this trumps merely metaphysical arguments like the Argument from Composition.

The underlying idea here is apparently that ontological reductionism is based "on a naïve and inaccurate conception of the practice of physics, and the relationship between physics and metaphysics."[19] Even putting the manifest Exclusivist bias and solitary focus on physics to one side, my work shows that such contentions find no application with the scientific reductionist. For the type of argument driving her ontologically reductive conclusion is not one offered by metaphysicians, but by working scientists. Furthermore, this reasoning is not narrowly, and hence perilously, rooted in physics alone, but is based upon a variety of scientific explanation found across the full range of successful sciences.

Parsimony reasoning like the Argument from Composition, or similar arguments, cannot therefore be dismissed as "mere metaphysics" or impositions of alien metaphysical doctrines or notions upon

---

[18] For instance, see the "defense manual" offered to cognitive scientists by Ross and Spurrett (2004) against Kim's "metaphysical" arguments.

[19] Ross and Spurrett (2004), p. 603.

the sciences. Such reasoning arises from within the sciences using the concepts used in successful scientific explanations. And we can consequently charitably interpret Kim's pioneering work as seeking to articulate this scientific approach, albeit through an unfortunate "functionalist" framework. Though previous attempts to articulate ontologically reductive arguments may have imported alien concepts, I have now shown that the ontologically reductive claim that "Wholes are nothing but their parts" can be given a scientifically grounded defense that scientists themselves have long appreciated even if philosophers continue to overlook it.

3.4.2 *Concerns about the Parsimony Principle*

Two kinds of objection focus on the use of the Parsimony Principle in driving compositional reduction. The first concern is that use of the Parsimony Principle is unjustified in philosophy and hence that one of the key premises of the Argument from Composition cannot be trusted. This type of worry about the use of the Parsimony Principle has recently been ably pressed by Michael Huemer (2009). However, regardless of our assessment of Huemer's success in challenging the use of the Parsimony Principle in philosophy, we can see a very simple response for scientific reductionism to make to such concerns. As I noted above, it is commonplace in the sciences to judge hypotheses using the Parsimony Principle. A reductionist such as Weinberg is thus simply *continuing* to apply the Parsimony Principles to assess scientific hypotheses in the ways the principle is often used in the sciences and where even critics like Huemer are happy to accept we are justified in using the principle given its past success. We can therefore see that it is very important to appreciate the nature of scientific reductionism and its core reasoning. For the mistake behind this type of worry is to misplace compositional reduction as a philosophical, rather than scientific, enterprise.

However, a second kind of objection focuses on the scientific reductionist's use of the Parsimony Principle as a part of a scientific project. This concern can take two forms. One variant of such a worry presses the claim that any use of the Parsimony Principle in the sciences is unjustified. The other variant focuses upon the scientific reductionist's particular use of the Parsimony Principle in reasoning like the Argument from Composition and challenges whether these particular applications are legitimate. Let us consider these versions in turn.

In response to the general concern about the justification of the Parsimony Principle in general, I should again note that we have overwhelming evidence from successful scientific practice about the utility of using the Parsimony Principle and this success provides the most obvious justification for its continued use. And let me reiterate that there are a number of other plausible justifications of the Parsimony Principle. It therefore appears plausible that the scientific reductionist is on firm ground in deploying the Parsimony Principle given its continuing role in scientific inquiry.

But what about the second variant of the worry focused on the reductionist's particular use of the Parsimony Principle? In response, we immediately need to ask what the precise worry is, since we cannot assess the problem until it is spelled out. It appears the objector is concerned that something is amiss with the Argument from Composition, so I should note that I will critically examine this reasoning at length in Parts III and IV. I ultimately suggest that the real problem is not that the Parsimony Principle is unjustified, but that the Argument from Composition in which it is deployed can be seen to be an invalid argument. Without an alternative understanding of the objection, it appears likely that this is the problem that the objection is getting at, so I defer further discussion of the concern until later chapters where I lay it out in detail.

3.4.3 Interventionist and Other Responses to the Problem of Mental Causation

Kim's arguments against mental causation have garnered a great deal of attention and many responses have been offered to them. Here I can only address the most pertinent responses and note why I put the others to one side. First, it bears emphasis once more that Kim's various arguments focus on causal explanation and causal overdetermination whilst the scientific reductionist's challenge is centered on scientific composition. And this difference is significant. For example, important responses using the interventionist approach to causal relations have been given to Kim's claims about causal explanation in the higher sciences by writers such as Larry Shapiro (2010),

(2011) and others, and plausibly show that Kim's reasoning fails to establish that causal explanations in the higher sciences are not true or legitimate.

But I should immediately note that the scientific reductionist plausibly accepts the conclusion of these interventionist arguments, for this chapter has outlined how the reductionist accepts there are many true causal explanations illuminated by higher sciences, though the reductionist takes such explanations to be true under her ultimate understanding of their predicates as referring to collective phenomena. Consequently, recent interventionist responses to Kim's arguments about causal explanation do not translate into rebuttals of scientific reductionism that accepts their conclusions. The scientific reductionist challenge starts with the truth of such explanations and focuses instead on the nature of the truthmakers for such claims.

In addition, there are many, many other responses to Kim using standard "functionalist" machinery that consequently translates into what I have now shown is an inadequate account of scientific composition that mischaracterizes its features. And such mischaracterizations of scientific composition mask the deeper nature, and extent, of the scientific reductionist challenge, which is driven by reflection upon compositional concepts in the sciences.

To see how the use of "functionalist" machinery mischaracterizes what is required to answer the reductionist, consider the writers, such as Kim himself (1998), who under the influence of the Flat/Subset model of realization wrongly suggest that because properties posited in sciences like chemistry contribute novel powers this protects their status from the ontological reductionist in contrast to "functional" properties whose powers overlap with those of their realizers. Unfortunately, we have now seen that the novel powers of realized properties are ubiquitous across cases of compositional explanation at all levels of the sciences and do not protect the relevant realized properties from the scientific reductionist's arguments. (For more on this point see section 3.4.6 below.) I therefore leave to one side the range of responses to Kim's arguments using "standard" functionalist machinery. Furthermore, I should also note that one or other of the scientific reductionist's responses to the objections I outline in this chapter can plausibly be extended to rebut such responses as well.

3.4.4 Ontological Arguments from the Sciences (I): the Multiple Realization Argument

If we look to the sciences for reasons to rebut scientific reductionism, then we find a range of arguments seeking to use ontological features of scientific explanations or findings to object to "reductionism." I divide these arguments into two groups. In this section I consider the most famous of these objections in the Multiple Realization Argument. Then in the next section I consider other objections. As we shall see, once we distinguish the ontological nature of scientific reductionism from the semantic reduction of the Nagelian, and also appreciate how scientific reductionism is rooted in compositional explanations and their features, then we can see how all of these concerns are comfortably handled.

Taken on the terms of antireductionists in the philosophy of science, such as Putnam, Fodor, Kitcher, or Wimsatt, the phenomenon of multiple realization grounds a powerful argument against the identities needed by the Nagelian model. And I should note that just this type of concern has been directly pressed against Weinberg's scientific reductionism, in one of the few discussions of it in philosophy, by the team of the prominent scientific emergentist Stuart Kauffman and the philosopher Phillip Clayton.[20] So let me briefly reprise where my earlier work leaves the Multiple Realization Argument as an objection to scientific reductionism.

Each of the examples of multiple realization is still a case in which we have a compositional explanation positing a realization relation – that is, each is an example where we commonly understand the situation as involving a team of component entities jointly filling the role of the composed entity, in this case a realized property instance. But, given this point, the scientific reductionist has an obvious response to the Argument from Multiple Realization. For ontological parsimony reasoning, like the Argument from Composition, can thus be mounted *using* the evidence about multiple realization supplied by compositional explanations, since multiple realization is built out of the application of the concept of realization that drives such reasoning. I should also note, as I outlined in

---

[20] Kauffmann and Clayton (2006), pp. 511–13, where they refer in their objection to the "multiple platforms" argument. It is worth noting that in the course of their argument, Kauffman and Clayton (p. 512) wrongly ascribe a "greedy" form of semantic reduction to Weinberg, thus once again missing the nature of his position.

3.2.2, that the scientific reductionist can also plausibly supply their own account of multiple realization through their Collectivist Ontology.

Although the Multiple Realization Argument of Fodor, Kitcher, and Wimsatt plausibly undermines the widespread existence of inter-level identities in the sciences, and hence the existence of Nagelian reductions, scientific reductionism relies on no such identities. Contrary to the claims of Kauffmann and Clayton, the Multiple Realization Argument thus poses no threat to scientific reductionism. In fact, the phenomenon of multiple realization, or any other form of multiple composition, simply supplies ammunition, in posited examples of the applicability compositional concepts, which the scientific reductionist can use to run her core arguments for compositional reduction.[21]

3.4.5 Ontological Arguments from the Sciences (II–X): Objections from Environmental Conditions, Reductions in Degrees of Freedom, Organization, Novel Powers/Properties, Higher Laws, Autopoiesis, Non-Linearity, or Feedback Loops

There are a welter of lesser objections from ontological features of higher sciences that have been broached as objections to "reductionism," so let me swiftly consider eight of these to show that scientific reductionism can handle them with aplomb. A number of writers have pressed the idea that the role of environmental or background conditions, i.e. entities often beyond the boundaries of the object phenomena, poses a problem for reductionism. Thus, for example, John Dupré (2008), (2009) persuasively argues that environmental conditions are crucial in many cases where a biological entity, Y, has been shown to be composed by certain biochemical entities, X1–Xn. And acceptance of this point flows from the features of the compositional explanations we find in these areas as our exemplars illustrate, since background conditions always play a crucial role in such explanations.

The difficulty for the objection is that we have now seen that the scientific reductionist plausibly accepts the role of such environmental

---

[21] And the multiple implementation that Kauffman and Clayton (2006) claim is found in certain biological entities is open to the same treatment as multiple realization, so too is the multiple constitution of individuals in biology and elsewhere (Gillett (2013)).

factors including the entities that are background conditions. First, we should note that the scientific reductionist will argue that the distal entities are also at the component level and productively interact with the components X1–Xn commonly understood to compose Y. Second, given the latter point, the scientific reductionist can plausibly press their compositional reduction of our commitment to Y and X1–Xn, down to X1–Xn, whilst accepting that other entities at the component level, although in the environment, also play a determinative role alongside X1–Xn. Thus scientific reductionism accepts the role of entities at the component level in the wider environment alongside components.

Similar points provide responses to objections that scientific reductionism faces a problem with reductions in degrees of freedom of components, or because "organization" is central in compositional explanations (Sperry (1986)), or that the latter show that we have novel powers or properties at higher levels (Kim (1998)). Once again, we have now seen that scientific reductionism plausibly accepts all of these phenomena, but still runs its core parsimony arguments so the concerns are unfounded. For reductions in degrees of freedom, we saw earlier the reductionist takes these to result from other component-level relations or properties. With regard to "organization," as I noted earlier in Chapter 2, this is charitably understood as making the point that the properties and especially the relations of component individuals are central in cases of scientific composition. But the scientific reductionist accepts this point, for the role of relations at the component level is central to their distinctive Collectivist Ontology. But this in no way impedes their core parsimony reasoning and so "organization", and reductions in degrees of freedom, by themselves pose no problem for scientific reductionism.

Similarly, we have also seen that the scientific reductionist accepts that composed entities, as commonly understood, are routinely taken to have novel powers and properties. But the compositional relations that go along with such Qualitative emergence drive the reductionist's parsimony arguments and hence fuel the position rather than posing a problem for it. Furthermore, we have also seen how the scientific reductionist's Collectivist Ontology accommodates higher levels, ultimately understood, involving novelty and the fact that "More is Different" – albeit under their own proprietary understanding of these phenomena. So the common understanding of compositional explanation as involving novel powers, properties, or other higher-level entities also poses no problem for the reductionist.

In an analogous fashion, it is also sometimes objected that the existence of higher laws, autopoietic systems, non-linear frameworks, or feedback loops in the sciences undermine "reductionism." With regard to higher laws, I show in the next chapter that the scientific reductionist embraces such laws and my comments on her stance toward higher sciences points to how she may do this. But I defer detailed discussion to the next chapter.

As to autopoiesis, it has been used by scientists and philosophers as a reason why reductionist arguments do not apply (Van Inwagen (1990)), but I should note that the scientific reductionist accepts that there are autopoietic collectives – for example the reductionist accepts that the collective of molecular-level individuals that is commonly understood to compose a eukaryotic cell is autopoietic. As a result, the question is why autopoiesis provides a reason to accept the existence of composed entities as well as the autopoietic collectives commonly understood to compose them. Unless more is added to such objections, it appears that the scientific reductionist's core reasoning again applies in these examples and the existence of autopoiesis by itself poses no obstacle to the scientific reductionist.

With regard to non-linear frameworks, as I noted above, philosophers such as C. D. Broad articulated reductionist positions committed to the linearity of aggregation in the early twentieth century. But the lower sciences have long since shown that aggregation is very often non-linear in nature and non-linear frameworks for components in collectives abound. Reflecting these empirical findings, the reductionist's Simple view of aggregation is thus solely committed to the same principles of composition (and laws) holding in complex collectives as hold in the simplest collectives and to aggregation being continuous, where this plausibly encompasses cases where non-linear frameworks apply.

Lastly, the existence of feedback loops with various phenomena in the higher sciences is often broached by scientists and some philosophers as a problem for reductionism. This objection appears to be founded, like the last one, on the assumption that the reductionist must take aggregation to be linear. But, again, the Simple view of aggregation entails no such feature. And feedback loops are often found with components and so, in principle, the existence of positive or negative feedback loops poses no problem for the scientific reductionist. The reductionist can happily accept that positive and/

or negative feedback between entities *at the component level* is central to many complex collectives, and the processes in which they are involved, and that application of non-linear frameworks is necessary for us to understand such cases.

Overall, we therefore see that once we appreciate that the scientific reductionist accepts the centrality of compositional explanations, and embraces their features and associated empirical findings about nature, then we find that the obvious ontological characteristics of these, and other, explanations in the sciences simply do not provide the quick knock-out objections that have been effective against semantic reductionism or the reductionist positions of earlier periods.

3.4.6 The Objection from the Indispensability of Higher Sciences

There is one last, and very important, kind of objection to reduction in the philosophy of science – namely, that higher sciences, and their predicates, explanations, laws, and theories, are indispensable. Thus, for instance, the famous Predicate Indispensability Argument pressed in different ways by Putnam, Fodor, Kitcher, Wimsatt, and others, is taken to show that reduction fails. In response, the differences between semantic and compositional reduction loom large.

It is true that semantically greedy Nagelian reduction entails the dispensability of higher sciences, so establishing indispensability cripples such a view. But we have seen that scientific reductionists such as Weinberg stress that scientific reductionism is thoroughly ontological in nature and claim to spurn such a greedy account of the role of lower sciences. And my budding account of the scientific reductionist stance on the sentences of higher sciences appears to support such an appraisal. Using her Collectivist Ontology, we have seen that the scientific reductionist can accept that higher science sentences are *true* when ultimately understood. Scientific reductionism therefore may well have the makings of a more plausible account of higher sciences and hence be able to respond to objections based on the indispensability of higher sciences.

However, despite these promising signs, I should note that I have yet to articulate a full response. Many readers will quite rightly ask how higher sciences could be indispensable if, as the scientific reductionist contends, all that exists are components and their collectives. In order to properly assess whether the scientific reductionist can answer this important question, we need to more carefully examine what scientific reductionism has to say about a range of wider ontological, semantic, and methodological issues that arise in the sciences. I therefore propose to defer a final judgment on this last objection until the next chapter, where I provide a wider treatment of these commitments of scientific reductionism.

3.5 Scientific Reductionism as a *Live* Form of Reductionism?

The work of this chapter outlines how the goals, and strategy, of scientific reductionism sharply contrast with those of the semantic reductionism that philosophers have often focused upon. Nagelian reductionism overlooks compositional explanations, seeks to establish the dispensability of higher sciences, and endorses the apparent non-existence of any macro-world. In contrast, scientific reductionism accepts the central role of compositional explanations, and embraces both a macro-world and higher sciences to study it, for the scientific reductionist is primarily engaged in a dispute over the *best understanding* of our compositional explanations, the macro-world, and higher sciences rather than seeking to do away with them. The focus of the scientific reductionist is instead to overthrow the account of compositional explanations, and the macro-world and higher sciences, offered in the Scientifically Manifest Image and to replace such views with what the reductionist claims is the more adequate account offered by her own Fundamentalist position.

The dialectical strategy of the scientific reductionist is therefore one that is unfamiliar to philosophers of science who are used to the failed approach of greedy semantic reductionism. Perhaps more surprising, however, is the success of the reductionist in prosecuting her case for these bold claims. I have now shown that the scientific reductionist has interesting negative arguments against the reigning Scientifically Manifest Image and associated views in philosophy. Perhaps more importantly, I have also shown that the scientific reductionist's Fundamentalist position has plausible positive accounts of both the macro-world and higher sciences.

Philosophers of science of an antireductionist leaning are likely to be unmoved by such reasoning or by the more detailed Fundamentalist framework I have supplied for scientific reductionism. However, we have begun to see that we need to be very careful in trusting such quick assessments, for scientific reductionism is obviously a very different beast than the semantic reductionism that these philosophers of science have so successfully critiqued. In the next chapter, I therefore further examine what the Fundamentalist framework of scientific reductionism entails about a range of wider issues, but looking especially closely at its stance on the higher sciences. For I have found surprisingly strong indications that the scientific reductionism of working scientists may well offer a *live* option, contrary to the received wisdom in philosophy.

4 Understanding Scientific Reductionism: Fundamentalist Views of Ontology, Laws, Sciences, and Methodology

> Science is the characteristic product of our culture. Similarly, understanding where science fits in … is our characteristic philosophical problem … As of now, the hardest part is to reconcile a physicalistic ontology with the apparently ineliminable multiplicity of discourses that we require when we try to say how things are.
>
> – Jerry Fodor[1]

Fodor rightly frames providing a plausible account of the sciences, and in a way that coheres with our ontological commitments, as one of the intellectual challenges of our times. And the main philosophical critique of "reductionism" has been its abject failure with this task. Focusing as they do upon the Nagelian model, or related accounts, it is not hard to understand why philosophers of science have adopted this stance. Nagelian reductionism allows no place for compositional explanation or its compositional concepts. And the greedy assumption of semantic reductionism is that higher sciences are ultimately replaceable by lower sciences that are implicitly taken to be explanatorily omnipotent. But such claims about the higher, *or* lower, sciences are deeply implausible when we consider actual scientific practice, including the facts that we find compositional explanations across the sciences and that higher sciences have not disappeared but continue to flourish and multiply. It is therefore unsurprising that many philosophers regard "reductionism" as a dead position.

Such philosophical judgments are obviously jarring to scientists such as Weinberg who see little relation between such semantic reduction and their own views. And Weinberg, as well as philosophers such as Kim, each independently locate a weakness in philosophical critiques, since Weinberg and Kim take the viable forms of scientific reductionism to be *ontological*, rather than *semantic*, in nature.

[1] Fodor (1998), p. 6.

One of my goals in this chapter is to show that Weinberg and Kim are rightly concerned, but in order establish this negative point I must first use the findings of the last chapter to illuminate the wider commitments of a scientific reductionism built around compositional reduction, the Collectivist Ontology, and the alternative Fundamentalist account of the structure of compositional explanations and nature that results. The positive work of the chapter therefore focuses on detailing the implications of scientific reductionism for ontological and nomological issues, semantic and explanatory matters, and methodology.

In reconstructing the commitments of scientific reductionism, I carefully link to Weinberg's claims to show that he is plausibly interpreted as holding just such a sophisticated view, but similar points hold for other scientific reductionists. And my most significant conclusion in this chapter is that the resulting scientific reductionism is indeed best understood as what Weinberg aptly terms a "compromising" reductionism – that is, a combination of ontological reductionism and semantic or theory *anti*-reductionism.[2] Why have philosophers of science overlooked such a view? The diagnosis I offer is that the long conditioning of philosophers of science to see issues through the lens of semantic models of reduction has led them to implicitly endorse a *false dichotomy* overlooking a compromising position like scientific reductionism.

4.1 The Nature of the Determinative Entities and Determinative Laws: Implications of Scientific Reductionism (I)

The appropriate starting point for any examination of scientific reductionism is obviously its ontological commitments. Using my work from the last chapter, in 4.1.1, I briefly review the ways in which we saw that a Collectivist Ontology allows scientific reductionism to endorse the existence, and significance, of the macro-world, but I also highlight the priority Fundamentalism still accords to component entities. I then turn to a fresh issue, in 4.1.2, by considering what scientific reductionism says about laws, and I outline why Fundamentalism accepts the truth, and hence existence, of higher scientific laws. But I again show that the view accords important forms of priority to the laws of lower sciences.

[2] See Weinberg (1992), p. 52 and (2001), p. 13.

4.1.1 *The Reality, and Importance, of the Macro-World and the Priority of Components*

My work in the last chapter illustrated that scientific reductionism does not reject the existence of a macro-world, with all its rich features, but only rejects the understanding of the macro-world pressed by the Scientifically Manifest Image and associated views. As we saw, scientific reductionism accepts the existence of a macro-world but understands its nature through its parsimony arguments and Collectivist Ontology. Scientific reductionism takes the macro-world to consist of collective phenomena. And the Fundamentalist accepts that isolated components behave differently than when these same entities bear the relations that they do in a collective. The Fundamentalist thus allows that "More is Different" in the macro-world and also endorses both "composition" and "levels" in nature understood through her Collectivist Ontology.

Despite endorsing the existence and significance of such a rich, complex macro-world, I need to reiterate how and why scientific reductionism still accepts the ontological priority of component entities. For Fundamentalism takes ontological parsimony reasoning, such as the Argument from Composition, or the reasoning outlined in later chapters, to show that only component entities exist – or, at least, to establish that only such component entities are determinative. Scientific reductionism therefore takes the determinative "go" of the universe, to use an older term, to reside *solely* with components that are alone the entities that make anything happen, in so far as it is determined to happen at all. Fundamentalism is thus committed to the component entities studied by lower sciences having an important form of *ontological priority* over the phenomena studied by higher sciences, for component entities are the only entities or at least the only determinative entities.

4.1.2 *The Existence of Higher-Level Laws and the Priority of Laws about Components*

I have not yet discussed what scientific reductionism says about the laws of higher sciences, but the detailed ontological commitments I have now laid out allow us to understand its commitments on this issue. For example, where we have collective propensities we have true, counterfactual supporting generalizations about the kinds of collective that have such propensities and the scientific reductionist will thus plausibly accept that there are many, many higher laws about such collective propensities. In addition, given that such macro-homogeneities are often found across micro-homogeneity, there is often no single lower-level law that corresponds to such a higher-level law. Rather there is usually a tangle of different lower laws applying to the collectives with different components that nonetheless all share the relevant collective propensity. The higher laws accepted by the scientific reductionist are thus triply "higher-level": they concern macro-phenomena, there is no corresponding law at lower levels, and as I show in the next section these laws are often only expressible, even in principle, using the predicates of higher sciences.[3]

Though scientific reductionism consequently accepts that there are many higher-level laws studied by higher sciences it is also important for us note how, and why, the Fundamentalist framework nonetheless again acknowledges a substantive form of priority for lower-level laws about components. Weinberg expresses the point vividly in reference to the biological and psychological levels when explaining why he endorses Anderson's claim that "More is Different." He tells us:

> ... phenomena like life and mind do *emerge*. The rules they obey are not independent truths, but follow from scientific principles at deeper levels; apart from historical accidents that by definition cannot be explained, [biological and psychological systems] have evolved to what they are *entirely* because of the principles of macroscopic physics and chemistry, which in turn are what they are *entirely* because of the principles of the Standard Model of elementary particles. (Weinberg (2001), p. 115. Original emphasis)

Here we find Weinberg accepting that there are higher scientific laws, but pressing the point that the laws of lower sciences have an important priority because they are the laws that *entirely*, and hence *solely*, make things happen. Let us explore what Weinberg is claiming here and the argument leading to such a conclusion.

It will help my discussion if I stipulatively define a "determinative law" about some phenomenon as a law that describes determinative entities involved with that phenomenon.[4] Given this terminology, we

---

[3] For an alternative defense of the existence of higher-level laws even under ontological reductionism see Loewer (2008).

[4] I try to remain neutral over whether such laws govern entities or merely describe the natures of entities.

can see that the scientific reductionist will argue that whenever we have lower laws about components, and higher laws about some entity commonly understood to be composed by these components, then the lower laws are the only determinative laws with regard to these component and composed entities. For the scientific reductionist applies her Argument from Composition, or analogous ontological parsimony arguments, to show that the components are the only entities that exist or, at the least, are the only determinative entities involved with the phenomenon. Thus the scientific reductionist takes lower laws to have priority over the higher laws, since they contend that only the former are determinative laws.

We can therefore see that although Fundamentalism accepts the existence and significance of the laws of higher sciences, the reductionist's claims about the determinative entities lead to a clear form of priority for the laws of lower sciences. Laws of higher sciences may be true, but under Fundamentalism such laws and everything else about nature is the way it is *solely* because of the lower laws about components. Taken in its global version, the Fundamentalism of the scientific reductionist thus leads to the stance on laws embodied in dreams of a Final Theory and the Completeness of Physics – namely, that the laws holding of the fundamental microphysical entities in the simplest collectives are the only determinative laws and that everything is the way it is solely because of such laws, at least to the degree that anything is determined at all.

4.2 The Truth, Indispensability, and Utility of Higher Sciences: Implications of Scientific Reductionism (II)

I can now turn to *semantic* issues, especially about the nature and status of higher sciences, and their predicates, explanations, theories, etc. As I have begun to sketch, scientific reductionism has a different character than the reduction studied by philosophers of science. And Weinberg has been careful to stress these differences, though very often to ears dead to the relevant distinctions. For instance, Weinberg tells us that:

Ernst Mayr informs me that what I mean by "objective reductionism" is what he means by "theory reductionism." Maybe so, but I prefer to keep these terms separate, because I wish to emphasize that what I am talking about here is not the future organization of the human scientific enterprise, but an order inherent in nature itself ... (Weinberg (2001), p. 18)

Weinberg thus emphasizes that his *ontological* reductionism differs from the semantic kinds of reduction, i.e. "theory reductionism," upon which philosophers of science have long focused and which Ernst Mayr clearly wants to apply to Weinberg, though somewhat ironically given Mayr's own concerns about philosophical frameworks.

I work in stages to draw out the important differences between semantic forms of reduction and scientific reductionism. I focus, in 4.2.1, on the most glaring contrast, for I show that scientific reductionism should endorse the *in principle* indispensability of higher science predicates, explanations, and laws – a claim that is fatal to semantic views of reduction. Against this background, in 4.2.2, I highlight other mistakes made by greedy semantic forms of reductionism and show why scientific reductionism avoids them.

4.2.1 The In Principle Indispensability of the Predicates of Higher Sciences

Many philosophers will be dubious that any reductionist does, let alone *ought* to, endorse the in principle indispensability of higher sciences. So let us start by marking what Weinberg tells us about the role of lower and higher sciences. For example, in the case of statistical mechanics, Weinberg contends that:

The study of statistical mechanics, the behavior of large numbers of particles, and its application in studying matter in general, like condensed matter, crystals, and liquids, is a separate science because when you deal with very large numbers of particles, new phenomena emerge ... even if you tried the reductionist approach and plotted out the motion of each molecule in a glass of water using the equations of molecular physics ... nowhere in the mountain of computer tape you produced would you find the things that interested you about water, things like turbulence, or temperature, or entropy. Each science deals with nature on its own terms because each science finds something in nature that is interesting. (Weinberg (2001), p. 40)

Here we see a commitment to the idea that the various sciences find aspects of nature that are significant and that we hence need a variety of sciences with their own proprietary predicates to capture all the truths about nature. But how can a *reductionist* accept that higher sciences are indispensable in this way?

That this question springs to mind so readily betrays the recent philosophical focus on semantic forms of reduction that cannot accept such indispensability by their very natures. Semantic reduction works by showing higher-level laws or explanations can be derived from lower-level laws or explanations and can thus, at least in principle, be dispensed with in favor of the lower-level laws or explanations. However, a scientific reductionist such as Weinberg embraces *ontological* reduction and our work now allows us to see why such reductionists can endorse the *in principle* indispensability of the predicates of higher sciences. The space for a scientific reductionist to make such a move arises because the argument she uses to drive a compositional reduction has no sub-conclusion about the semantic dispensability of higher explanations or laws. Instead, her ontological parsimony arguments solely focus upon the dispensability of our commitments to composed *entities*, thus allowing such a reductionist to potentially endorse the indispensability of higher *predicates*.

It is therefore plausible that the scientific reductionist can endorse the in principle indispensability of higher predicates, but *should* they do so? At this point, I should note what some philosophical reductionists have said on these issues, especially since philosophers such as Bickle (2003), (2006) who have engaged the types of reduction espoused in the sciences have endorsed the dispensability of higher sciences. In contrast to Bickle, Kim (1998) has stressed that reductionism leaves us with higher-level predicates and no higher-level properties, but Kim provides no reason why such higher-level predicates would be in principle indispensable and apparently rejects this claim. Under his brand of ontological reductionism, John Heil (2003) has suggested that sentences involving higher-level predicates might express truths. For Heil raises the possibility that such higher-level predicates might actually refer to similar lower-level component properties. Heil thus comes close to endorsing something like the scientific reductionist's ultimate understanding of sentences of higher sciences, although Heil uses the flawed Flat/Subset model of realization to frame his points. However, Heil also provides no reason why such higher science sentences, and their predicates, might be indispensable.

Previous ontological reductionist accounts in philosophy have thus left it a mystery why higher sciences and their predicates simply will not go away, hence bolstering the implication of philosophical antireductionists, such as Fodor or Kitcher, that such indispensability must be rooted in an ontologically non-reductionist structure to nature. Given this background, appreciating the scientific reductionist's Collectivist Ontology, and her sophisticated treatment of the macro-world, is of some moment. For the latter shows why the scientific reductionist can, and should, endorse in principle indispensability about the higher sciences. To appreciate this point, let me start by giving a more detailed outline of the antireductionist Argument for Predicate Indispensability using work by Kitcher, though the same approach allows scientific reductionism to respond to other versions of the objection offered by writers such as Putnam, Fodor, and Wimsatt.[5]

To underpin his objection, Kitcher focuses on the case of molecular biology and various higher biological sciences, such as classical genetics, cytology, and embryology. Kitcher convincingly argues that properties and individuals commonly understood to be referred to by predicates in the higher biological sciences are taken to be multiply realized in the case of properties, and multiply constituted in the case of individuals, by heterogeneous molecular properties and individuals. Furthermore, Kitcher argues at length that we very often fail in attempts to replace key explanations using the predicates of higher-level sciences, whether of classical genetics, cytology, embryology, or others, with explanations using the predicates of molecular biology.

To illustrate the thrust of these arguments for my purposes it suffices to consider the example that Kitcher uses to summarize his wider conclusions. Kitcher tells us:

The distribution of genes to gametes is to be explained, not by rehearsing the gory details of the reshuffling of the molecules, but through the observation that chromosomes are aligned in pairs just prior to the meiotic division; and that one chromosome from each matched pair is transmitted to each gamete. We may formulate this point in the biologists' preferred idiom by saying that the assortment of alleles is to be understood at the cytological level. What is meant by this description is that there is a pattern of reasoning ... which

---

[5] Kitcher (1984). Other versions of such objections are found in Putnam (1975a) and (1975b), Fodor (1974) and (1975), and Wimsatt (1976), and (2007).

involves predicates that characterize cells and their large scale structures. That pattern of reasoning is to be objectively preferred to the molecular pattern which would be instantiated by the derivation that charts that complicated rearrangement of individual molecules because it can be applied across a range of cases which would look heterogeneous from a molecular perspective. Intuitively, the cytological pattern makes connections that are lost at the molecular level, and is thus to be preferred.[6]

Here Kitcher presses the indispensability, in framing certain true explanations, of higher science predicates referring to macro-homogeneities involving collectives of individuals that are molecularly and hence microphysically heterogeneous in their properties and relations. For example, if we solely used the predicates of molecular biology (or fundamental physics), then Kitcher convincingly argues that we would fail to refer to the macro-homogeneity shared between the molecularly and microphysically heterogeneous collectives that are central in truly explaining the shared behavior of collectives such as meiotic processes. As a result, higher science predicates referring to the relevant macro-homogeneity are, in principle, indispensable in formulating this and many other true explanations.

Kitcher summarizes the underlying basis of his argument by referring to what he terms the "organization of nature," which he suggests entails that higher science predicates are indispensable in the ways he outlines.[7] And Kitcher's emphasis on the "organization of nature" points us to what the scientific reductionist ought to say about such cases given her Collectivist account of the macro-world. Where we have such macro-homogeneity across micro-heterogeneity of the type Kitcher points to, the scientific reductionist *agrees* that predicates of lower sciences fail to capture these macro-homogeneities and also *agrees* that only predicates of higher sciences, at least when ultimately understood, suffice to express the relevant truths about these macro-homogeneities that we find where we have micro-heterogeneity. But the rub is that scientific reductionism consequently defends the in principle indispensability of higher science predicates whilst only accepting the existence of component entities and their collectives.

[6] Kitcher (1984), pp. 22–3.
[7] Kitcher (1984), p. 22.

Crucially, the scientific reductionist disagrees with "antireductionists" such as Kitcher over how we should understand such macrophenomena. Recall the example of the hardness of the tooth and filling. Under the reductionist's proprietary account of multiple composition, the predicates of lower-level sciences simply serve to express the differences at the level of component entities across the relevant collectives, i.e. the micro-heterogeneity, thus missing the truths about shared collective propensities and the macro-homogeneities in such an example. Consequently, in these kinds of example, Fundamentalism claims the predicates of the higher sciences, such as "hard," "turbulent," or the predicates of cytology in Kitcher's example, allow us to express the truths, under the ultimate understanding, that are not captured using predicates of lower sciences. Such truths concern collective propensities of collectives, but they still express objective, mind-independent truths about nature nonetheless.

The scientific reductionist should therefore accept that the predicates of higher sciences, when used with the ultimate understanding of them, are *in principle* indispensable for expressing truths about nature. And I want to mark that there are likely other reasons that can be given for why the scientific reductionist should accept the in principle indispensability of higher sciences and their predicates. For example, the existence of underivability or uncomputability of various kinds of higher-level explanations, laws, or theories may well offer another reason to accept such in principle indispensability. Similarly, issues concerning the level of abstraction of various predicates and explanations may provide another reason.[8]

Scientific reductionism thus plausibly endorses a representational *pluralist* position, which accepts that many different representational schemes, in the predicates of many sciences, are needed to express all the truths about nature. Once again, this is far from surprising once

---

[8] There have recently been other positions combining ontological reductionism and some form of semantic antireductionism such as Heil's (2003) and Loewer's (2008) in philosophy of mind or science. In metaphysics, such positions have been defended by Van Inwagen (1990) and Merricks (2001), amongst others. We should therefore note that the machinery of these writers *potentially* provides a resource for the scientific reductionist in articulating her position, so long as she is careful about the rather different commitments of these writers on a range of issues such as the nature of scientific composition, commitment to a Collectivist Ontology, and the character of the arguments driving scientific reductionism.

we distinguish compositional and semantic reduction. Regardless of how many phenomena explain this situation, we have thus found that the Fundamentalist position of scientific reductionism is indeed best understood, as Weinberg suggests, as a *compromising* combination of ontological reductionism and semantic *anti*-reductionism. In a robust sense, scientific reductionism should accept, to use Weinberg's words, that "each science deals with nature on its own terms" and discovers "something in nature that is interesting."[9]

Though scientific reductionism accords important roles to higher sciences, I should finally emphasize the ways that this position takes the lower sciences studying components, and their predicates, laws, explanations, and theories, to still have an important form of priority over higher sciences and their predicates, laws, explanations, and theories. For example, when discussing the situation with regard to chemistry and scientific defenses of its indispensability, Weinberg highlights the compromising stance of scientific reductionism, but also the priority it still accords to lower sciences. Weinberg tells us:

> In an attack on those who seek to reduce chemistry to physics, Hans Primas listed some of the useful concepts of chemistry in danger of being lost in this reduction: valence, bond structure, localized orbitals ... I see no reason why chemists should stop speaking of such things as long as they find it useful or interesting. But the fact that they continue to do so does not cast doubt on the fact that all of these notions of chemistry work the way that they do because of the underlying quantum mechanics of electrons, protons and neutrons. (Weinberg (1992), pp. 43–4)

We have now seen why the scientific reductionist agrees with antireductionists, either in the sciences such as Primas (1983), or in philosophy of science such as Fodor or Kitcher, that higher sciences like chemistry, and their predicates and laws, are indispensable – even in principle indispensable. But the scientific reductionist nonetheless still argues that lower sciences, and their predicates and laws, have a robust form of priority over such higher-level sciences and their predicates and laws. For using the Argument from Composition or other parsimony reasoning, the scientific reductionist argues that lower sciences are the only disciplines studying the determinative entities, and

---

[9] Weinberg (2001), p. 40.

determinative laws, involved in the phenomena that we truly describe as "covalently bonded," "turbulent," or "happy." Though accepting a rich macro-world, and the indispensability of the higher sciences in understanding it, scientific reductionism thus still endorses an important form of priority for lower sciences, and their predicates, explanations, and theories.

4.2.2 Compromising versus Greedy Reductionism: Moving Beyond Necessary Derivability, Semantic Dispensability, Explanatory Omnipotence, and Identities

Scientific reductionism, like Weinberg's, provides a far more plausible account of higher and lower sciences than earlier greedy accounts of reduction, so I want to tease out some further contrasts between these positions and highlight some common philosophical mistakes that derive from the focus on greedy semantic reductionism. I just raised the possibility of underivability or uncomputability, so let me start by exploring the contrasting commitments on this issue.

Once more, Weinberg highlights the salient points in discussing an actual case of such underivability or uncomputability when he emphasizes that:

We may even say that something is explained even where we have no assurance that we will ever be able to deduce it. Right now we do not know how to use our standard model of elementary particles to calculate the detailed properties of atomic nuclei, and we are not certain that we will ever know how to do these calculations, even with unlimited computer power at our disposal. (This is because the forces in nuclei are too strong to allow the sort of calculational techniques that work for atoms or molecules.) Nevertheless, we have no doubt that the properties of atomic nuclei are what they are because of the known principles of the standard model. This "because" does not have to do with our ability to actually deduce anything but reflects our view of the order of nature. (Weinberg (1992), p. 28)

Here Weinberg is actually implicitly noting a feature of compositional explanations in contrast to the deductive-nomological model of explanation favored by the Positivists.[10] And Weinberg then draws out how

---

[10] See Craver (2007) for a detailed articulation of this point.

this distinctive feature of compositional explanation leads compositional reductions to have an equally novel feature.

The initial point is that compositional explanations, and their supporting evidence, often establish compositional relations between higher- and lower-level entities *without* also sufficing for detailed derivations of higher-level explanations or laws from lower-level laws or explanations. It is the natural necessitation of the compositional relation that underpins the explanation, rather than the logical relation in a derivation. A qualitative, rather than quantitative and derivational, account of the relevant compositional relation can suffice to drive a compositional explanation. But if we have evidence in this way that certain entities studied by a higher science are composed, even without being able to deduce many claims about them, then the Argument from Composition, or related parsimony reasoning, can still be applied to the composed entities. So long as we have independent evidence for compositional relations between entities a successful compositional reduction therefore *does not require* that we be able to derive or compute the relevant higher laws, explanations, and theories, from the laws, explanations, and theories of the lower sciences. I earlier reflected this point in my account of the requirements for a compositional reduction, and the same point holds for the alternative parsimony argument I outline in Chapter 8, for neither argument requires derivational or other semantic relations in seeking to establish a compositional reduction.

Again, such claims about the relations of higher and lower sciences are often befuddling to those focused on semantic reductionism that cannot operate without derivational or computational relations. But the still deeper point concerns what a scientific reductionist should, and should not, expect from *lower* sciences. Semantic reductionism leads to a bloated, utterly unrealistic, view of the capabilities of lower sciences. Weinberg apparently seeks to express this point in discussing implausibly strong views of what a Final Theory can yield, but the underlying point actually holds generally. Let us look at what Weinberg says about this case before drawing out the wider lesson. Weinberg notes with regard to fundamental physics that:

various silly things … might be meant by a final theory, as for instance that the discovery of a final theory in physics would mark the end of science. Of course a final theory would not end scientific research, not even pure scientific research, nor even pure research in physics. Wonderful phenomena, from turbulence to thought, will still need explanation whatever final theory is discovered. The discovery of a final theory in physics will not necessarily even help very much in making progress in understanding these phenomena ... (Weinberg (1992), p. 18)

Here Weinberg makes the point that even if lower sciences complete a Final Theory that describes the only determinative entities and the only determinative laws, then Fundamentalism *still* accepts that lower sciences do not supply predicates, explanations, and theories that suffice to express all the truths about nature, including all the true explanations and laws.

To finish, it is worth briefly marking one final vestige of semantic reductionism still widely endorsed by many philosophical sympathizers of reduction that is again rejected by scientific reductionism. For a common driving idea of many reductionists in philosophy is that all true scientific statements, whether framed in the vocabulary of physics or some other lower-level science, or using the different predicates of the higher sciences, all have the same referents – the entities of physics or the relevant component entities.

Apparently swayed by the Nagelian approach and its focus on identities, philosophers have long assumed that the latter point entails that if reductionism is true, then there must be true identity statements showing that the predicates of physics, or some other lower-level science, and those of the higher sciences have *exactly the same* referents. Thus the claim is that for every predicate of the higher sciences there must be a predicate of the lower sciences that can be used to frame such a true identity claim.

After directly engaging the nature of the relevant scientific explanations in more detail, we see that this claim is also plausibly false under the Fundamentalist account of the scientific reductionist. The Collectivist Ontology that underlies the scientific reductionist's view of the macro-world, and the ultimate understanding of higher scientific predicates it underpins, both imply that such an assumption is mistaken. To see these points, we need to be careful of the common and ultimate understandings of higher scientific predicates when framing the relevant identities. To highlight the relevant points, I use the case most favorable to the philosophical reductionist where certain higher-level predicates always go along with the same kind of lower-level entities in the case of "diamonds" and their "hardness," i.e. a case of univocal realization. However, the reader should note that the problems outlined below are even more striking if we look at cases of multiple composition.

To begin, let me note the identity claims that should not be entertained, since they are obviously false:

"A diamond, as commonly understood, is identical to a certain collective of carbon atoms."

Or:

"The hardness of a diamond, as commonly understood, is identical to the relations of bonding and alignment of certain carbon atoms."

Under the common understanding of either a diamond or an instance of hardness these are a single, unified entity and cannot be identical to a many, whether in the plurality of individuals or property/relation instances found with a collective. Furthermore, the persistence conditions of a diamond or instance of hardness, so understood, are different from those of a collective.

However, philosophical reductionists are not usually attracted to the latter identities. Instead, they press the truth of claims like these:

"A diamond, as ultimately understood, is identical to a certain collective of carbon atoms."

Or:

"The hardness of a diamond, as ultimately understood, is identical to the relations of bonding and alignment of certain carbon atoms."

Here we have pluralities of entities on either side of the identity, so we avoid obvious falsehood. But the scientific reductionist *still* rejects the truth of even these putative identities using *ultimately understood* higher scientific predicates and lower-level predicates. And, again, her Collectivist Ontology is central to the scientific reductionist's claims.

In this case the problem is not that we are trying to relate a one to a many. Instead, the reductionist notes that "diamond," ultimately understood, is best understood as referring to a collective falling under a *collectivist kind* – that is, a collective individuated by certain collective propensities. As a result, we cannot identify "diamond," ultimately understood, to any *particular* collective of carbon atoms because we find "diamond" with a range of differing collectives of carbon atoms. Similarly, a particular "diamond" can persist across time whilst the starting collective of carbon atoms is destroyed by losing one carbon atom. Thus we see that a "diamond," ultimately understood, is not identical to any collective of carbon atoms.

Similar points hold for "hardness," ultimately understood, since the scientific reductionist takes this predicate to be best understood as referring to a collective propensity. And "hardness," ultimately understood, is found with many arrays of instances of bonding and alignment in greater and small collectives of carbon atoms. And, again, a particular example of "hardness," ultimately understood, can persist whilst the starting array of instances of bonding and alignment is destroyed by losing the instances of bonding and alignment that are lost when a carbon atom is lost from the relevant collective.

Under scientific reductionism we therefore do not get true identities between the predicates of higher sciences and lower sciences. When ultimately understood, the predicates of true sentences in the higher sciences refer to collectives of component entities, or collective propensities of them, whilst the predicates of lower sciences refer to particular component entities. So we thus still usually fail to get true identity statements. Using her Collectivist Ontology, the scientific reductionist can and should therefore press the point that it is no accident that we cannot dispense with the predicates of lower *or* higher sciences if we wish to express all the truths about nature.[11]

4.3 Scientific Methodology and the Relations of the Sciences: Implications of Scientific Reductionism (III)

I now want to briefly turn to methodological issues in the sciences about which I want to highlight how the Fundamentalist framework

---

[11] The latter provides a still further set of reasons, on top of those offered in Chapter 2, to reject the claims of Kim (1998), as well as Esfeld and Sachse (2011) and others, that a viable ontological reductionism grounds true identity claims involving predicates of higher and lower sciences.

of scientific reductionism again entails a range of fresh claims. I begin, in 4.3.1, by looking at what scientific reductionism says about using the common understanding of higher predicates in everyday science after carefully noting the complex ways in which this view is, or is not, "eliminativist" and "conservative." I then examine broader methodologies, in 4.3.2, showing that scientific reductionism endorses a co-evolutionary research strategy, but I also lay out the argument that Fundamentalists give for why lower sciences always have one reason for priority in funding lacked by higher sciences.

4.3.1 Eliminativism, Scientific Discovery, and the Role of the Common Understanding of Scientific Predicates

I have already highlighted how working scientists use the common understanding of the predicates of higher sciences in the everyday practices. A pressing question is therefore whether scientific reductionism should be "eliminativist" or "conservative" about such methodological practices. However, in making claims about "eliminativism" or "conservatism," one might be talking about higher-level *entities*, such as higher-level composed properties like being a voltage-sensitive gate. Or one could be asking about higher scientific *predicates* such as the term "voltage-sensitive gate," which is used in neurophysiology. Alternatively, one could be asking about higher scientific *concepts* such as the idea of a composed higher-level entity, for example the concept of a higher-level property such as being a voltage-sensitive gate. So let me carefully distinguish whether scientific reductionism is "eliminativist" or "conservative" about entities, predicates, and concepts because the position gives different answers with regard to these issues.

For example, with higher-level composed entities, as commonly understood, we have seen that scientific reductionism is a form of *higher entity eliminativism*. Its central claim is that when we reflect more carefully on the implications of compositional concepts, then we should eliminate higher-level entities, as commonly understood, from our ontology. In contrast, with regard to higher scientific predicates we have now seen that scientific reductionism is a version of *higher predicate conservativism*, since as a compromising position it claims that higher predicates are in principle indispensable for expressing certain truths when these predicates are ultimately understood as referring to collective phenomena.

This brings us to whether scientific reductionism should be eliminativist or conservative about higher scientific concepts such as those underlying higher scientific predicates as commonly understood. With regard to this question, it is important to emphasize that in everyday "practices of discovery" working scientists routinely use the common understanding of higher science predicates as they investigate various phenomena and construct their explanations and theories about them. Such an approach has been awesomely successful, since this practice basically produced all of our compositional explanations in the whole range of higher sciences. Against this background, and given the plausible principle that we should continue to pursue practices in the sciences that are known to have been successful in the past, scientific reductionism should therefore plausibly endorse the continued use of the common understanding of higher science predicates in scientific practices of discovery, although not in providing accounts of the ultimate structure of nature. Scientific reductionism will thus plausibly be a form of higher concept conservatism with regard to practices of discovery.

Once again, such a stance may appear odd, since the scientific reductionist takes sentences of the higher sciences as commonly understood to be false or meaningless. But how could it be a successful methodology for scientists to understand their sentences in a manner that strictly renders them false or meaningless? As I began to outline in the last chapter, scientific reductionism has ready answers about the utility of such a practice.

First, the common understanding of higher predicates is one that scientists are competent and comfortable with. Cognitive science provides good evidence that everyday concepts are very much like those involved in what I am terming the common understanding of higher predicates. Most of us find it natural to see the world as falling under determinative higher-level kinds of entity, which are composed of other, qualitatively different, entities. Furthermore, the sciences train researchers to tacitly apply higher concepts and hence to use the common understanding of higher predicates as referring to determinative, composed entities.

Second, and more importantly, I have already illuminated why the scientific reductionist takes such practices to reveal *truths* about the natural world. Higher science sentences are discovered by scientists using a common understanding of them and are indeed false under such an understanding. But under the ultimate understanding, articulated by the scientific reductionist, these sentences are very often *true*, and even express truths that cannot be captured using other predicates, because these predicates refer to the collective phenomena that, under Fundamentalism, form the macro-world.

Under scientific reductionism there is consequently no mystery about how the common understanding of higher predicates continues to be a successful part of ongoing scientific practices of discovery that reveal truths about the natural world. Within the context of practices of scientific discovery, scientific reductionism thus endorses *higher concept conservatism* about the concepts that underpin the common understanding of higher scientific predicates, although scientific reductionism embraces *higher concept eliminativism* with regard to accounts articulating truths about the ultimate structure of nature.

4.3.2 Co-Evolutionary Research Strategies and the Methodological Priority of Component Sciences

Given her Collectivist Ontology, the scientific reductionist takes cases commonly understood as involving realization to be about components forming collectives where the properties and relations of these components result in a certain collective propensity in the collective. This ontological point about the relation of components and collectives means that there are strong reasons to expect that, under certain circumstances (such as having sufficiently well-confirmed theories), there will be inter-theoretic constraints between the disciplines studying such components and their collectives respectively. For example, if one has a very well-confirmed theory of the collective propensities of some collective, then this theory can be used "top-down" to guide and even constrain research about the natures of the components forming the collective. For these components must have properties and relations that acting together result in the relevant collective propensities, so one can exclude certain hypotheses about components, or prioritize others, depending upon whether these hypotheses make claims about these components that allow them to act together in the relevant ways.

In the reverse direction, working "bottom-up," if one has a well-confirmed account of the nature of the components found in some collective, then this constrains theories about the collective propensities of collectives involving these components in various ways. For instance, precise knowledge of the nature of the components can exclude, or prioritize, certain hypotheses about what collective propensities are had by collectives. How does this work? Basically, such theories of the collective propensities of aggregates are, in part, plausible to the degree to which we can see that such propensities can result from properties and relations that our well-confirmed accounts tell us constituent individuals possess.

Scientific reductionism is thus plausibly committed to what writers such as William Wimsatt (2007) and Patricia Churchland (1986) term a "co-evolutionary" research strategy – lower and higher sciences evolve together in a dynamic flow of insight, and constraint, in both directions over time and across the ever-changing circumstances in these disciplines.

Although scientific reductionism accepts important methodological roles for the higher sciences, it accords lower sciences an important form of methodological priority over higher sciences to go along with the ontological, nomological, and semantic priority outlined in earlier sections. Weinberg articulates this form of such methodological priority that generalizes to all lower sciences, though Weinberg focuses on contrasting research in particle physics with the work on superconductivity in condensed matter physics that inspires scientific emergentists such as Anderson and Laughlin. Weinberg argues that:

> when condensed matter physicists finally explain high-temperature super-conductivity – whatever brilliant new ideas have to be invented along the way – in the end the explanation will take the form of a mathematical demonstration that deduces the existence of this phenomenon from *known* properties of electrons, protons and atomic nuclei ... Elementary particle physics thus represents a frontier of our knowledge in a way that condensed matter physics does not. (Weinberg (1992), p. 59. Original emphasis)

The deeper point here is that scientific reductionism argues that components are the only determinative entities (and the laws outlined by lower sciences are the only determinative laws), so the lower sciences that study components (and the laws about them) always work closer to the "frontier" of fundamental scientific research than higher sciences do.[12] Thus Fundamentalism implies that neurophysiology is closer to the "frontier" than psychology, biochemistry is closer than the biological sciences, and so on. In its "global" form, scientific reductionism implies that particle physics is the closest of all sciences to the "frontier" of fundamental research, including work in condensed matter physics.

More importantly, the latter point underpins an argument that scientific reductionists such as Weinberg offer about funding priority. To start, it is only fair to mark the nuanced nature of this claim, which is *not* that lower sciences always have funding priority over higher sciences. Weinberg carefully emphasizes the range of relevant factors involved in deciding the relative importance, and hence funding priority, of various scientific disciplines. Scientific reductionists such as Weinberg are clear that there are many factors to be weighed when allocating funding where these various factors may sometimes favor either lower or higher sciences depending on the circumstances.[13] In addition, Weinberg also acknowledges that decisions about how to weight these various factors are always difficult ones about which there are rarely easy answers. But we see Weinberg making the latter points alongside his claim that, amongst the various reasons we need to weigh, lower sciences always have one claim to funding priority lacked by higher sciences. Weinberg tells us that:

All I have intended to argue for here is that when the various scientists present their credentials for public support, credentials like practical value, spinoff, and so on, there is one special credential of elementary particle physics that should be taken into account and treated with respect, and that is that it deals with nature on a level closer to the source of the arrows of explanation in other areas of physics. But how much do you weight this? That's a matter of taste and judgment ... (Weinberg (2001), p. 23)

---

[12] One can read Rosenberg's (2006) contention that higher sciences must always be completed, corrected, deepened, or otherwise be made more precise, by lower sciences as having similar implications. In contrast, we need to leave aside Weinberg's apparent claim here that we need a derivation in favor of the more plausible contention that a compositional explanation would suffice to ground this claim.

[13] Weinberg (2001), pp. 120–1.

Here we see Weinberg acknowledging the range of factors in funding decisions and also the difficulty of such decisions. However, he also defends a special claim to funding priority of fundamental physics.

I can generalize the underlying point here so it applies to any lower science studying components if I roughly reconstruct a general version of such reasoning. This is what I term the "Methodological Priority Argument," which proceeds as follows. All else being equal, investigating the nature of the determinative entities involved with some phenomenon of interest (and/or the determinative laws associated with this phenomenon) usually yields greater insight than investigating the nature of other entities. Given the Argument from Composition, or similar parsimony reasoning, we can see that with what is commonly understood to be a composed, higher-level phenomenon the component entities studied by the relevant lower science(s) are the sole determinative entities whilst the composed, higher-level entities studied by the relevant higher science(s) are not determinative. (And/or we can also see that the laws illuminated by the lower science(s) are the only determinative laws associated with this phenomenon.) Therefore we can conclude that, all else being equal, with the study of a composed, higher-level phenomenon the investigations of the relevant lower-level science(s) into the nature of the relevant component entities (and the laws describing these component entities) will usually yield more insight than the investigations of the relevant higher science(s). But it is also plausibly true that we should prioritize, in funding, status, or other ways, the investigations into the object phenomena that yield greater insight, over investigations that yield less insight. Consequently, we can therefore further conclude that, all else being equal, with the study of a composed higher-level phenomenon we should prioritize, in funding, status, or other ways, the investigations of lower science(s) studying the relevant components over those of higher science(s).

The "all else being equal" rider in the premises, and conclusion, of the Argument for Methodological Priority is intended to capture the role of the other factors that scientific reductionism accepts are involved in assessing the funding claims of various areas of research. One may apply the Methodological Priority Argument to defend the "local" methodological priority of molecular neuroscience over the psychological sciences, or biochemistry over the biological sciences, and so on, as well as the "global" priority of particle physics over all

sciences. The scientific reductionist thus follows through the implications of her ontological conclusions for methodology and, contrary
to the claims of some philosophers of science, far from being mere
rhetoric I have shown that the scientific reductionist offers *principled*
arguments why lower sciences always have one special type of claim
to funding priority lacked by higher sciences.

4.4 Appreciating the Dialectical Implications of Scientific Reductionism

Now that I have articulated its wider commitments, I want to step
back and take a look at the implications of scientific reductionism for
wider debates. I begin, in 4.4.1, by summarizing the various commitments of scientific reductionism that I have illuminated and the rich
arsenal of resources they provide for answering objections based upon
scientific practice, including those from the philosophy of science.
Given its apparent viability, an obvious question is why philosophers
of science have overlooked scientific reductionism. I offer a diagnosis,
in 4.4.2, by highlighting the false dichotomy that philosophers of science have apparently fallen into about their options with regard to
reduction. Overall, I conclude, in 4.4.3, that Weinberg is right to question whether we all ought to be scientific reductionists and I examine
responses from philosophical antireductionists that claim they can
embrace this view at no cost.

4.4.1 A Scientifically Plausible Reductionism: Compromising Reductionism and its Nuanced Account of Scientific Practice

I have now shown that the novel features of scientific reductionism
leave it with a wide-ranging set of commitments in marked contrast to those of Nagelian reductionism or even Kim's reduction-by-
functionalization. It is worth briefly summarizing these claims, since
we have seen that scientific reductionism accepts at least the following
theses (the sections where these claims were defended are noted in
parentheses):

(a) True identity statements involving the predicates of higher sci-
    ences, whether commonly or ultimately understood, and predi-
    cates of lower sciences are rare in the sciences. (4.2.2)

(b) Working scientists illuminate, and endorse, many true sentences that they commonly understand to be about the realization and multiple realization of higher science properties, as well as true sentences commonly understood to be about examples of multiple composition involving other categories of entity such as the multiple constitution of individuals. (2.3, 3.2.2)

(c) The predicates of higher sciences, including those commonly understood to be about multiple realized properties, are projectible. (2.3)

(d) The higher sciences and their predicates provide true, law-like generalizations. (4.1.2)

(e) The higher sciences and their predicates provide true explanations. (3.3, 4.2.1)

(f) The best explanations in many contexts, given our purposes, are often those using predicates of higher sciences. (4.2.1, 4.2.2)

(g) Many true explanations and/or true law-like generalizations can, in principle, only be properly expressed using the predicates of the higher sciences. (4.2.1, 4.2.2)

(h) Many true explanations and/or true law-like generalizations properly expressed using the predicates of the higher sciences cannot be deduced or computed from true explanations and/or true law-like generations using predicates of lower sciences. (4.2.2, 5.2.3)

(i) Many higher-level sciences use non-linear frameworks, feedback loops, and/or can only explain their object phenomena using simulations. (3.2.2, 5.2.3)

(j) An important part of successful scientific practices of discovery is that working scientists use our common understanding of higher science predicates and the sentences based upon them, thus taking these higher predicates to concern composed entities or relations of composition as commonly understood. (4.3.1)

(k) The higher and lower sciences, which respectively study realized and realizer properties (and other composed and component entities), as commonly understood, inter-theoretically constrain each other under appropriate conditions (such as sufficient confirmation of the relevant theories, etc.) in such a way that these higher and lower sciences consequently co-evolve, through a process of mutual fit and adjustment, in a variety of beneficial ways. (4.3.2)

This is obviously something of a laundry list, but I hope it is dialectically useful in being shocking to philosophers. For amongst the dirty secrets revealed about scientific reductionism is that it endorses many of the claims that philosophers and scientists have long thought undermine "reductionism."

As a consequence, unlike Nagelian reduction, scientific reductionism does not simply fall to objections pointing to this or that feature of scientific practice, the macro-world, or higher sciences since the Fundamentalist framework embraces these phenomena and gives accounts of them. And I have now shown that the two main philosophical arguments against "reduction" both fail against scientific reductionism. I dispatched the Argument from Multiple Realization in the last chapter and I have now also shown that the Predicate Indispensability Argument fails to find its mark with scientific reductionism. I have thus provided a rebuttal of the last objection considered in Chapter 4 that I deferred for discussion.

Unlike the easy target of greedy Nagelian reductionism, with its panoply of implausible claims about the higher and lower sciences, scientific reductionism is thus far harder to attack, since it is founded in the compositional explanations we find in actual practice and thus has commitments such as (a)–(k) to use to defend itself alongside its underlying Collectivist Ontology and compromising approach. Overall, scientific practice therefore does not so easily undermine scientific reductionism in the ways the last generation of philosophers of science undercut Nagelian reduction. Properly understood, scientific reductionism can also meet the challenge from Fodor with which I started the chapter. Scientific reductionism provides a plausible account of the sciences, both lower and higher, in a way that coheres with our ontological commitments – albeit with a distinctive twist on what those commitments are – and presents a resilient, and apparently viable, form of reductionism.

4.4.2 The False Dichotomy in Philosophical Discussions of "Reductionism" and "Antireductionism"

I have confirmed that there has indeed been a radical dislocation between scientific and philosophical understandings of reductionism in the sciences. We have seen that working scientists have articulated a form of reductionism that is unlike the semantic reductionism upon which philosophers have focused and which is untouched by the objections philosophers have offered to such a reductionism. Scientists have thus come closer to a viable form of reductionism than the received wisdom in philosophy has registered.

My findings also allow me to diagnose what has gone awry in philosophical discussions about reductionism, since they bring two facts to center stage. Unfortunately, the focus on the dominant Nagelian approach, and its descendants, has apparently led the ontological strand in philosophical discussions of reduction to collapse into a solitary focus on semantic issues. The reason for this situation is not hard to find.

The semantic approaches to reduction center upon showing that lower sciences can do the explanatory work of higher sciences. Under such an approach there is thus a natural tendency to think that if reduction is successful, then the theories, explanations, and predicates of higher sciences must be, at least in principle, dispensable. But if higher sciences are unnecessary it is natural to assume the entities of lower sciences are all that exist. The ontological strand in discussions of "reduction" thus collapses into the semantic one and many philosophers have consequently assumed that they only have a dichotomy of options: *Either* be a greedy reductionist who endorses some brand of semantic reduction for the higher sciences as well as a reduction in our ontology, often to the entities of microphysics alone; *Or* be an antireductionist who accepts some brand of semantic antireductionism about higher sciences and an unreduced ontology encompassing both the entities of physics and levels of composed, determinative higher scientific entities as commonly understood.

My work illuminating the nature, and coherence, of scientific reductionism shows that the received wisdom in philosophical debates embraces a *false dichotomy*. The Fundamentalist framework of scientific reductionists such as Weinberg, and other working scientists, offers a philosophically overlooked option combining ontological reduction and semantic antireductionism. Our options are not thus limited to two positions, but actually encompass a third view. And philosophers of science have offered no response to the challenge of scientific reductionism for they have effectively wagered that their critiques of the Nagelian model apply to all "reduction." But I have shown that this bet is a failure, since I have established that the famous objections to semantic reduction fail to apply to the compositional reduction of scientific reductionism.

4.4.3 Are We All (Compromising) Scientific Reductionists?

Given the false dichotomy that underlies recent debates in the philosophy of science, many writers need to re-evaluate which theses they should actually endorse and why. Pressing questions arise for philosophers of science who label themselves either "antireductionists" or "reductionists." Let me briefly lay out the situation.

Self-proclaimed "reductionists" in philosophy need to carefully assess their stance on semantic reductionism because this group includes many who have apparently been primarily committed to establishing the truth of ontological reductionism. For example, do these writers focused on ontological reduction still wish to continue to make their previously strong claims about identities, derivability, or the dispensability of higher sciences when we see that no such claims are required to defend a thoroughgoing ontological reductionism – especially given the apparent conflict of these claims with features of actual scientific practice?

Similarly, self-identified "antireductionist" philosophers of science include many researchers who were often concerned primarily with defending versions of semantic antireductionism, and hence the continuing significance of higher sciences. Such "antireductionists" thus need to consider whether they really do endorse ontological antireductionism now that we see that it is *not* required for semantic antireductionism. For instance, should these writers actually accept, or at least remain agnostic about, the truth of a position like scientific reductionism, since this is consistent with acceptance of semantic antireductionism? Similarly, should such "antireductionists" continue a knee-jerk adherence to ontologically non-reductive views mirroring the Scientifically Manifest Image once we see this is not necessary for higher sciences to be in principle indispensable?

In the spirit of such re-evaluation, "antireductionist" philosophers of science quite reasonably float the suggestion that scientific reductionism actually poses no challenge to their positions.[14] Certainly, they allow, there may be problems in *philosophy*, or *metaphysics*, about whether in the face of the challenge of ontological reductionism we can still defend a layered view of the world properly regarded as a form

---

[14] Thanks to Colin Allen and Jerry Fodor for independently pressing me on
    this point.

of ontologically non-reductive physicalism. But, such philosophers of science protest, all they ever *really* wanted was to establish the *scientifically* important conclusion that higher sciences are significant and/ or, in principle, indispensable. And they note that scientific reductionism cedes these points. So why, ask these philosophers, does scientific reductionism pose any problem for antireductionism in the philosophy of science or, still more importantly, in the sciences themselves?

These are just the types of issue that need to be thoroughly explored. And we should recall that scientific reductionists such as Weinberg have repeatedly pressed just this kind of question – while contending that once the position is properly understood then *all* sides can see that they are, and always were, scientific reductionists.

It bears emphasis that the suggestion under discussion is *not* that we all embrace *everyday* reductionism and its search for compositional explanations – that has always clearly been accepted by "antireductionist" writers such as Putnam, Fodor, Kitcher, and Wimsatt, who use such explanations in critiques of Nagelian reduction. The suggestion is that everyone, including the philosophical "antireductionists," should endorse the more substantive position of scientific reductionism. And, in response, I can now use my outline of the commitments of scientific reductionism to show why it will not be so easy for "antireductionists" to embrace scientific reductionism.

Consider just four of the claims of scientific reductionism I have sketched so far and one further implication of scientific reductionism that I will illuminate in Part IV. First, there is the ontological implication of scientific reductionism that the only entities, or at least the only determinative entities, are components. Second, consider the connected claim that the only determinative laws are those solely concerning component entities – and that consequently higher sciences never examine either determinative or fundamental laws. Third, again built around its ontological claims, there is the conclusion of the Methodological Priority Argument – namely, that lower sciences always have a claim to methodological priority lacked by higher sciences. Fourth, consider scientific reductionism's claim that lower sciences are always closer to the "frontier" of fundamental research than higher sciences. Lastly, in Part IV, I show that the Fundamentalist framework of scientific reductionism entails that in any example of a successful compositional explanation in the sciences we must accept a scientific hypothesis about the structure of such an example which takes it to have just one level of determinative entities and laws in the component level.

There are still further substantive commitments of scientific reductionism, but just these five will trouble many "antireductionists" since they are not simply claims about abstract metaphysics or philosophy. Scientific reductionism uses its ontological conclusions to provide principled arguments for conclusions with very real practical implications in the sciences. Thus philosophical "antireductionists" face far harder choices than it might initially appear.

My guess from published work, and conversation, is that in re-evaluating their commitments a large group of "antireductionist" philosophers of science will baulk at endorsing one or more of these integrated wider commitments of scientific reductionism. If my guess is correct, then it is important to emphasize what is needed to rebut the compromising embrace of scientific reductionism. To stave off scientific reductionism, philosophical "antireductionists" plausibly need to show how it is even *possible* to have a situation in which we ought to accept that a composed entity is determinative. And the scientific reductionist's challenge to such "antireductionists" is especially pressing, since compositional reductions are driven by just the compositional explanations whose significance "antireductionists" have long championed. Weinberg thus really is owed a response to his question: *Why* are we not *all* scientific reductionists now?

4.5 Scientific Reductionism Redux? The Failure of Mainstream Philosophy and the Crying Need for Critical Assessment

My work in this chapter shows that confusions over "reductionism" have led philosophers badly astray. We have seen that Kim is correct that the philosophical focus on semantic approaches to reduction have led to unfortunate debates that are indeed "beside the point for issues of real philosophical significance" (Kim (1999), pp. 12–13), since philosophers have been left endorsing a false dichotomy about "reductionism" that leads them to overlook the compromising view of scientific reductionism.

Scientific reductionists such as Weinberg are therefore perfectly justified in pressing the importance of their wider conclusions. Although the Scientifically Manifest Image initially appears to be highly

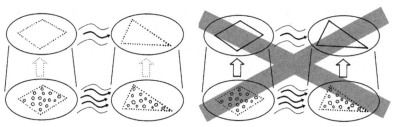

Figure 4A The scientific reductionist's view of our options about the structure of nature.

plausible, I have outlined how scientific reductionism supplies strong reasons supporting its conclusion that we must reject the Scientifically Manifest Image and instead adopt its Fundamentalist framework for compositional explanations and nature itself. And I have shown that scientific reductionism is immune to the extant critiques of reduction in the philosophy of science because it embraces the wider conclusions of such objections about nature and the sciences. Overall, Figure 4A thus frames the conclusions that scientific reductionists press upon us: We have two choices – the Scientifically Manifest Image or Fundamentalism – but the scientific reductionist claims she has shown that only the Fundamentalist picture of compositional explanation is acceptable.

However, before we can proclaim scientific reductionism the best account either in a local scientific example, or about the whole of nature, a difficulty immediately confronts us. It is a proven heuristic that positions are only properly assessed by sustained critical scrutiny and engagement with opposing positions. Unfortunately, as this chapter has laid out, the philosophy of science has failed to engage scientific reductionism since it mistakenly wagered that objections to semantic reduction would suffice, but to properly assess scientific reductionism we need to critically engage its claims.

We therefore need to look *beyond* the philosophy of science for resources to critically engage scientific reductionism. So, in Part III, I bring my discussion full circle by examining whether any emergentist views, in the sciences and philosophy, can ground more substantive critical engagement with scientific reductionism and its arguments. For my conclusion so far is that, as writers such as Weinberg and Kim have long pressed, scientific reductionism and its

Fundamentalist framework are arguably not just a live position, but may very well be the view of nature and the sciences, at both higher and lower levels, best supported by our scientific explanations, findings, and practices.

PART III

The Fruits of Emergence

5 | The Varieties of Emergence: On Their Natures, Purposes, and Obligations

Lately the emergence concept has, it seems, created a bandwagon effect, engendering high enthusiasms and expectations. What is needed at this point, I believe, is a healthy dose of realism about the prospects of the emergence concept. What are the chances that it will turn out to be the kind of magical philosophical or scientific tool some of its most ardent friends expect it to be ...?

– Jaegwon Kim[1]

As Kim makes clear, although now intellectually engaged with emergence many philosophers are concerned that it is incoherent, flawed, or does not live up to the claims made for it by its scientific proponents. Despite these philosophical qualms, over the last four decades there has been a huge wave of interest in emergence in the sciences driven by the need to grapple with empirical findings. Thus we find Philip Anderson suggesting that:

"emergence" is in fact the key to the structure of twentieth century science, on all scales ... the fundamental philosophical insight of twentieth century science [is]: everything we observe emerges from a more primitive substrate, in the precise meaning of the term "emergent," which is to say obedient to the laws of the more primitive level, but not conceptually consequent from that level ... To me it seems to argue effectively that the "God Principle" of emergence at every level is far more pervasive in our understanding of the universe than any possible "God Particle" representing a hypothetical milestone in the reduction to ever simpler and more abstract laws ... (Anderson (1995), pp. 2019–20)

As Anderson illustrates, diverse scientific proponents of emergence claim it is of central importance in the sciences, perhaps their most important focus, in ways that challenge scientific reductionism and its Fundamentalist picture.

[1] Kim (2006), p. 191.

These two passages illustrate the longstanding, and obviously very different, reactions to emergence of working scientists and philosophers. We need to carefully engage, and assess, these conflicting claims about the implications of emergence for scientific reductionism. My strategy in pursuing this task is again to leverage my framework for scientific composition, along with our better understanding of scientific reductionism, to fruitfully engage claims about emergence in two steps. Initially, I distinguish four distinct notions of "emergence," outlined by both scientists and philosophers, which are potentially relevant to debates over reduction and emergence.[2] I provide a clearer account of the features of each of these concepts of emergence as well as providing a broad criterion for the application of each of them. Then, in the second step, I examine whether the concept helps in meeting the challenge of scientific reductionism.

One immediate conclusion that my work establishes is that even a careful critic such as Kim is plausibly mistaken in his widely shared talk of "the" concept of emergence, for there are a plethora of distinct notions of emergence springing both from philosophical debates as well as the rich array of scientific usages.[3] And I show that there are hidden battles even over how many tenable notions of emergence there actually are, and such disputes are masked by talk of "the" concept. We therefore need to be aware that the contested nature of various notions of emergence means there is a very tangled terminology. For example, the terms "Weak" and "Strong" are used in existing debates to refer to differing (and sometimes even incompatible) notions.[4] To

---

[2] I should also note that I again focus primarily on the case of property emergence, since this is the most developed area of research. However, the kinds of emergence I distinguish, and my arguments about them, can also easily be extended to the emergence of powers, individuals, and processes. And given the coadunative character of such entities outlined in Chapter 2, one may expect emergence in one category to be accompanied by emergence in other categories. The notions I examine cover a very large number of writers on emergence, but I should note that a few anomalous notions do fall outside my net. Thus, for instance, Humphreys (1997) and Schaffer (2007), (2010) frame notions of emergence in quantum mechanical cases that do not easily fall under the broad kinds of concept I distinguish. However, I do examine Schaffer's position further in Chapter 8.

[3] See Corning (2002), pp. 21–2 for the wide range of uses of emergence in the sciences that easily outstrips even the numerous ideas recently considered by philosophers.

[4] See Gillett (2002a) and (2006b).

facilitate this chapter's initial work of clearly distinguishing various concepts by avoiding terminological confusion as much as possible I refer more neutrally to "W-," "O-," and "S-emergence."

What does my survey establish? Overall, I show that Kim is correct that we need a sobering realism because there are a number of concepts of emergence that scientific emergentists and philosophers have thought undermine reduction, but which we can now see fail to work against scientific reductionism once we properly appreciate its nature. These problems might be thought to result from something flawed in notions of emergence, but in almost every case I show that the problems *also* involve confusions about *reduction*. In concluding the chapter, I therefore outline three "rules" for more productive uses of notions of both emergence and reduction in future discussions.

In addition, I also defend an important positive conclusion. I use my theoretical frameworks to highlight how scientific emergentists such as Anderson, Freeman, Laughlin, Prigogine, and others have used concrete scientific cases to express novel, and philosophically overlooked, ontological insights about the form of emergence they champion. My final conclusion is that this sophisticated notion, in what I term "Strong" or "S-emergence," presents us with a neglected position with real promise for critically engaging scientific reductionism. Just as importantly, I also suggest that this overlooked scientific notion of emergence has real potential for articulating an alternative ontological account of nature compatible with our core scientific evidence.

5.1 Qualitative Emergence, Scientific Composition, and the Danger of Vacuity

We encountered our first concept of emergence in Chapter 2 when I noted that *qualitative novelty* accompanies compositional relations. Every compositional level is commonly understood to have novel powers, novel properties, novel kinds of individuals, and hence novel processes. Diamonds are hard, though carbon atoms are not; diamonds base the process of scratching glass, but carbon atoms never do. Similarly, ion channels are voltage-sensitive gates, but protein sub-units are not; ion channels base a process of opening, whilst it makes no sense to talk of a protein sub-unit opening. And the same features can be found for almost any pair of compositionally related levels in the sciences.

This common characteristic of composed entities as qualitatively distinct from their components underpins what I earlier termed "*Qualitative* emergence." Using property instances as our focus, I can frame a criterion for this notion as follows:

(Qualitative Criterion) A property instance X, instantiated in an individual *s**, is Qualitatively emergent *if* (i) *s** is constituted of lower-level individuals *s1–sn* which instantiate properties P1–Pn, and (ii) X is a property not had by any of *s1–sn*, i.e. X is not identical to P1 or P2 or P3 or … etc.

I should carefully mark that this only gives a *criterion* for this type of emergence, rather than being an account of what is constitutive for such emergence. My goal in offering this criterion, and those for the other concepts of emergence examined below, is thus simply to highlight distinctive features of the relevant notion of emergence and to bring some order to the various accounts given for specific concepts of emergence.

All kinds of writers have noticed that higher-level properties in the sciences are Qualitatively emergent and have suggested this is a significant fact about nature posing problems for scientific reductionism. For example, Gilbert and Sarkar ((2000), p. 2) use this notion of emergence in defending a new "organicism" and, in a book appropriately entitled *The Emergence of Everything*, Harold Morowitz (2002) details twenty-eight levels of emergent properties where the relevant notion is clearly Qualitative emergence. However, we can now see there are serious questions about whether Qualitative emergence, *by itself*, poses any problem for the scientific reductionist.

In fact the reverse is true. Compositional explanations in the sciences routinely explain Qualitatively emergent entities using lower-level entities taken to compose them. But, as we saw in Chapter 3, the scientific reductionist uses compositional relations, through the Argument from Composition or similar parsimony reasoning, to show that we should only accept component entities. Consequently, scientific reductionism is far from worried by Qualitative emergence, for such emergence goes along with the compositional relations that drive the reductionist's own arguments for ontological reduction!

Qualitative emergence, and the composition it accompanies, only starts the fight with scientific reductionism, rather than finishes it. Confusions about Qualitative emergence posing problems for scientific reductionism may be one reason why critics, such as Kim or Weinberg, have been concerned by claims about the significance of emergence in the sciences and in this case their worries are warranted.

5.2 Weak or W-Emergence: "Apologetics" for Scientific Reductionism

With the rise of the new sciences of complexity, with their distinctive modes of explanation, what is commonly termed "Weak" emergence, or what I term in this chapter "W-emergence," has recently garnered much attention. After outlining its nature, I critically examine the utility of W-emergence in responding to scientific reductionism.

5.2.1 The Nature of W-Emergence

Basically, a W-emergent property instance is realized, but also has certain *semantic* and/or *epistemic* features. In particular, the law statements, and/or explanations, and/or theories, that hold of a W-emergent property cannot be derived, computed, or predicted from the laws, explanations, and/or theories that hold of its realizers. With this core idea in mind, the following account provides a useful criterion for W-emergence:

(W-Criterion) A property instance X, instantiated in an individual s^*, is W-emergent *if* (i) X is realized by property instances P1–Pn of individuals $s1$–sn (and s^* is constituted by $s1$–sn), and (ii) the higher scientific law statements and/or theories and/or explanations taken to be true of X cannot be derived and/or computed and/or predicted from the lower-level scientific law statements and/or theories and/or explanations holding of the property instances P1–Pn that realize X in s^*.

Although ontologically a W-emergent property instance is understood to be intimately related to lower-level properties, being realized by them, the higher scientific law statements, explanations and/or theories concerning this property nonetheless in practice *float free*, in various ways, from the lower scientific law statements, explanations, and theories concerning its realizer properties.

Again, I should mark that this only gives a *criterion* for a property to be W-emergent. However, it does allow us to fix our target concept of emergence and explore its implications. For our criterion plausibly applies to the differing accounts of W-emergence offered by philosophers such as Mark Bedau (1997), (2002), (2008), Paul Humphreys (2008a), (2008b), Cyrille Imbert (2007), David Newman (1996), Alexander Rueger (2000), and William Wimsatt (1997), (2000), (2007), amongst others.[5]

5.2.2 W-Emergence as a Failed Route to Blocking Scientific Reductionism

Concepts of W-emergence have been widely used in recent debates over emergence and reduction in both philosophy and the sciences. But given their lack of detailed theoretical frameworks it is often hard to tell whether emergentists in the sciences *only* endorse W-emergence or whether they also endorse a more ontologically committed form of emergence. For example, one can read Anderson (1972) as endorsing W-emergence given certain things he says, but one can also plausibly read him as endorsing a still more ontologically committed notion of the kind I examine in 5.4 below. However, many emergentists in the sciences, such as Anderson, are very clear in opposing scientific reductionism, so I leave the thorny exegetical questions to one side and simply examine whether W-emergence alone would suffice to accomplish this.

To this end, I examine the work of a philosopher who does explicitly address why W-emergence putatively undermines reduction. The philosopher of science David Newman (1996) focuses on chaotic systems to argue that the W-emergence found in such systems blocks reduction. In particular, Newman considers a form of emergence where he claims that each instance of the emergent property is identical to an instance of a combination of microphysical properties/relations, i.e. a structural property (Newman (1996), p. 251). Thus, since structural properties are plausibly realized properties (if they exist at all), then Newman's emergent properties are plausibly composed by lower-level properties.

---

[5] The criterion has the merit of covering the array of differing accounts of W-emergence, since there is now a nascent debate over how to define W-emergence with a number of competing views. See, for example, the recent discussion amongst Bedau (1997), (2008), Imbert (2007), and Humphreys (2008a) and (2008b).

Yet Newman notes that the theories and/or laws about this kind of emergent property cannot be derived or predicted from the theories and/or laws, and information, holding about the relevant microphysical realizers. Newman argues at length that chaotic systems of a certain kind have properties of this variety that are obvious examples of W-emergence. More importantly, Newman also argues that such W-emergence results in higher-level properties that are "irreducible" and that this hence protects the causal efficacy of higher-level properties manifesting such W-emergence. Given these points, and especially the latter one, it appears Newman implies that W-emergence does, or would, block scientific reductionism.

Unfortunately, in arguing for such "irreducibility" Newman takes the Nagelian model, based on derivational relations between theories and/or laws, as his target notion of reduction. And Newman is plausibly correct that his kind of W-emergence precludes the derivational relations necessary for a successful semantic reduction. However, the wider problem is that Newman has not appreciated, or spoken to, the challenge raised by scientific reductionism and the Argument from Composition or similar ontological parsimony reasoning. For such ontological reductionism is based on compositional relations between properties and, as we saw in the last chapter, does not require derivational relations between laws, theories, and/or explanations. The features of W-emergent properties noted by Newman thus do not prevent the Argument from Composition, or similar parsimony reasoning, from applying to such properties, whether of chaotic or any other system, since W-emergent properties are realized and compositional concepts apply to them. For all Newman has shown about the interestingly different features of the explanations we find with chaotic systems, the Argument from Composition, or similar reasoning, *still* establishes that these W-emergent properties, given their realized nature, are *not* ontologically "irreducible."[6] Such W-emergence therefore does not by itself block scientific reductionism.

---

[6] Once again, I argue in Chapter 8 that the relevant parsimony argument is actually what I term the Argument from Composition and Completeness, rather than the Argument from Composition.

5.2.3 W-Emergence as Reductionist Apologetics and the Obligations of W-Emergence

Finding W-emergence in nature, or the sciences, is not news to scientific reductionism, which, as I outlined in the last chapter, endorses W-emergence of various kinds *throughout* the higher sciences. Scientific reductionism claims we often have *semantic* and *epistemic* dislocations between the predicates, theories, explanations, and/or laws of higher- and lower-level sciences even though we have established compositional relations, including realization between property instances, between the entities of these sciences. So the W-emergence found with the sciences of complexity, and elsewhere in the higher sciences, actually conforms to scientific reductionism's picture of nature and these sciences.

Overall, W-emergence thus plausibly provides what we might term a powerful "apologetic" tool for scientific reductionism. To see this point, consider what Mark Bedau tells us in one of the papers that first drew the attention of philosophers to W-emergence in the sciences of complexity. Bedau argues that his account of W-emergence:

meets these three goals: it is metaphysically innocent, consistent with materialism, and scientifically useful in the sciences of complexity that deal with life and mind. (Bedau (1997), p. 376)

We can now clearly see why W-emergence is taken to be innocent, because it avoids a commitment to the determinative composed entities that the ontological parsimony arguments of scientific reductionism putatively show we should never accept. W-emergence is consistent with "materialism," i.e. a world where compositional concepts apply comprehensively, because it concerns higher-level properties that are realized.

Against this background, Bedau's third feature illuminates why W-emergence is apologetically useful for the scientific reductionist, because it can be used to interpret the emergence common in findings in the sciences of complexity, which have the epistemic features associated with W-emergence. For example, after discussing cases where we can only explain via simulation, Bedau argues that:

In cases of weak emergence, the macro depends on the micro because, in principle, each instance of the macro ontologically and causally depends on the micro because, in principle, each instance of the macro ontologically and causally is nothing more than the aggregation of the micro-causal elements … At the same time, weak emergence exhibits a kind of macro autonomy because of the incompressibility of the micro-causal generative explanation of the macro-structure. Because the explanation is incompressible, it is useless in practice (except in so far as its serves as the basis for a good simulation of the system). (Bedau (2008), p. 449)

Bedau does not outline the basis for his claim that the macro composed entities are "nothing more" than the collective of their micro-level components. However, we are taken to have compositional relations in such cases, so Bedau is charitably interpreted as taking the reductionist's ontological parsimony reasoning to apply and this underwrites such "nothing but" claims. But the necessity of explaining by simulation, and the impossibility of using the components to explain, results in the *semantic* and/or *epistemic* autonomy that characterizes W-emergent properties. And the reductionist may use such W-emergence to explain why defenders of the Scientifically Manifest Image mistakenly accept the existence of composed, determinative properties. The latter is an illusion generated by the fact that the law statements and/or theories and/or explanations concerning W-emergent properties are semantically or epistemically autonomous from lower-level predicates or law statements, etc. Thus W-emergence helps the scientific reductionist explain away why so many scientists endorse determinative composed entities.

Since one of the primary obligations of scientific reductionism is to explain away the appearance of evidence for composed determinative entities, we can see how W-emergence helps the ontological reductionist. But this immediately suggests that difficult questions face the many writers who claim the existence of W-emergence blocks reductionism in the sciences. What does my work suggest about the obligations of these writers? It appears such "antireductionist" W-emergentists have two options with very different obligations.

On one hand, we can now see that such writers can laud their important victory over "greedy" semantic reductionism with the conclusion that higher sciences are in principle indispensable in various ways. But under this option my work also shows that the primary obligation of these W-emergentists is again that of learning to live with Fundamentalism and its implications. For we have now seen that by endorsing W-emergence *alone* these emergentists have no response to the underlying arguments of scientific reductionism. Whether endorsing the commitments of scientific reductionism is a price that these emergentists in the sciences are willing to pay is a real question. And considering just the five features of scientific reductionism I highlighted at the end of the last chapter shows that the price is indeed a steep one.

My bet is again that there will not be many takers for this first option, so let us turn to the second option, which focuses on continuing to resist the conclusions of scientific reductionism. Under this option, given my findings about W-emergence, the obligation of scientific defenders of W-emergence is consequently to illuminate *another* kind of emergence, which is more ontologically robust in character than W-emergence. Crucially, this alternative notion of emergence must have a nature that provides an ontological reason why we may accept the existence of composed entities that are still determinative, hence undermining the Argument from Composition. Whether this can be achieved is obviously a substantive question, but its chances of success do not look as remote as that of using W-emergence alone to block scientific reductionism.

5.3 Ontological or O-Emergence: Articulating the Anti-Physicalist Position in Philosophy

An increasing group of philosophers, though virtually no scientists, has recently focused on an ontologically robust concept of emergence, so one may wonder whether it could help emergentists in the sciences to block scientific reductionism. This concept is based around higher-level properties that are *unrealized* by any lower-level properties in what is often called "Ontological emergence" and which I refer to in this chapter as "O-emergence."

5.3.1 The Nature of O-Emergence

An O-emergent property instance is, first, a property instantiated in a higher-level individual (whether constituted by other individuals or a fundamental non-physical individual). Second, an O-emergent property instance is an unrealized property. And, third, an O-emergent property is a productive, and hence determinative, property instance.

We can therefore capture the heart of this notion in the following criterion:

(O-Criterion) A property instance X, instantiated in an individual s^*, is O-emergent *if* (i) s^* is an individual which is either constituted by other individuals or is an unconstituted, non-physical individual; (ii) X is an unrealized property instance; and (iii) X is productive and hence determinative.

The complexity of condition (i) is produced by O-emergentists differing on an important issue. One group takes *both* properties *and* individuals to be uncomposed and hence O-emergent, thus generating unrealized properties in unconstituted, non-physical individuals (Hasker (1999), Lowe (2008)). But another group claims that properties *alone* emerge (O'Connor (1994), Clayton (2004)), leaving us with unrealized properties of *constituted* individuals. However, for my purposes the key claim of both groups is arguably claim (ii), that O-emergent properties are not realized, since we shall see in the next section that this claim may be thought to protect the truth of (iii) in the face of the scientific reductionist's parsimony arguments.

5.3.2 O-Emergence and the Response of Anti-Physicalist Philosophers to Compositional Reduction

The philosophy of mind is the main area in which philosophers have recently embraced O-emergence and with respect to phenomenal consciousness in particular. Although the new sciences of the mind and brain are making progress at an astonishing rate, these disciplines are only a few decades old in their present guise and have made very few inroads with regard to explaining phenomenal consciousness. Furthermore, philosophers are well acquainted with the battery of venerable, *a priori* arguments, which they have retooled using modern philosophical machinery, that phenomenal consciousness, or even mental individuals like minds, cannot be composed by other entities.[7] Furthermore, a range of other arguments, with *a*

---

[7] See Chalmers (1997) for the most high-profile use of such *a priori* arguments based around phenomenal consciousness and Lowe (2008) for a review of arguments about minds and/or persons.

posteriori elements, have also recently been offered to bolster such claims.[8]

In response to such difficulties understanding how consciousness or the mind is composed, a bevy of philosophers have sought to use O-emergence to provide what they take to be a more adequate account of the mental and the mind–body relation. As I noted earlier, there is an important division between the resulting forms of O-emergentism concerning how far they depart from traditional substance dualism about the kinds of individuals we find in nature. Our O-criterion has the merit of covering the views of the philosophers who have been drawn to both kinds of position based on O-emergence, including the accounts of Philip Clayton (1997), (2006), Tim O'Connor (1994), Tim Crane (2001), William Hasker (1999), E. J. Lowe (1996), (2008), Michael Silberstein and John McGeever (1999), and others.

However, rather than the philosophy of mind, my focus here is on whether O-emergent properties provide any response to the arguments of scientific reductionism about cases of compositional explanation. And in this respect we can see why someone might think O-emergence has attractions. To appreciate the point, consider the early work of Clayton, who accepts O'Connor's (1994) account of O-emergent properties as unrealized by lower-level properties (Clayton (1997), p. 248). Clayton consequently suggests that as a result we do:

not need to embrace *either* a radical dualism of mind/soul and body *or* the physicalism that is widespread among scientists and philosophers today. The theory of *emergent properties* forms an attractive *via media* between these two poles of the discussion. (Clayton (1997), p. 247. Original emphasis)

Clayton rightly juxtaposes O-emergence to physicalism, since we have seen the latter is the claim that higher-level entities in space-time are comprehensively composed by lower-level entities in the ways illuminated by compositional explanations in the sciences. And commitment to *unrealized* O-emergent properties found at higher levels of nature gives us a form of what we may term "Anti-Physicalism," since O-emergent properties are not composed. Note, however, that rejecting

---

[8] See, for example, Hasker (1999) for arguments focused on the unity of consciousness, but the philosophical debates over consciousness contain a range of other arguments.

"a radical dualism" is optional, since O-emergentists, such as Lowe and Hasker, argue we need to situate O-emergent properties in spatially located, but still unconstituted, higher-level individuals, which are a second basic, and mental, substance, albeit a located one in contrast to that of Descartes.

It is important to briefly articulate some of the implications of O-emergence. If we take a force to be an entity that changes the motions of masses, then I should note that O-emergent properties are plausibly fundamental, non-physical forces since they do change the motions of masses, are unrealized, and are non-physical. In addition, it appears that new fundamental, non-physical energies are also therefore implied by the existence of O-emergent properties and so too are new kinds of conversion from, and to, existing types of energy.

Below I briefly explore the import of the latter implications, but let me now examine why O-emergentism might be thought to avoid the arguments of scientific reductionism. The reason is blindingly simple. O-emergent properties are not realized by lower-level properties and thus parsimony reasoning like the Argument from Composition cannot apply to these properties – an entity cannot be "nothing but its parts" if it has no "parts." Positing O-emergent properties, whether in unconstituted individuals or not, might thus be thought to allow one to endorse the determinativity, and existence, of such higher-level properties whilst being untouched by the arguments of scientific reductionism.

5.3.3 The Obligations of O-Emergentism: Ameliorating Conflicts with Core Scientific Findings

O-emergence might be thought to provide a shield against scientific reductionism, but two critical points are worth noting about this use of O-emergence that also highlight the key obligations of O-emergentists. First, and most importantly, it is true that if we have O-emergence, then the arguments of the scientific reductionist do not apply. But in a case of compositional explanation we do have comprehensive applicability of compositional concepts, so we do not have O-emergence. Any insights about the nature of O-emergence therefore do *not* show anything defective about the Argument from Composition, or related parsimony reasoning, about the implications of compositional explanations. So O-emergence leaves the scientific reductionist's central argument untouched. And this means that the potential applicability of the use of O-emergence as a bulwark against scientific reductionism is actually very narrow indeed. We have successful compositional explanations across vast swathes of scientific levels and phenomena, so a resort to O-emergence is therefore unavailable in these examples.

Second, the latter points begin to illuminate why rejection of O-emergence by scientists is widespread and often colorfully put. For example, we find reductionists such as Weinberg stating that:

Henri Bergson and Darth Vader notwithstanding, there is no life force. This is the invaluable negative perspective that is provided by reductionism. (Weinberg (2001), pp. 115–16)

Here we could just as easily substitute fundamental "mental force," "chemical force," or "social force" instead of "life force," since as I noted above we are left with new fundamental, non-physical forces in whatever scientific domain one posits O-emergence. And it is important to mark that scientific emergentists such as Anderson, Freeman, Prigogine, and Laughlin, amongst many others, agree wholeheartedly with Weinberg in rejecting O-emergence. In our opening passage in this chapter, for example, Anderson makes it clear that emergent entities are covered by the fundamental physical laws and are thus presumably composed entities. Similarly, we find Laughlin stating:

The term emergence has unfortunately come to mean a number of different things, including supernatural phenomena not regulated by physical law. I do not mean this. (Laughlin (2005), p. 5. Original emphasis)

Working scientists thus almost universally reject O-emergence and the nature of our core empirical evidence from compositional explanations plausibly underlies their stance.

The wider problem facing the O-emergentist is that the evidence of *everyday* reductionism, in the form of our pervasive compositional explanations, makes a strong case that there are no fundamental – i.e. uncomposed – forces except the four (possibly three) physical forces in the weak and strong nuclear forces, the electromagnetic force, and the gravitational force. (Similar points hold about the kinds of fundamental energy.) So our core scientific evidence provides general reasons to deny the existence of O-emergence. In philosophy, Brian McLaughlin

(1992) in his seminal work on the earlier philosophical movement labeled "British Emergentism" has foregrounded this argument against O-emergence and its commitment to fundamental forces and forms of energy. And McLaughlin's objection has led many philosophers to dismiss robust "emergentism" as a coherent but empirically disconfirmed position because philosophers routinely equate robust "emergentism" with O-emergentism.[9]

To conclude, it appears the main obligation of proponents of O-emergence is to explain away its apparent conflict with our core empirical evidence in our pervasive compositional explanations and wider evidence that there are only four (or possibly) three fundamental forces that are all physical. Returning to my more narrow focus here on cases of compositional explanation, I have also highlighted why O-emergence in no way undercuts the scientific reductionist's arguments about the implications of a compositional explanation and its compositional concepts. Consequently, if there is a more robust notion than W-emergence that can provide emergentists in the sciences with an ontological reason to reject scientific reductionism in cases of successful compositional explanation, then O-emergence simply is not that concept.

5.4 Strong or S-Emergence: Condemned by the Unholy Philosophical Alliance, Defended by Scientific Emergentists

Scientists such as Anderson, Freeman, Laughlin, Prigogine, and others have apparently pressed what appears to be a further notion of emergence that is distinct from those I have so far surveyed. This notion is a sophisticated species of what is sometimes termed "*Strong* emergence" (Gillett (2002a), (2003a), (2003b), (2006a), (2011b)), but that term has also often been used by other writers to refer to O-emergence, for reasons that I outline below. To avoid confusion, I therefore use the term "S-emergence" for the notion in this chapter.

---

[9] For interesting recent work on O-emergentism see Hasker (1999), O'Connor and Jacobs (2003), O'Connor and Wong (2003), and Lowe (2008), amongst others. In Chapter 8, I also use my work to broach a still further innovation and a novel version of O-emergentism that results – though I mark that this innovation fails to address the key conflict with empirical evidence.

5.4.1 The Basic Nature of S-Emergence, the Unholy Alliance, and Two Kinds of S-Emergentist

The core idea of S-emergence is easy to frame, since an S-emergent property is simply a higher-level, realized property that is determinative. We thus get this criterion:

(S-Criterion) A property instance X, instantiated in an individual s^*, is an S-emergent property instance *if* (i) X is realized by property instances P1–Pn of individuals $s1$–sn (and s^* is constituted by $s1$–sn), and (ii) X is determinative.

S-emergent properties just are the properties posited by working scientists in intra-level productive and inter-level compositional explanations and are the heart of the Scientifically Manifest Image – that is, composed properties that are nonetheless taken to be determinative. Thus, for example, researchers take the ion channel to instantiate the realized, but productive, property of being a voltage-sensitive gate in order to explain why the ion channel opens when the surrounding charge changes. Or posit the realized, but productive, property of being extremely hard in the diamond in order to explain its scratching the glass, amongst many other effects.

A large number of philosophers deny that S-emergence even represents a viable option. Philosophical reductionists such as Kim (1999), as well as Anti-Physicalist philosophers such as Clayton (2004) or Chalmers (2006), are all *united* in being what I term "dichotomists" who take W- or O-emergence to *exhaust* the substantive kinds of emergence that are possible in addition to Qualitative emergence.[10] This odd coupling of reductionists and Anti-Physicalists is aptly termed the "Unholy Alliance" not simply because it unites vehement opponents but also due to the unfortunate actions I show below that it prompts in philosophers.

We can now understand why the Unholy Alliance denies we should ever accept the existence of S-emergent properties if we see its members as implicitly endorsing reasoning like the Argument from Composition, which is taken to establish that we should *never* accept that a property instance is *both* realized *and* determinative – that is, an S-emergent

---

[10] See also Bedau (1997) and Wilson (Forthcoming).

property instance. The members of the Unholy Alliance thus take *two* theses to be such that they should not be held together: Namely, that a higher-level property instance is realized and that this property instance is determinative.

The disparate members of the Unholy Alliance agree that we consequently have a *dichotomy* of options and corresponding notions of emergence. Either you accept that the Argument from Composition is both sound and valid, thus only accepting W-emergence and rejecting the existence of determinative, realized property instances. Or one can endorse the Argument from Composition as valid, but deny that it is sound by rejecting the claim that certain higher-level properties are actually realized properties. This option involves abandoning the universal applicability of compositional concepts in nature to endorse Anti-Physicalism, thus leaving one accepting O-emergent properties and fundamental, non-physical forces. Driven by reasoning like the Argument from Composition, the Unholy Philosophical Alliance thus takes W- and O-emergence to exhaust the substantive kinds of emergence and denies that S-emergence should ever be accepted or is perhaps even possible at all.[11]

This philosophical opposition to S-emergence as a viable theoretical option can thus be seen to line up with the scientific reductionist's long campaign against positions endorsing what we can now see is S-emergence. And the confluence of positions appears to result from the endorsement of a common argument in the Argument from Composition.

There is consequently profound opposition to acceptance of S-emergence and appreciating this feature of recent debates allows us to see that we need to distinguish amongst two kinds of defenders of S-emergence whose positions leave them in rather different situations. On one side, we have what I term "Naked" views, which are committed to S-emergence but offer no further ontological commitments that might illuminate how we can have S-emergence – thus providing no resources for rebutting the Argument from Composition or similar reasoning. In this group fall the many scientists who endorse the Scientifically Manifest Image. And many (most?) so-called

---

[11] Kim (1999) offers arguments that conclude that S-emergence is not even *possible*, rather than being simply a phenomenon that we should never accept, as parsimony reasoning like the Argument from Composition establishes.

"non-reductive physicalists" in philosophy who also appear to be Naked S-emergentists, including the antireductionists such as Fodor, Wimsatt, and Kitcher, who, as I outlined in Part II, mistakenly take their semantic/epistemic points about predicate indispensability, or multiple realization, to protect S-emergence.[12]

In contrast, what I term "Enriched" views endorse the existence of S-emergence, but then go on to outline *further ontological commitments* that are explicitly aimed at illuminating how we can have property instances satisfying the S-Criterion and which hence supply a response to the Argument from Composition. I show in the next section, and following two chapters, that we find scientific emergentists in this group, such as Freeman or Laughlin, since these scientists offer a rich trove of ontological claims about S-emergence *beyond* features (i) and (ii). And there has been a long tradition of theoretical work by scientists, as well as those working in cybernetics or systems theory, that falls squarely into espousing Enriched views, such as Bertalanffy (1950), Weiner (1948), Pattee (1970), (1973a), and Salthe (1985), amongst many others. In philosophy, we find Enriched views of S-emergence in the work of Atmanspacher and Bishop (2006), Dupré (2001), (2008), (2009), Dyke (1988), Juarrero (1999), Mitchell (2009), (2012), Richardson and Stephan (Boogerd et al. (2007)), Stump (2012), and Van Gulick (1993), amongst others.[13] I have also defended an Enriched position about S-emergence in a series of papers.[14] In the area of science and religion, Enriched views are found in the work of a number of writers, but the most prominent examples are Arthur Peacocke's view about the natural world (1990), (1995), (1999) and Murphy and Brown (2007).

There is a very wide range of ideas offered by proponents of Enriched views concerning the further ontological commitments that underpin

---

[12] I do not take commitment to multiple realization to be a further ontological commitment relevant to protect S-emergence, since if it exists then it simply falls out of a commitment to ubiquitous scientific composition. Similarly, I also will not take commitment to the Completeness of Physics to be an ontological commitment intended to defend S-emergence, since it is not adopted for that reason and we shall see in Chapter 9 that it has exactly the reverse effect.

[13] I leave aside views that seek to use features of flawed views of scientific composition to defend S-emergence such as Shoemaker (2001) or perhaps Melnyk (2003).

[14] For my papers, see Gillett (2002a), (2003a), (2003b), (2006a), (2006b), and (2011b).

the existence of S-emergence, including downward causation; the role of boundary, environmental, or background conditions; reduction in degrees of freedom and/or higher-level constraints; entrainment or enslavement; unpredictability or undeducibility of certain kinds; and more. Rather than delve into all of these ideas, my approach is once again to engage the recent scientific emergentist accounts. And after articulating their claims, I return to other ideas offered by Enriched theorists in later chapters. But before I turn to the scientific emergentist's positive ideas about S-emergence, let me briefly highlight the kind of treatment meted out to defenders of such views in the sciences by the members of the Unholy Alliance as a result of reasoning like the Argument from Composition.

5.4.2 Why the Unholy Alliance Claims Scientific Emergentists Cannot Really Mean What They Say

Understanding the Unholy Alliance's view of the problematic, and the viable options, is important because it configures the wider actions of its members. For example, these philosophers consequently "charitably reinterpret" the claims of past, and present, proponents of S-emergence whether in philosophy or the sciences. The dichotomists contend that philosophers or scientists cannot charitably be taken to espouse S-emergence, so they are reinterpreted as holding either W- or O-emergence even when they explicitly endorse S-emergence.

To see how this works, let us consider a couple of examples of such charitable reinterpretations of scientists who espouse S-emergence. For instance, consider this passage from Kim, where see how acceptance of the dichotomy of options can be important. Kim asks:

are there emergent properties? Many scientists have argued that certain "self-organizing" phenomena of organic, living systems are emergent. But it is not clear that these are emergent in our sense of nonfunctionalizability. (Kim (1999), p. 18)

Here Kim suggests that the test of "robust" emergence is being non-functionalizable and hence non-realizable. For Kim endorses the dichotomy of options. On one hand, we "functionalize," apply compositional concepts, and run the Argument from Composition to leave us with at most W-emergence. Or, on the other hand, the relevant entities are "nonfunctionalizable," are not composed, Anti-Physicalism is true and we are left with the empirically implausible O-emergence it entails. Thus Kim implies, *really*, the emergence pointed out by working scientists, and crucially the underlying empirical evidence driving their claims, can be nothing more than W-emergence after all, since these researchers accept universal composition given our core scientific evidence.

Kim shows how charitable reinterpretation proceeds in the abstract, but let us now consider concrete cases. As we saw earlier, Bedau's (1997) work on W-emergence in the sciences of complexity can be interpreted as embodying this type of approach, for he suggests that complex systems can *only* be instances of W-emergence given our evidence for the comprehensive applicability of compositional concepts in such cases. And another example in the same area is the response that the philosopher of science Stephen Kellert gives to a team of researchers from the sciences of complexity. In a prominent piece in *Scientific American*, Crutchfield et al. (1986) claim that the undeducibility of laws, and changes of behavior, due to "global" features in chaotic systems undermines "reductionism." In response, Kellert states (where he uses the term "holistic" instead of "S-emergent"):

> it is important to clarify that chaos theory argues against the universal applicability of the method of microreductionism, but not against the philosophical doctrine of reductionism. That doctrine states that all properties of a system are reducible to properties of its parts ... Chaos theory gives no examples of "holistic" properties which could serve as counterexamples to such a claim. No researcher is likely to say, for instance, that the positive Lyapunov exponent of a system is a property of the system as a whole which is ontologically distinct from the properties of all its parts. (Kellert (1993), pp. 89–90)

Here we see Kellert arguing using the received dichotomy: If higher-level properties are composed, then they cannot be "holistic" or S-emergent properties that are "ontologically distinct" from lower-level entities. But all the higher-level properties of a chaotic system are composed. Therefore such systems, regardless of what scientists apparently say about them, do not, *really*, have "holistic" or S-emergent properties after all. But what kind of emergence is there? Apparently, just like Kim and Bedau, Kellert implies that we can only have W-emergent properties after all, whatever working scientists suggest to the contrary.

5.4.3 Scientific Emergentism and its Claims about S-Emergence: Overlooked Ontological Insights?

Interestingly, as I noted above, many scientific emergentists appear to hold sophisticated versions of Enriched positions that seek to articulate ontological reasons why S-emergence can exist. Unfortunately, philosophical critics have largely ignored what scientific emergentists say about the ontological features of the scientific cases they claim to involve S-emergence. I therefore want to briefly explore some of these ideas to see whether they have any promise as defenses of S-emergence and hence also as responses to scientific reductionism and its arguments.

Given the complexity and novelty of their ideas, I examine scientific emergentism in stages. In this section, my goal is simply to give a broad-brush picture of key ideas and their potential implications. I then spend the following two chapters in a more detailed articulation, and assessment, of scientific emergentism and its ontological commitments. If the reader finds some of the ideas outlined in this section puzzling, then they can be assured that a more detailed examination of them follows in the subsequent chapters.

We get a feel for some of the most significant ontological claims of scientific emergentism when Laughlin tells us that:

I think primitive organizational phenomena such as weather have something of lasting importance to tell us about more complex ones, including ourselves: their primitiveness enables us to demonstrate with certainty that they are ruled by microscopic laws but also, paradoxically, that some of their more sophisticated aspects are insensitive to the details of these laws. In other words, we are able to *prove* in these simple cases that the organization can ... begin to transcend the parts from which it is made. What physical science thus has to tell us is that the whole being more than the sum of the parts is not merely a concept but a physical phenomenon. Nature is regulated not only by a microscopic rule base but by powerful and general principles of organization. (Laughlin (2005), p. xiv. Original emphasis)

Here we see the claim that the existence of emergence is not based upon ignorance, but putatively derives from our detailed quantitative understanding of the behavior of certain components in both simple and complex collectives, and the laws that are needed to fully explain the behavior of components in such complex collectives. Laughlin argues that we know the "microscopic laws" in simpler systems, but in certain complex collectives we now see that further laws must also come into play given what our quantitative understanding illuminates about the behavior and powers of the relevant components in these complex collectives. We also see Laughlin emphasizing that the result of these laws is not merely an epistemic phenomenon like W-emergence, but an ontologically robust form of emergence. When combined with Laughlin's claim that all higher-level properties are composed, which I noted earlier, then we see Laughlin is not simply asserting the existence of S-emergence. Instead, Laughlin is offering an Enriched account of S-emergence built upon a novel, and apparently sophisticated, set of ontological commitments drawn from empirical cases.

Central to Laughlin's position is the claim that our empirical research, and crucially the quantitative accounts of components in collectives that it supplies, has shown that components behave differently in complex collectives. But how are these ideas about "parts" behaving differently in "wholes" supposed to underpin S-emergence? Unpacking the claims here obviously takes some work given the complexity of the view, and its various commitments about determination, parts/wholes, and laws, but I can offer a crude first pass.

Laughlin is apparently rejecting what I earlier termed the Simple view of aggregation, which we saw is an ontological cornerstone of the dominant strain of scientific reductionism. Recall that the Simple view claims the aggregation of components is continuous and only involves determination by other components. In contrast, Laughlin espouses what I shall roughly term the "*Conditioned* view of aggregation." For Laughlin is suggesting that our empirical findings show that certain components sometimes contribute different powers, and hence behave differently, under the condition of composing a certain higher-level entity, but where the component would not contribute these powers if the laws applying in simpler collectives exhausted the laws applying in the complex collective. Let us call such powers contributed by a component "differential powers."[15] Finally, Laughlin apparently contends

---

[15] I should note that I have defined differential powers to leave it open whether their contribution is determined by a composed entity or other component

that the composed entity determines that the relevant component contributes its differential powers.

All manner of issues and questions arise about such claims and I will seek to engage those concerns in coming chapters. However, we can begin to grasp the purported significance of these ideas for S-emergence. For these commitments may allow an option that is missed by reasoning like the Argument from Composition. The basic point is that when the Conditioned view of aggregation holds true, then a realizer property instance, i.e. a component entity, may only contribute certain differential powers when realizing a certain higher scientific property. And scientific emergentists take the *realized instance* to determine the contribution of differential powers by its realizers. As a result, it appears that such a property instance can be determinative not simply by contributing powers itself, but also by determining the contributions of powers by *other* property instances including its own realizers. Thus we have an overlooked way, or set of ways, to potentially defend the further claim that "Wholes are more than the sum of their parts" because "Parts behave differently in wholes." In such a situation we would have properties that are realized, but which still play a determinative role in nature both by determining the powers of their realizers, their "parts," and also by their own contributions of powers producing effects at their own level. When the Conditioned view of aggregation holds, we would thus potentially have composed properties that we ought to accept as determinative despite being realized – and hence have S-emergent properties!

5.4.4 The Obligations of S-Emergentism: Pursuing a Theoretical Task

Both the scientific and philosophical proponents of reasoning like the Argument from Composition will contest the coherence of the latter ideas of scientific emergentism, but the novel claims of the emergentist at least have potential as a response to scientific reductionism. However, there is much work to be done to give scientific

---

level entities. I also intend differential powers to include not only extra powers that add to the powers contributed in simpler collectives, but also contracted sets of powers excluding powers contributed in simpler collectives. There can also be mixed cases where differential powers are both added and subtracted.

emergentism's key ideas and arguments a clearer theoretical formulation and then to assess whether the broached type of position is even coherent, let alone a successful response to scientific reductionism and the Argument from Composition. Nonetheless, my initial survey of the ideas of the scientific emergentist raises the suspicion that exploring her views may be worthwhile. For scientific emergentists like Laughlin do not appear to be so confused and in need of "charitable reinterpretation" after all. Far from being unaware of the concerns that drive the scientific reductionist's reasoning, whether framed as the Argument from Composition or in some other manner, the core ontological ideas of the scientific emergentist just outlined suggest she may have offered interesting ontological reasons, driven by empirical findings, that both seek to illuminate why we sometimes ought to accept S-emergence and also why the arguments of scientific reductionism may be in error.

The latter points, and my earlier work, illustrate the most pressing obligation of scientific emergentism. As I noted in the Introduction, one strand of thought, what I termed the More Data strategy, contends that the most pressing obligation of scientific emergentism is to gather more *empirical* evidence about cases of S-emergence. But my work highlighting the charitable reinterpretations that critics offer of scientific emergentism, and the reasoning behind them, allows us to diagnose why the reams of empirical evidence about scientific cases of emergence, and hence the More Data strategy, have made such little headway with critics whether in philosophy *or* the sciences. Even given more data, the Argument from Composition, or similar reasoning, in combination with our empirical evidence for the applicability of concepts of composition, still leads to the conclusion that emergence simply cannot be anything more robust than W-emergence.

Supplying more empirical evidence about putative cases of S-emergence thus makes no headway against reasoning like the Argument from Composition, and associated charitable reinterpretation. So we can see that the most pressing obligation of S-emergentists presently is to provide sharper *theoretical* frameworks for their ontological ideas and then use this work, if they can, to rebut the Argument from Composition and related reasoning. Otherwise the empirical findings, the "data," offered by scientific emergentists in concrete scientific cases will continue to be dialectically ineffective however remarkable this may seem.

5.5 Jobs for Philosophers and Three Rules by Which to Live or Die in an Age of Emergence or Reduction

Overall, my survey has highlighted both the real problems, but also substantive insights, of recent discussions of emergence. But we can fruitfully harvest the insights only if we do something to systematically resolve the problems. In concluding, I therefore want to use my findings, both about the debates over reduction as well as emergence, to suggest some guidelines that, if followed, would increase the productivity of future discussions by avoiding some of the problematic patterns running through recent discussions. To this end, consider these three simple "rules" for future discussions about both emergence and also reduction:

> *First Rule*: Remember that evidence for Qualitative emergence and/ or scientific composition starts the fight over "emergence" and "reduction," rather than finishing it.
>
> *Second Rule*: Assume that there is no such thing as "the" concept of "emergence" or "reduction."
>
> *Third Rule*: Guard against using theoretically unarticulated assertions of "emergence" or "reduction" that are either vacuous or damaging or both.

My goal in offering these rules is to spell out some of the dangers I have illuminated and ways to avoid them in the future, so let me briefly sketch a rationale for each rule.

Anyone who has read recent work inspired by, excited by, or generally focused upon emergence has come across writers, of the kind I noted earlier, who point to Qualitatively emergent composed phenomena and then conclude that such emergence leads to all kinds of substantive and surprising conclusions. Equally common are writers who point to other features of scientific composition, such as the novel powers or processes of composed entities, and claim they also support important claims. However, my work has highlighted, in a range of ways, why acknowledging the evidence for the existence of Qualitative emergence, and scientific notions of composition more generally, is the *beginning* of the battles over the structure of nature, and the sciences, in which we are now involved, since these phenomena drive the arguments of scientific reductionism. Respecting the First Rule thus avoids the initial mistake of overlooking the challenge of scientific reductionism.

A second problem that I have confronted is that all too often in recent debates different notions of emergence are not distinguished and general conclusions about "emergence" are drawn from a specific concept when these conclusions do not apply to other notions of emergence. Part II showed that similar points hold about notions of reduction. Embracing the Second Rule forces us to acknowledge that when we find someone discussing emergence or reduction, or when we confront a phenomenon that we think involves emergence or reduction, there are a range of notions that this person could be using or which we can use to understand the relevant phenomenon. Rather than simply assuming one is using "the" concept of emergence or reduction in understanding other researchers, or various phenomena, we need to respect the Second Rule and remember that there are a range of notions of emergence and reduction, and that reasons need to be given about why a certain notion is the most suitable for the work at hand.

Lastly, by following the Third Rule we provide detailed articulations of the conception of emergence or reduction that we are positing, or opposing, and the reasons supporting our contentions. What name or term one uses for one's favored concept is basically unimportant, so long as one articulates the underlying claims that one is making through one's assertion of emergence or reduction and the support one gives for them. If everyone follows the Third Rule, then researchers can consequently more clearly see who they agree/disagree with on various issues, and why, hence also promoting applications of the Second and First Rules and allowing us to move the debate forward.

When one follows these three rules, the substantive concepts of emergence and reduction I have now articulated plausibly have a central, and significant, role to play in a number of debates in the sciences and philosophy. In particular, my work in this chapter shows that concepts of W- and S-emergence have potentially important contributions to make to scientific debates over reduction and emergence by articulating opposing interpretations of cases where we have successful compositional explanations – W-emergence provides a useful tool for scientific reductionism and a substantive species of S-emergence does so for scientific emergentism.

However, my primary focus in this chapter was to assess which concepts of emergence, if any, might provide aid in assessing scientific reductionism. So let me summarize my findings about this question. If we are interested in the *critical* assessment of scientific reductionism, then I showed that exploring W-emergence is ineffective because its existence is actually a key element of scientific reductionism. And with regard to O-emergence I concluded that it offers little help in assessing the key argument of scientific reductionism, since O-emergence does not exist in the relevant cases where we have successful compositional explanations and leaves the scientific reductionist's arguments untouched.

In contrast, I highlighted how looking further at the ideas of the scientific emergentists who defend Enriched views of S-emergence has two connected virtues. First, and most importantly for my present purposes, the most pressing obligation facing proponents of S-emergence is to show that the Argument from Composition is mistaken and that there can be situations in which we ought to accept S-emergence. Exploring the scientific emergentist's complex view of S-emergence is thus indirectly also a critical examination of the core reasoning, and ontological assumptions, of the scientific reductionist. Second, scientific emergentism is also an interesting position in its own right not least because it is potentially at an evidential "sweet spot." In contrast to O-emergence, S-emergence has the capacity to accommodate the scientific evidence for compositional relations between higher- and lower-level entities apparently supplied by compositional explanations. Whilst in contrast to W-emergence, S-emergence has the potential to accommodate the apparent evidence supporting the existence, and determinative role, of composed entities in nature. S-emergentism thus potentially accommodates a far wider range of scientific evidence than rival notions of emergence – *if* a position endorsing S-emergence can successfully address its theoretical challenges.

The final conclusion of my survey of emergence is that pursuing the theoretical project of exploring, and more carefully articulating, the novel ontological ideas offered by scientific emergentism about the existence of S-emergence (or what I from hereon term "Strong" emergence) offers the most promising vehicle with which to critically engage the Fundamentalism of the scientific reductionist. Happily, this is a job for philosophers, or other theoreticians, and my frameworks for scientific composition and reductionism provide a platform for just such a project. So, in the remainder of Part III, I focus on reconstructing the key ideas of scientific emergentism and assessing whether there is any situation in which we ought to accept Strong emergence, contrary to the key argument of scientific and philosophical reductionists.

6 | A Whole Lot More from "Nothing But": Conditioned Aggregation, Machresis, and the Possibility of Strong Emergence

> ... "emergence" does not mean mysteries popping out of the undergrowth; it means that with a sufficient understanding of interactive processes, we should come to understand why a complex whole has properties its parts lack on their own, and how the parts are modified by the context in which they lie.
>
> – Richard Gregory[1]

Scientific examples of emergence from superconductivity to slime mold, and beyond, have driven the views of scientific emergentists. As the psychologist Gregory makes clear, these cases do not involve mysterious uncomposed entities, like souls, élan vital, or entelechies. Instead, natural phenomena have led researchers to suggest, as Gregory highlights, that components behave differently, and obey new fundamental laws, in emergent composed entities. However, such claims of scientific emergentists are yet to be given a more precise articulation and, as we have now seen, critics question whether there is any coherent position to be clarified at all. Our most pressing business with scientific emergentism, as it was with scientific reductionism, is thus to provide a framework for its core ontological claims and assess whether they are coherent.

I start with what scientific emergentists themselves say by surveying the claims of researchers from a range of higher and lower sciences to draw out their array of sophisticated claims about Strong emergence. To more carefully explore and then evaluate the promise of these new ideas, I follow a common scientific procedure for assessing theoretical claims by constructing a thought experiment using scientific concepts to show that Strong emergence is possible in a case of compositional explanation. Furthermore, my scientific thought experiment also establishes that the scientific reductionist's Argument from

---

[1] Gregory (1994), p. 360.

Composition is *invalid* and that the Unholy Alliance of philosophers has endorsed a number of *false dichotomies* about the viable positions and coherent forms of emergence.

Overall, I establish that, driven by their insights about real cases, scientific emergentists have appreciated that mutually determining, and interdependent, composed and component entities can make it true that "Parts behave differently in wholes." There are thus no mysteries with Strong emergence – no strange new fundamental forces or kooky élan vital – for the scientific emergentist simply claims that emergent composed entities non-productively determine new behaviors and powers of their components. Scientific emergentism thus illuminates how composed entities can be "nothing but" their components, in the sense of being composed by them, and yet *still* play a determinative role in nature.

6.1 The Ontological Insights of Scientific Emergentism: Parts, Wholes, Aggregation, Emergence, and More

To give the flavor of these researchers' core ideas, in 6.1.1, I examine passages from a number of scientific emergentists. I highlight the sophisticated array of interrelated commitments endorsed by these writers, in 6.1.2, but I foreground their Conditioned view of aggregation and commitment to a novel type of non-productive, but non-compositional, determination between composed and component entities. Finally, in 6.1.3, I outline why, once we have a clearer grip upon its character, the overall package of ontological claims presented by these scientific researchers offers a potential insight into the nature and possibility of Strong emergence.

6.1.1 *A Different Universe: The Emergentism of Couzin and Krause, Freeman, and Laughlin*

We find scientific emergentists in many disciplines, but let me begin with two researchers working in a very high-level science. Iain Couzin and Jens Krause describe what they take to determine the flocking behaviors in birds as follows:

[I]t is usually not possible to predict how interactions among a large number of components within a system result in population-level properties. Such systems often exhibit a recursive, non-linear relationship between the individual behavior and the collective ("higher order") properties generated by these interactions; the individual interactions create a larger-scale structure, which influences the behavior of the individuals, which changes the higher-order structure, and so on. (Couzin and Krause (2003), p. 2)

Here we see very much to the fore the notion that "Parts behave differently in wholes" and the further claim that the composed entity "influences" the component's altered behaviors and powers. If we descend down the levels of the sciences for our next example, we find the neuroscientist Walter Freeman making similar claims about the relations of neurons and neuronal populations. Freeman tells us:

An elementary example is the self-organization of a neural population by its component neurons. The neuropil in each area of cortex contains millions of neurons interacting by synaptic transmission. The density of action is low, diffuse and widespread. Under the impact of sensory stimulation, by the release from other parts of the brain of neuromodulatory chemicals ... all the neurons come together and form a mesoscopic pattern of activity. This pattern simultaneously constrains the activities of the neurons that support it. The microscopic activity flows in one direction, upward in the hierarchy, and simultaneously the macroscopic activity flows in the other direction, downward. (Freeman (2000a), pp. 131–2)

Notice that once again we have no mysteries here in the sense of uncomposed entities at the higher level that "jump from the bushes," as Gregory puts it, since the neural population, like the flock of birds, is completely composed. Instead, Freeman's claim is rather that the *components* behave differently, and have different powers, when they compose a certain higher-level entity – and that this composed entity is responsible for the altered behavior and powers of its components. Like Couzin and Krause, Freeman thus posits "upward" compositional determination, but also some kind of "constraint" and "downward" determination by the relevant composed entity on its components.

Moving toward the lower levels of nature and the sciences we this time find Laughlin, like Anderson (1972) before him, focused on examples of symmetry breaking in physics. Even at this low level, Laughlin is making strikingly similar ontological claims to these other scientists when he tells us:

The idea of symmetry breaking is simple: matter collectively and spontaneously acquires a property or preference not present in the underlying rules themselves. For example, when atoms order into a crystal, they acquire preferred positions, even though there was nothing preferred about these positions before the crystal formed. When a piece of iron becomes magnetic, the magnetism spontaneously selects a direction in which to point. These effects are important because they prove that organizational principles can give primitive matter a mind of its own and empower it to make decisions ... Symmetry breaking provides a simple convincing example of how nature can become richly complex all on its own despite having underlying rules that are simple. (Laughlin (2005), p. 44)

There is again a lot to unpack in this passage. As well as making similar claims about the relations of the composed and component entities, Laughlin also adds important observations about the laws involved in such cases.

We should note that Laughlin is committed, like the three researchers from the higher sciences, to his cases being ones where "Parts behave differently in wholes." Thus the atoms have preferred positions in the crystal, which they lack elsewhere. And Laughlin uses his observations about laws to more precisely articulate how such differences arise and to imply they are not the kinds of different powers of aggregated components posited under scientific reductionism. For Laughlin, repeating points I highlighted in the last chapter, claims that concrete cases of emergence are to be found where we now know that the laws that hold of components in simpler collectives cannot *alone* exhaustively determine the behaviors and powers of the components in more complex aggregations. The behaviors and powers involved with the "preferences" in the more complex structures are not completely determined by the laws holding in simpler collectives. Thus components in these complex aggregations have what I dubbed differential powers and "Parts behave differently in wholes" in a stronger sense than that endorsed by reductionists such as Weinberg. As a consequence, Laughlin concludes that "Nature is regulated not only by a microscopic rule base but by powerful and general principles of organization."[2] These new laws, or "principles of organization," cover the novel behaviors and differential powers

---

[2] Laughlin (2005), p. xi

of the components that compose specific higher-level entities, such as superconductors, magnets, or crystals, which are emergent along with their properties.

Again, Laughlin thus accepts that the relevant higher-level entities are fully composed and is at pains to explicitly reject dualism, vitalism, or any other kind of Anti-Physicalism.[3] Laughlin thus concludes that we do not have Ontological emergence. But Laughlin claims our better understanding of concrete scientific cases establishes that composed entities determine the contribution of differential powers by component entities. Consequently, Laughlin is clear that the emergence we get is not the conceptual or epistemic variety we find with Weak emergence. And Laughlin also tells us:

... we are able to *prove* in these simple cases that the organization can ... begin to transcend the parts from which it is made. What physical science thus has to tell us is that the whole being more than the sum of the parts is not merely a concept but a physical phenomenon. (Laughlin (2005), p. xiv. Original emphasis)

Here we see Laughlin linking the claim that "Parts behave differently in wholes" to the more famous emergentist thesis that "Wholes are more than the sum of their parts." For based upon our quantitative findings about components across simpler and more complex collectives, Laughlin, and the other researchers, are claiming that composed entities have been shown to play a determinative role in the universe by determining the powers and behaviors of their *components*.

As a result, composed entities play a determinative role in the universe by determining the powers of their components, as well as through their own productive roles, and hence such composed entities *are* determinative *in addition* to their components. Similar points apply to Freeman, and Couzin and Krause, where these scientific researchers are all plausibly taken to endorse Strong emergence, rather than merely Weak (or as Laughlin puts it "conceptual") emergence or the Ontological emergence of Anti-Physicalist philosophers.

---

[3] Laughlin (2005), p. 6.

6.1.2 An Initial Assay of Scientific Emergentist Commitments: The Conditioned View of Aggregation, Machretic Determination, and other Claims

Rather than merely spouting empty slogans, as we have seen some philosophers of science would have us believe, Couzin and Krause, Freeman, and Laughlin are like other scientific emergentists in making a rich array of substantive, and interrelated, ontological claims. I cannot examine all of these claims at once, so let me therefore briefly summarize some of the central ontological commitments of scientific emergentists and then focus in most detail on a couple that are especially relevant to assessing the coherence of Strong emergence.

First, as I noted in Chapter 1, our scientific emergentists accept both compositional relations between emergent and lower-level entities, and also the ubiquitous existence of the compositional explanations that illuminate and utilize such relations. Second, scientific emergentists like Laughlin claim that new fundamental "emergent" or "organizational" laws, indexed to "organizational" phenomena like composed entities, apply to components in the complex collectives that compose certain higher-level entities. Such organizational laws concern the novel differential powers of components had as a result of the determinative influence of the emergent entities they compose and such laws thus refer to these emergent entities. More attention needs to be given to the character of these "organizational" laws and I pursue this task in the next chapter.

Third, and connected to the latter points about laws, we have seen that Couzin and Krause, Freeman, and Laughlin all apparently endorse a very different account of aggregation than the reductionist's Simple view. These writers deny that the powers contributed by component properties to individuals as they aggregate are continuous in character, thus denying that we have a common principle of composition for such components across all collectives. In certain complex collectives, these emergentists contend, we have discontinuities where components contribute differential powers not accounted for by the laws and/or principles of composition holding of such components in simpler collectives. Crucially, these writers contend that the new powers and behaviors of the components are *not* those that they would have if the laws applying in the simpler collectives were the *only* laws applying to the complex, composite systems – hence leading to differential powers and Laughlin's connected claim that further "organizational" laws also hold in these cases.

The result is the Conditioned view of aggregation under which aggregation is more complex than reductionism has long assumed. But this commitment to extra complexity in the structure of nature is not arbitrary, nor is it based upon our ignorance. Instead, we have already seen that our scientific emergentist researchers contend that our better, usually quantitative, understanding of components in simpler and more complex collectives, and/or the laws that hold of them, is what forces acceptance of the Conditioned view of aggregation in certain cases of compositional explanation.

Fourth, our emergentists take composed entities to be responsible for the altered contributions of powers by components. Abandoning the Simple view of aggregation opens the space for this novel kind of determination. If aggregation is always continuous, and the laws holding in the simplest collectives (which only concern components) exhaustively determine components under all conditions, then components are plausibly only ever determined, in so far as they are determined at all, by *other components*. In contrast, when the Conditioned view of aggregation holds, then components have novel differential powers, under certain conditions, which are *not* determined in the ways described by the laws holding in the simplest collectives that only involve other components. We thus have the space for a novel variety of determination between composed and component entities.

My examination of the writings of scientific emergentists thus confirms their commitment to the novel kind of "downward" determination that I highlighted in Chapter 1. With regard to such relations, we saw Couzin and Krause talk of "influence," Freeman uses the term "constraint" in the passage we looked at above though elsewhere he refers to "circular causality" (Freeman (1999), p. 159), and Laughlin also uses the term "cause." However, even where these researchers use the term "cause" it is charitable to interpret them as simply meaning "determine in some way," rather than using "cause" in the very specific meaning that philosophers give to this term. For example, all of these writers take the relevant determination to hold between entities that compose each other and which are hence in some sense the same. In addition, Laughlin explicitly states that the determination of components by "organizational" entities is "spontaneous" and hence apparently synchronous in nature. These features plausibly mean such determination is *not* a causal relation or a productive relation based around the triggering and manifestation of powers.[4]

Although its relata bear compositional relations, this novel non-causal, non-productive determination relation is also plausibly not itself a compositional relation. This relation is not based upon joint productive role-filling, since the processes and powers of the composed entity are not such that they can fill the roles of their components. Instead, as their remarks suggest, these scientists assume we have non-causal, non-productive determination of components by composed entities through something like productive role-*shaping* or role-*constraining*, rather than role-*filling*.

Our scientific researchers have thus apparently been concerned with a kind of non-productive, but also *non*-compositional, determination that has been overlooked by philosophers. We thus need a new term to mark this kind of determination relation. Combining the Greek words "macro" and "chresis," where the latter is roughly the Greek for "use," we get the terms "machresis," and "machretic determination," for the general phenomenon of composed, or "macro," entities that non-productively, and non-compositionally, determine the nature of their components through productive role-shaping. This type of determination again deserves more attention and I examine it further in the next chapter, but let me press on to other commitments.

Fifth, given the latter, our scientists endorse the existence of *mutually determinative*, and *interdependent*, composed and component entities. As we saw Freeman, Laughlin, and Couzin and Krause all explicitly contend, we have non-productive determination in "upward" compositional relations, whilst also having "downward" non-productive machretic determination as well, thus leaving us with *mutual* determination between the relevant composed and component entities. I am therefore going to refer to the resulting overall kind of position we find in scientific emergentists as "Mutualism," or a "Mutualist" view, to contrast with the Fundamentalism of scientific reductionists.

Furthermore, the altered components only exist because of the machretic determination of the Strongly emergent composed entity, which in turn only exists due to its components, resulting

---

[4] In the next chapter I outline more detailed reasons why this has to be a non-productive and non-causal relation if it is to allow Strong emergence.

in interdependent entities. We should carefully note that it is only the differential powers contributed, and hence the *altered* components in the realizer property instances that contribute them, that are dependent, since in simpler collectives, or in isolation, we can often have such realizer instances when they are not contributing the differential powers and without the relevant composed entity. And it is of course true that the composed entities are also dependent upon their components. But, under Mutualism, such altered components mean we have *dependent components* of a strikingly novel kind that Fundamentalists, like scientific reductionists and their philosophical allies, have rarely envisioned.

Sixth, and finally, as a result of the foregoing commitments we have seen that Couzin and Krause, Freeman, and Laughlin take emergent entities to be determinative, rather than being examples of Weak emergence involving merely epistemic phenomena. However, this robust emergence is not a form of Ontological emergence, since all of the relevant emergent entities are taken to be composed. Since they endorse composed entities that are nonetheless determinative it is most plausible to interpret these writers as defending the existence of Strong emergence in the cases they are focused upon, contrary to the "charitable reinterpretations" often pressed by philosophers. My brief survey of their ontological commitments thus allows us to confirm that our researchers do *not* simply *assert* that we have Strong emergence in what I earlier termed Naked positions. Instead, scientific emergentists hold Enriched views of Strong emergence offering a sophisticated and novel set of ontological commitments to putatively illuminate *how* we may have Strongly emergent entities.

6.1.3 A Deep Insight? Glimmers of a Possibility for Strong Emergence

Scientific emergentism rejects a couple of the reductionist's epochal assumptions, including the Simple view of aggregation and the Completeness of Physics, so unsurprisingly scientific emergentism has rather different implications than Fundamentalism. As a result of appreciating these different implications, scientific emergentism claims to teach us a lesson about the number of ways an entity can be determinative at a time. (Again, I state the point for property instances, but one can generalize it to individuals, processes, and powers.)

Although all of these options may only be open when the Conditioned view of aggregation holds, nonetheless if scientific emergentism is correct then it appears that there are *three* ways a property instance can make a difference to the powers of individuals at a time t. *Either* (i) the property instance solely contributes powers to individuals at t; *or* (ii) the property instance solely determines the powers contributed to individuals by other property instances at t; *or* (iii) at time t the property instance both contributes powers to individuals itself and also determines the powers contributed to individuals by other property instances.

The upshot of the latter points is that the Argument from Composition may embody a flawed account of the ways in which a property instance can determine the powers of individuals at a time, for this reasoning is focused solely on the powers contributed by a property instance itself – hence neglecting (ii) and (iii). For when we have Conditioned aggregation, then *alongside* the powers contributed by the emergent realized property that produce certain higher-level effects, scientific emergentists contend it is *also* the case that realizers only contribute some of their powers as a result of the machretic determination of the emergent realized property instance. Thus a realized property instance may be determinative *both* by determining the contributions of differential powers by its realizers, *and also* through its own contributions of powers.[5]

6.2 Exploring the Possibility of Strong Emergence

Little explicit philosophical attention has been given to a world in which the Conditioned view of aggregation holds, let alone alongside the machretic "whole"–"part" determination posited by the scientific emergentist. I therefore propose to explore a detailed scenario for just such a situation. I begin, in 6.2.1, by describing my methodological approach based around a common use of thought experiments in the sciences. Against this background, I then apply my earlier framework

---

[5] It can actually be still more complex, since the realized property instance my also itself be a realizer of still higher-level property instances and hence compositionally determine their powers as well. But I leave this further complexity to one side in this chapter, where I focus on machretic relations to bring out some of the key insights of scientific emergentism.

for scientific composition, in 6.2.2, in a scientific thought experiment built around a concretely described scenario in which the Conditioned view holds true and where we also have machresis. I show that in my scenario, first, all of the higher-level property instances are indeed realized and then, second, that these realized properties are nonetheless determinative. In 6.2.3, I briefly summarize my main findings from my thought experiment.

6.2.1 Thought Experiments in the Sciences and the Bounds of Composition

Scientists such as Einstein and Galileo, to name just two of the most famous, often use thought experiments to establish conclusions, especially when a theoretical issue is in dispute. Central to one kind of thought experiment is the assumption of well-understood scientific concepts: One considers a representation where these well-confirmed concepts apply and then one considers what can also be the case in order to show what is, or is not, allowed by these concepts – that is, what is, and is not, possible when these concepts comprehensively apply.[6] This type of thought experiment is useful for my purposes for three reasons.

First, we have seen that scientific and philosophical reductionists, in effect, claim that concepts of scientific composition, when they comprehensively hold of some situation, preclude acceptance of that Strong emergence. Scientific emergentists defend the opposite claim. A thought experiment focused on concepts of scientific composition would thus speak to the key theoretical dispute between the two sides over what concepts of composition do, and do not, allow us to accept. Second, I have now supplied a framework for scientific concepts of composition, so I am in a position to mount just this kind of thought experiment using such concepts. And, third, I have illuminated substantive ideas drawn from scientific emergentist researchers, in the

---

[6] There are obviously other kinds of scientific thought experiment and also competing accounts of their nature. See Brown (1991) for a survey of a variety of different thought experiments in the sciences. The two most prominent competing accounts of the nature of thought experiments in the sciences are nicely contrasted in Brown (2004) and Norton (2004). Our thought experiment, given its structure, can be charitably adapted to fit the constraints of either view.

Conditioned view and machretic determination, which I can explore in such an exercise.

I am therefore nicely situated to pursue such a scientific thought experiment. But it is important to remember that I am assessing claims that Strongly emergent entities are akin to square-circles. This is significant because it constrains what we are, and are not, allowed to assume in our thought experiment. Consider the situation with someone seeking to establish the possibility of a square-circle through a thought experiment. The critic allows that anything may be assumed in the thought-experiment, for the critic claims that nothing we imagine can coherently be a square-circle. Of course, since this is a scientific thought experiment, we do need to constrain our scenario so that it does not conflict with well-confirmed background assumptions, but otherwise we are free in our assumptions.[7] Thus, in the case at hand, I am plausibly allowed to assume both the Conditioned view and machretic determination in my scenario. For the scientific reductionist is committed to such assumptions being of no use – nothing we imagine, claims the scientific reductionist, results in a coherent representation of a composed entity that is still determinative.

The latter points should flag for the reader that although superficially similar to the thought experiments commonly used amongst philosophers, I am pursuing the far more constrained procedure common in the sciences. Writers such as Kathleen Wilkes (1988) have raised a number of important worries about whether the unconstrained thought experiments in philosophy can actually be trusted to establish interesting conclusions. In contrast to such philosophical exercises, my thought experiment is based upon well-confirmed scientific concepts, and is compatible with well-confirmed background assumptions, in the ways that Wilkes and others suggest is necessary for a more productive use of thought experiments.

6.2.2 *A Scenario for Conditioned Aggregation with Machresis: Blobs and their Collectives, Revisited*

Although the reader will see features in my scenario that mirror those of the concrete cases that Couzin and Krause, Freeman, Laughlin, and

---

[7] For example, the scenario should not posit anything known to be nomologically impossible, or which conflicts with well-confirmed theories, and so on.

other scientists all highlight, I am *not* looking at a real case. As we shall see in later chapters, interpreting the ontological structure of concrete cases is a difficult and detailed process over which the various parties are presently battling. A large part of the utility of using a thought experiment is that it avoids these difficulties and provides us with a simple, imagined case that cleanly and directly focuses on the heart of the theoretical issue in dispute.

To pursue my scenario, I am therefore going to focus once again on my fancifully named "blobs" and "globules" examined in Chapter 3, but with a couple of additions. I will assume that the Conditioned view of aggregation holds of blobs in the complex collectives that constitute globules. And I also assume that in such complex collectives the properties of globules machretically determine some of the differential powers contributed by the properties of the blobs that are their constituents. However, for simple collectives of blobs the scenario is the same as that used to illuminate scientific reductionism, so the example is also useful in easily contrasting the implications of the various theses held by competing positions.

To reprise the situation, I am assuming that blobs are individuals that can exist in simple, as well as complex, collectives. In simple collectives, blobs bond with each other in a certain way and also have instances of the property, call it "P1," which exerts a weak type of attractive force on other blobs in a random direction, amongst many other features and powers. In simple collectives, instances of the property P1 thus contribute to blobs the power of weakly attracting other blobs in some random direction in addition to a range of other powers that we can leave to one side. I also again assume that when aggregated in the right way in simple collectives blobs constitute globules that have a range of realized properties. Amongst these properties, globules have the realized property, call it "F," of exerting a weak attractive force on other globules in all directions, for F is simply the attractive force that results from the powers together of all the instances of property P1 found in the blobs that constitute such a simple globule. The left side of Figure 6A illustrates the familiar situation in such simple collectives.

In addition, of course, I am now also stipulating that in the new scenario the Conditioned view of aggregation holds true in certain complex collectives. So I assume that at a certain point we get complex collectives of blobs that constitute a globule that as well as shape, and

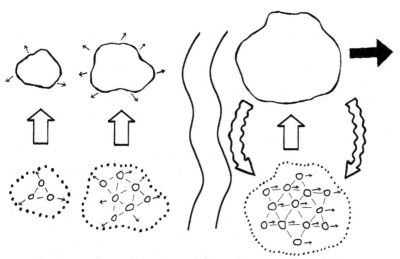

Figure 6A The scenario in our thought experiment. On the left we again have the blobs and globules found with simple aggregations, whilst on the right we have the nature and relations of the blobs and globules found with complex aggregations of blobs. (Again, the large circles at the top are globules, small circles at bottom are blobs; the enclosed upward arrows are relations of composition; the enclosed downward wavy arrows are machretic relations; straight solid lines are bonds; solid arrows are forces where the size of arrow equals the magnitude of the force.)

other properties, also has the distinctive property of exerting a strong attractive force on other globules in a particular direction. Let me call this property of globules "G." Furthermore, I also stipulate that when a globule has an instance of the property G, then the instances of property P1 in its constituent blobs continue to contribute the power of attracting blobs with the same magnitude they have in simple collectives, but *also* contribute the further power, "C1," of attracting blobs only in the same preferred direction as the other constituent blobs in the relevant globule. In addition, I assume the instances of P1 contribute all the other powers they contribute in simpler collectives aside from the power of attracting in a random direction. That is, the instances of property P1 of the blobs contribute the powers they previously contributed, but also one more power that is different from that which they contribute in simpler collectives. It will help if I dub "Pz" the lower property instances that are produced as effects of the power C1 and also assume that G contributes powers to produce events involving instances of a higher-level property "H" that is realized by various instances of Pz together. Let me also stipulate, what should be obvious from the description of the simple collectives, that the power C1 is a differential power that would not be contributed if the laws applying to simpler collectives accounted for the contributions of powers by instances of P1 in such complex collectives.

In the globules found with complex collectives, blobs thus continue to bond with each other and continue to exert a weak force with the same magnitude, amongst other powers, but now the force of each blob is exerted only in the same preferred direction as that of other blobs constituting the globule. Finally, following the idea of scientific emergentists, let me also stipulate that the instance of property G of a complex globule *machretically* determines, i.e. non-productively and non-compositionally determines, that instances of property P1, in the constituent blobs, contribute their differential power C1. My full scenario is outlined in Figure 6A, with the simple collectives of blobs and resulting globules on the left, and the complex collectives of blobs and associated globules on the right.

With a better grip on its features, let me now examine what holds true in our scenario for Conditioned aggregation and machresis.[8] First, we should note that all of the powerful properties/relations of the globules like the property G, such as attracting other globules, or its mass or shape, are plausibly realized by properties/relations of blobs. For example, consider the instance of property G and its power to strongly attract globules in a certain direction. This power is comprised by the many powers of attracting blobs in a certain direction had by each of the blobs that constitute the globule. And the processes based by the globule are plausibly implemented by processes based by the blobs. For instance, the process of attracting globules is implemented by the processes of attracting blobs. Thus the instance of G is realized and the processes it grounds are implemented. Similar points hold for the other powerful properties/relations and processes of the globule, including more mundane properties like shape or mass and the processes they ground. It thus appears that concepts of scientific composition *comprehensively* apply to the higher-level property instances in my scenario.

---

[8] Abstract characterizations of this kind of scenario are offered in Gillett (2003a), (2003b), (2006a), (2006b) and (2011b).

What about the determinativity of the instance of G in the globule? The properties of the constituent blobs contribute to them the power to attract blobs in a preferred direction. And this power would *not* be contributed if G were absent from the globule in which the blobs having these realizer properties are to be found. Furthermore, I have stipulated that G machretically determines the novel powers contributed by its realizers. Thus the property G does make a difference to the powers of individuals – in this case the powers of *component* individuals in the blobs. (In fact, as I outline in the next chapter, Strongly emergent properties like G also contribute powers that produce higher-level effects, in this case an event with property H, and are also joint causes of lower-level effects, but I have focused upon the machretic determination of G highlighted by working scientists.)

6.2.3 *The Possibility of Strong Emergence, Machretic Determination, and the Invalidity of the Argument from Composition*

Some readers may be wondering whether the scenario is too clear, or too simple, to establish any conclusions of substance. However, one may establish that a set of statements is logically coherent by trying to conceive of a situation, using a conceptual representation, in which these statements are all true. And with my scenario for the Conditioned view of aggregation, alongside machresis, I have successfully outlined just such a representation in which it is true that a property instance is realized, but nonetheless also true that this realized instance is determinative. My scenario also plausibly satisfies the conditions for an adequate scientific thought experiment. Consequently, the scenario provides good reasons to think that Strongly emergent property instances can indeed exist when scientific concepts of composition comprehensively apply. And this finding has important implications about the core ontological claims of scientific emergentism and the assessment of the arguments of scientific reductionism.

If we start with the latter, my thought experiment plausibly shows that in drawing its crucial sub-conclusion the Argument from Composition is *invalid*, since the sub-conclusion of this reasoning is false in our scenario whilst all of the relevant premises are true. Recall that the sub-conclusion is that in cases of comprehensive composition using components accounts for all the powers of individuals at the higher and lower level. However, in my scenario even though G is realized, and is the subject of successful compositional explanation, we still ought to accept its existence in order to account for certain powers of its components, amongst other features of the example. For in my scenario we *cannot* account for all the powers of individuals solely using realizer properties. In this type of case, we therefore ought to accept a Strongly emergent realized property instance in order to account for differential powers contributed by the realizers themselves. My thought experiment thus confirms my earlier suggestion that the Argument from Composition misses options by which properties may be determinative at a time – crucially, by determining the powers contributed by *other* properties including its own realizers.

If we turn to the illumination of the core claims of scientific emergentism, then my thought experiment helps to support the coherence of the machretic "downward" determination posited by scientists like Laughlin, Freeman, Couzin, Krause, and others. Concepts of composition apply comprehensively in my scenario, but the posited relations of machretic determination result in no obvious incoherence. We therefore have support for the coherence of machresis, but this novel relation deserves further investigation, which will be one focus of the next chapter.

6.3 Far Too Much from "Nothing But"? Objections to My Defense of the Coherence of Strong Emergence[9]

To better assess my thought experiment, and the conclusions I have drawn from it, it is useful to examine objections. It is important to distinguish two kinds of worry. The criticisms I am going to consider in this section seek to show that my use of the thought experiment to defend the coherence of Strong emergence is defective. Another type of objection to scientific emergentism attacks the actual existence of Strong emergence, but I defer consideration of such concerns to later chapters after we have a better grip on the detailed commitments of scientific emergentism.

6.3.1 The Argument from Inelegance

One objection to my thought experiment is that its structure is too inelegant or irregular to be possible. I start with this blunt objection

---

[9] The credit for the wonderful title of this section goes to Ernan McMullin.

because it is instructive for other worries. The obvious difficulty with the objection is that, once we put aside theologically motivated premises, we have no good reason to suppose that nature *must* be maximally elegant or regular in either its ontological structure and/or fundamental laws. That so many researchers appear to be attracted to such claims appears to result from our failure to be explicit about the ontological assumptions of scientific reductionism. Failing to articulate the Simple view of aggregation or the Completeness of Physics, or assess their status, apparently leaves many implicitly taking such claims to be *necessary* truths, rather than contingent foundational assumptions. And reflection suggests there is no justification for taking these claims to be necessary truths.

We thus have no reason to think that because the Mutualist picture of nature endorsed by scientific emergentism is complex, inelegant, or downright messy it means that it *could* not hold of our world. It may be that we have empirical evidence that leads us to believe that nature happens to be simple, elegant, or regular in ways that establish Strong emergence does not *actually* exist. But that evidence, even if it can be supplied, still would not support the very strong claim made in the Argument from Inelegance.

6.3.2 The New Force Objection

One of the most common objections to robust "emergence" is that it always involves new fundamental non-physical forces. This type of worry is pertinent to my thought experiment, for it may be argued that my scenario must involve a new fundamental non-physical force with any emergent property like G. But if G is a new fundamental force, concludes the objection, then G is not a realized property after all. Hence the requirements for G to be Strongly emergent are not satisfied, since it is not realized, and the thought experiment fails to show what it claims to. The latter is what I term the "New Force Objection."

This objection draws from McLaughlin's pioneering work on British Emergentism noted in the last chapter.[10] McLaughlin offers an account of an "idealized" British emergentist as endorsing the existence of fundamental, non-physical, *configurational* forces, i.e. forces

---

[10] The reader should note that I am not ascribing this objection to McLaughlin (1992).

exerted by higher-level individuals that are not composed by the fundamental forces exerted by microphysical entities. Working from McLaughlin's notion of a configurational force, the sophisticated New Force Objection first notes that, crudely put, a force is simply something that changes the motions of masses. But in my scenario the property instance G is therefore a "force," since G's causal efficacy means it plausibly changes the motions of masses. And in my scenario such "forces" only arise in certain collectives or "configurations." Thus even in my scenario, concludes the objection, we are still therefore committed to non-physical "configurational forces" of just the type McLaughlin claims we always find with such "emergentism." But since such non-physical "configurational forces" are uncomposed, then G fails to be realized in our scenario and we again fail to satisfy the criterion for a Strongly emergent property – and hence fail to show we can have such properties when scientific concepts of composition apply comprehensively.[11]

Unfortunately, the sophisticated New Force Objection is based upon a slide from a mundane use of "configurational force" as basically synonymous with "higher-level cause" to a far stronger use as meaning "ontologically fundamental force." To appreciate this point, we should first mark that if one accepts that universal composition exists, then all causal or productive effects will involve microphysical changes and very often changes in the motions of masses. Consequently, if one holds that a realized property like G is causally efficacious or productive, then it will be likely that one will take events that instantiate G to change the motions of masses in virtue of this property. Hence G is a "force" in the weak sense, but we should carefully mark that G is also realized.

The second point to note is that being a "force" in the weak sense just outlined does *not* entail that such a higher-level property is itself an ontologically *fundamental* force. When the Conditioned view of aggregation holds, and we also have machretic determination, then a realized property instance may produce higher-level effects, and be a joint cause of some change in the motions of masses, through the

---

[11] Similar arguments to the same conclusion can also easily be made about the existence of non-physical energies, as well as about what Kim (1992a) terms "non-physical intrinsic powers," in addition to forces. The response outlined below is easily adapted to rebut these objections about energies and powers in a similar fashion.

mediation of the fundamental microphysical forces and their contributions of powers that this Strongly emergent property machretically determines. However, all of the powers of such a Strongly emergent property instance are completely comprised by the powers contributed by microphysical properties and all of the processes resulting from such an emergent instance are implemented by processes of microphysical individuals. Consequently, although G may thus be termed a "force" in the weak way that any higher-level cause in a physicalist world will be a "force," G does *not* therefore have to be an ontologically *fundamental* property. Once we are careful about the weaker and stronger senses of "force," we can therefore see that the New Force Objection fails to pose any problem for my conclusions from my scientific thought experiment.

6.3.3 Objections from the Primacy or Completeness of Physics

A rather different objection argues from the primacy of physics and is articulated by David Papineau in the case of the mental, but can easily be generalized to other levels of nature. Papineau tells us that:

> It is one thing to hold that the current categories of energy, field and space-time structure leave a gap in the determination of certain physical effects. It is another to hold that this gap cannot be filled without bringing in the mental. If that were true, after all, then the obvious moral would be that physicists needn't build expensive particle accelerators to generate theoretically anomalous physical phenomena; instead they could find plenty of currently inexplicable physical phenomena simply by looking inside people's heads. I think we have good empirical reason to reject this possibility as absurd. (Papineau (1993), p. 32)

Though directly focused upon the mental, such an objection obviously translates to all higher-level phenomena. If superconductors, or populations of neurons, or other higher-level natural phenomena, are Strongly emergent, then this would imply that physicists need not build supercolliders to explore the fundamental structure of nature. Instead, we could further our fundamental understanding of nature by investigating superconductors, or neural populations, or whatever higher-level phenomenon is Strongly emergent. But, the generalized version of Papineau's objection concludes, this is an "absurd" suggestion.

We obviously need to do more work to illuminate the exact structure of this objection, though it appears either the Simple view of aggregation, or at least the Completeness of Physics, is involved in this argument. However, two quick observations are useful. First, what Papineau is claiming to be "absurd" is just what the researchers who are scientific emergentists, including three Nobel Prize winners in the physicists Anderson and Laughlin, and the chemist Prigogine, amongst others, are presently arguing is our best scientific approach! This ought to give one pause. Second, if Papineau were offering an objection to my thought experiment, and if as it appears plausible Papineau is using either the Simple view or the Completeness of Physics as premises, then Papineau would have to be taking these claims to be *necessarily*, rather than just *contingently*, true in order to press such an objection. But, again, though some writers may feel that such theses *must* be true we have no good reasons to support such contentions.

In fact, Papineau is apparently clear that our empirical evidence drives the objection, plausibly by justifying us in accepting either the Simple view or the Completeness of Physics that are consequently only contingently true claims that entail that Strong emergence does not actually exist. As we have begun to see, scientific emergentists contest the claim that empirical evidence supports either of these theses and argue our scientific evidence actually supports very different claims. I therefore defer consideration of this type of objection until Part IV, after we have more carefully articulated the claims of scientific emergentism.[12]

6.3.4 The Argument from Compositional/Mechanistic Explanation

Another interesting kind of objection to Strong emergence is based around the ubiquity of compositional explanations, primarily mechanistic

---

[12] Given their contested status, where scientific emergentists claim empirical evidence undercuts their justification, it is preemptory to take a global version of the Simple view, or the Completeness of Physics, to be well-confirmed theses that should be taken as background assumptions in our thought experiment. And, in Part IV, I show that we are not presently justified in accepting these theses are true.

explanations involving the composition of processes. The most detailed example of this important kind of objection is offered by Megan Delehanty, who uses what she terms "extensions" of such explanations to critique recent claims about the existence of "robust" emergence in actual scientific cases like that of slime mold.[13] Delehanty is focused on actual cases, so it is again unclear whether what I term the "Argument from Compositional/Mechanistic Explanation" is best framed as an objection against the *actuality* of Strong emergence, or as also attacking our thought experiment. However, what is obvious is that the concern is an important one that we can only address *after* more carefully examining the implications of scientific emergentism for laws and explanations. Given this point, I am going to respond to this worry in detail in the next chapter after considering what the Mutualism of scientific emergentism implies about these topics.

6.3.5 *The Transitivity and Temporal Priority Objections*

Objections to our scenario that are more metaphysical in nature focus on the machretic determination between instances of the putatively Strongly emergent property G and instances of its realizer properties like P1. Kim has recently raised such concerns against similar positions and about the various scientists presently espousing the existence of robust "self-organization" (Kim (1999), pp. 28–31). Kim expresses the concern about such cases, a worry not limited to card-carrying reductionists, as follows:

... how is it possible for the whole to causally affect its constituent parts on which its very existence and nature depend? If causation or determination is transitive, doesn't this ultimately imply a kind of self-causation, or self-determination – an apparent absurdity? It seems to me that there is reason to worry about the coherence of the whole idea. (Kim (1999), p. 28)

These general concerns about the coherence of "self-organization" obviously apply to our scenario through a couple of objections.[14]

---

[13] See Delehanty (2005).

[14] Kim (1999) has a more complex version of a related argument, which he elucidates with a complex principle, but the response outlined to the simpler Transitivity and Temporal Priority Objections illuminate points that also work against this more complicated reasoning. I therefore focus on the simpler arguments here.

The first argument, what I dub the "Transitivity Objection," can be pressed whether we take the Strongly emergent property G to be causally, productively, or non-productively related to its altered realizer P1, though Kim obviously favors a causal or productive relation. However, the key point is that "determination is transitive" so we get "a kind of self-causation, or self-determination – an apparent absurdity." For we have seen that scientific emergentism endorses mutually determinativity. The instance of G non-productively determines that the instance of P1 contribute power C1. And P1, together with other realizers, non-productively determines that the instance of G exists. But if we assume with Kim that determination is transitive, then when we put such "upward" and "downward" non-productive determination together we see that G determines itself!

Whether or not it is an absurdity, Kim is surely right that literal self-determination is problematic. We can also accept that compositional determination is transitive. We can even accept, for the sake of the argument, that machretic determination is transitive as well, although this is an issue we have not examined or hence resolved. However, even granting these assumptions, this still does not get us to a problematic form of self-determination. For the Strongly emergent property instance G *machretically* determines P1, whilst P1 together with the other realizers *compositionally* determines that an instance of G exists. And although chains solely of machretic relations, or chains solely of compositional relations, may each be transitive, it does not follow that chains *mixing* machretic and compositional relations should therefore also be transitive. In fact, we have reason to think that these species of determination are distinct, giving rise to skepticism about such transitivity. I have highlighted how composition involves joint productive role-filling, whilst machresis is apparently based around productive role-shaping *without* role-filling. Consequently, we have good reasons to deny the required transitivity of determination in the scenario and the Transitivity Objection fails.

That the Objection finds little purchase is perhaps unsurprising, since we have seen that Kim's favored interpretation of those endorsing "robust" emergence takes them to embrace just one kind of determination in *causal*, or *productive*, relations – thus bringing his Transitivity Objection into play. But Kim has not properly appreciated, first, that the scientific emergentist takes the relevant species of determination to be *non-causal* and *non-productive*. And, second, that the scientific emergentist assumes that there are *two* distinct forms of non-productive determination in play. Given these rather different commitments, the notion of Strong emergence in question thus avoids the claim that realized and realizer properties bear causal or productive relations to each other, or bear the same kind of non-productive determination to each other in both directions, and thus neatly steps around the Transitivity Objection.

A related concern articulates similar worries through looking at the temporal sequence by which various entities come into being in what I therefore dub the "Temporal Priority Objection." This reasoning applied to our scenario proceeds roughly as follows: Can one explain, asks the objector, *how* the property instance G could determine the nature of the altered instance of the property P1, i.e. an instance contributing C1, that is a necessary component of G, or vice versa? For, the objector argues, it appears that G needs to exist *prior* to the transformation of P1 to determine its nature, and yet G is only brought into existence *after* this transformation has occurred. Similarly, the transformed P1 would have to exist *prior* to the instantiation of G to realize an instance of this property, and yet the transformed P1 again would only be brought into existence *after* the instantiation of G. Whether the instance of G precedes the transformed instance of P1, or P1 precedes G, the situation is equally impossible. Since these are the only options, concludes the objector, this type of emergence is not logically possible.

It should be clear that a crucial assumption of this type of objection is again that a *causal* or *productive* determination relation holds between the instances of P1 and G in our scenario, since the objection assumes this determinative relation is temporally extended. However, this assumption is highly contentious. First, scientific composition presents a well-confirmed type of non-causal and non-productive determination in cases of realized and realizer properties. As we saw in Chapter 2, compositional relations all involve non-causal, non-productive determination. So the assumption that determination between "parts" and "wholes" *must* be causal or productive is dubious on general grounds.

Second, and building upon this point, we have seen that scientific emergentists posit another kind of non-productive determination relation from composed entities to their components in what I dubbed machretic determination. The latter shares with compositional relations the various features noted earlier, including being synchronous, occurring between entities that are in some sense the same, being mass-energy neutral, does not involve the transfer of energy and/or mediation of some force, and is not identical to the triggering and manifestation of powers. Consequently, a scientific emergentist may thus respond to the Temporal Priority Objection that it is either question-begging, in assuming causal or productive relations are involved, or based on something close to a category mistake in asking whether the relevant instance of P1, or that of H, exists first. For machretic relations between property instances, like other non-productive relations between instances, are synchronous relations. We can therefore see that the Temporal Priority Objection, like the Transitivity Objection, also fails to undermine the conclusions of my thought experiment once we appreciate the features of machresis as a non-causal and non-productive determination relation.

6.3.6 The Objection from Conditioned Fundamentalism

A more substantive challenge seeks to undermine the conclusions of my thought experiment by arguing that the best interpretation, or best explanation, of the contribution of the new powers by the realizers in our scenario, or any other, is *always* that *other component entities* determine their contribution. This alternative interpretation of the scenario is shown in Figure 6B and putatively allows us to make do with a more parsimonious ontology, so the objector concludes we should reject the coherence or acceptability of the representation I have defended – or at least reject the interpretation of it that includes a determinative role for the realized property instance G. Hence, concludes the objector, the criterion for Strong emergence should not be taken to be satisfied in my scenario.

This criticism is what I dub the "Objection from Conditioned Fundamentalism," since the worry assumes the position I term "Conditioned Fundamentalism" in a novel species of ontological reductionism about cases of compositional explanation driven by the arguments of scientific reductionism. The Conditioned Fundamentalist position overlaps with the Mutualism of scientific emergentism in endorsing the Conditioned view of aggregation, and accepting the existence of differential powers of components, but crucially Conditioned Fundamentalism only takes *other components* to

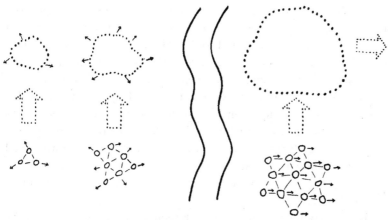

Figure 6B A diagram representing the interpretation that Conditioned reductionism provides of our thought experiment. Although accepting we have Conditioned aggregation and differential powers in the complex aggregation, the Conditioned reductionist claims the differential powers are *always* best interpreted as being determined by other components alone. (The large dotted circles at the top are collectives; the upward dotted straight arrows are collective relations; the horizontal dotted arrow at the top is the collective force of the components together; smaller circles at the base are blobs; straight solid lines are bonds; solid arrows are forces where the size of arrow equals the magnitude of the force.)

determine the contributions of such differential powers. Conditioned Fundamentalism consequently presses a parsimony argument to drive compositional reductions, putatively establishing that only components should be accepted as existent or at least determinative.[15] Conditioned Fundamentalism thus offers an important variant of scientific or philosophical reductionism, but I defer a detailed examination of its features to Part IV. My present concern is simply whether deployment of such a view undermines the conclusions I have drawn from my thought experiment, for the objector claims it always provides the best account of a case like my scenario regardless of the details.

---

[15] Rather than the invalid Argument from Composition, we can now see the Conditioned Fundamentalist needs to press the more complex parsimony argument I outline in Chapter 8.

One blunt response to this objection reminds us of the purpose of the thought experiment and the rules governing it. We are not looking at real cases where talk of the "best interpretation" or "best explanation" might make sense. Rather, we are stipulating a scenario in the course of a thought experiment to assess whether application of certain scientific concepts is compatible with specific features also holding in the broached scenario. In this context, one is allowed to stipulate the scenario and its conditions. That the objector might be able to imagine *other* scenarios is beside the point, for if the stipulated scenario is coherent then it establishes there can be a Strongly emergent property even when scientific concepts of composition apply comprehensively.

Even if this blunt response is correct, then it is unsatisfying. So for the sake of argument let me put aside concerns about applying parsimony criteria in "interpretations" of a thought experiment. Instead, I want to consider whether the kinds of method we use in the sciences to confirm hypotheses could provide us with reasons in a case of compositional explanation to favor the Mutualist hypothesis positing existence of a Strongly emergent property over a Conditioned Fundamentalist account. Pursuing this question is also a useful preliminary to my discussion in Part IV of how scientific evidence can confirm/disconfirm Mutualist, or Conditioned Fundamentalist, hypotheses about cases of compositional explanation. However, it is also important to be clear that my present goal in exploring this issue is obviously whether our actual practices, which we routinely take to establish various conclusions in everyday science, could suffice in *some* example to support the existence of a Strongly emergent property over a competing Conditioned Fundamentalist hypothesis positing no such property.

The ways in which we confirm/disconfirm compositional hypotheses in the sciences is actually an ongoing area of research.[16] And little work, to my knowledge, has been done on how we can confirm/disconfirm hypotheses about machretic relations. However, some obvious facts about compositional and machretic relations as determination relations highlight the broad ways to engage these issues. Generally, and ignoring many complexities, if entities X1–Xn compositionally

---

[16] As I detailed in Chapter 1, Craver (2007) adapts interventionist frameworks to compositional cases. However, as I outlined in Chapter 2, there are difficulties with the resulting account, so I leave it to one side here.

determine Y, under condition $, then we can expect that the absence of X1–Xn to accompany the absence of Y all else being equal. For similar reasons, if Y machretically determines that entities X1–Xn contribute certain differential powers, then we can expect that the absence of Y means we lack X1–Xn contributing such differential powers. Even lacking a more precise account of how to confirm/disconfirm hypotheses about compositional or machretic relations, these general points allow us to discern the broad outlines of the ways in which we can adapt common scientific methods to confirm/disconfirm competing Mutualist scientific emergentist and Conditioned Fundamentalist hypotheses.

Given the objector's claims we can obviously assume any scenario, but let me simply limit myself to building into my earlier scenario the new feature that, over time, we have a common form of multiple realization of the Strongly emergent property. To this end, let me assume that the scenario is as I have described it earlier, but I am further assuming the multiple realization of the instance of G over time and the multiple constitution of the globule s^* over time as well. Basically, the idea is that over time an ever-changing array of constituents, with an ever-changing array of properties and relations, realize and constitute G and s^* at different times. Consequently, let me also assume that although, at time t, $s2$–sn, and their properties P2–Pm, are constituents and realizers, at some later time t* $s2$–sn, and their properties P2–Pm, are not amongst the constituents and realizers of s^* and G, and that $s2$–sn, and their properties P2–Pm, do not in fact exist at t*, even though s^* instantiates the relevant instance of G at t*. As I noted in Chapter 2, we have considerable evidence that such cases are common in nature, for example with the molecular constituents of cells or our own bodies over time (Gillett (2013), Aizawa and Gillett (2009a, 2009b)).

In addition, let me also stipulate that the constituent $s1$ remains stable as a part of s^* at the lower level over the period of time in question by continuing to have the property instance P1, which contributes its differential powers to $s1$ throughout this period. Finally, let me assume that when s^* does lose G, then the instance of P1 in $s1$ no longer contributes its differential powers. And let me also stipulate that if the instance P1 of $s1$ does not contribute its differential powers, then s^* does not have G. Remember that I am plausibly allowed to stipulate we have a case of this variety because the objection is that in *any* case, whatever its features, we must always favor the Conditioned Fundamentalist hypothesis over the Mutualist scientific emergentist account of the scenario.

In this extension of my scenario, I want to consider competing Mutualist and Conditioned Fundamentalist hypotheses. For example, consider a Conditioned Fundamentalist hypothesis that claims that the differential powers of property P1 in *s1* are determined by property instance P2 of some other constituent *s2*. Call this hypothesis "H_{CF1}." On the other side, consider a Mutualist hypothesis, "H_{M1}," that contends that the differential powers of P1 in *s1* are determined by the realized property instance G of *s**.

Our goal is to see whether adaptations of our usual scientific methods could supply evidence cutting between H_{CF1} and H_{M1} in any scenario. To get a better grip on this question, consider some of the predictions made by these hypotheses about my extended scenario. The scientific emergentist's Mutualist hypothesis H_{M1} is plausibly committed to the following: If we can have situations lacking the Strongly emergent property instance G of the globule, *s**, under the relevant conditions, then we can expect that the differential powers of an instance of P1 in *s1* will also be absent. In addition, if we can have situations where the instance of P1 in *s1* does not contribute certain differential powers, under the conditions relevant to our scenario, then we may expect the Strongly emergent property instance G to be absent.

On the other side, the Conditioned Fundamentalist hypothesis H_{CF1} implies the following: If we can have situations lacking instance P2 of *s2*, under the relevant conditions, then the differential powers of the instance of P1 in *s1* will also be absent. In addition, if we can have situations lacking the differential powers of this instance of P1 in *s1*, while the other relevant conditions remain the same, then property instance P2 of *s2* will be absent as well.

With these differing predictions of the opposing hypotheses in mind consider whether we should accept H_{M1} over H_{CF1} in my extended scenario. As the reader has probably already discerned, when we assess which of the two hypotheses should be favored in this extension of our thought experiment then we get some obvious results.

First, notice that we must plausibly *reject* the Conditioned Fundamentalist hypothesis H_{CF1}, since its predictions are false. P1 *still* contributes its differential powers to *s1* in the absence of P2 and we have reason to deny that P2 determines P1's contribution of differential powers contrary to the Conditioned Fundamentalist's claims. Second, we have evidence *supporting* the scientific emergentist hypothesis H_{M1}, since its predictions are borne out. When G is absent from s^*, then P1 does not contribute differential powers to $s1$. And when P1 does not contribute differential powers, then the globule s^* does not instantiate G.

In my extended scenario, adaptations of our usual methodologies imply that we should accept the emergentist's Mutualist hypothesis H_{M1}, positing machresis and Strong emergence, over the Conditioned Fundamentalist's hypothesis H_{CF1}. And we consequently have reason to reject the Objection from Conditioned Fundamentalism since, as we saw earlier, its basic contention was that it is *always* a better interpretation, or the best explanation, of a thought experiment to take other realizers, or other entities at the component level, to determine P1, or any other component's, contribution of differential powers. But, depending on the details of the case, my extended scenario shows that this is not true.

At this point, the immediate response will be that I have only shown a Mutualist hypothesis should be favored over a *simplistic* Conditioned Fundamentalist hypothesis and that more complex hypotheses should be considered that would still be favored over the Mutualist account in this more complex scenario. In response, I should mark that whether the new Conditioned Fundamentalist hypothesis is that just P2 and P3, or all of the other realizers of G, P2–Pn, or some other group of component properties/relations, is chosen, then we can plausibly pursue the same procedure of constructing a scenario where the implications of the Conditioned Fundamentalist hypothesis will not be borne out, but those of a Mutualist hypothesis are. We thus plausibly have a procedure by which we can rebut iterations of the Objection through appropriately constructed extensions to my scenario where adaptations of our usual methodologies will once again support accepting a Mutualist account over the relevant Conditioned Fundamentalist hypothesis.

We can therefore see that the Objection from Conditioned Fundamentalism fails as a criticism of my thought experiment and the conclusions I have drawn from it. But I should emphasize that the latter point in no way undermines the fact, which I lay out in later chapters, that Conditioned Fundamentalism is a position that should be taken seriously as an account of various concrete scientific cases where we have successful compositional explanations. Which hypothesis we endorse about such cases must be resolved case by case using the evidence we have about it. All that I have argued at this point is that Conditioned Fundamentalism is not the best interpretation of *every* example involving differential powers of components *regardless* of the specific features of the case and hence our empirical evidence about it.[17]

6.4 Why Good Thinkers Have Gone Bad with Strong Emergence: Illuminating a *Trio* of Incompatible Theses and a *Trichotomy* of Options

All of the objections to my thought experiment plausibly fail. We have thus confirmed that reasoning seeking to show we should never accept Strong emergence in cases of compositional explanation, such as the Argument from Composition, is indeed flawed. And so too are the received views of the viable positions, or possible forms of emergence, built around such arguments. But a puzzle immediately confronts us: Why have so many adept theoreticians gone so wrong with Strong emergence?

In answering this question, we need to attend to Whitehead's observations about the role of epochal assumptions – that is, the deeper ontological assumptions underlying an era that, although widely shared, are rarely articulated. For example, I have begun to highlight how both philosophers, and also scientists, have failed to articulate the Simple view of aggregation that is apparently central to scientific reductionism, associated "dreams" of a Final Theory, and a precondition of the truth of the Completeness of Physics. Leaving this epochal assumption unarticulated has apparently left this thesis *implicitly* working its influence upon our thinking without any check or explicit critical assessment. And we can now see how this has apparently led theoreticians astray about Strong emergence.

---

[17] Highlighting Conditioned Fundamentalism frames the differences between scientific emergentism and the type of emergence broached in Shoemaker (2003b) and (2007). Shoemaker does not take his kind of "emergent" property to machretically determine the powers contributed by its realizers, thus leaving his "emergentism" looking more like the Conditioned Fundamentalism that I consider at more length in Part IV.

To start, I need to mark that the existence of Strong emergence is incompatible with the Simple view of aggregation and hence also the Completeness of Physics. For example, in my scenario, a composed entity machretically determines some of the powers of its components, so the Simple view of aggregation is false and so too is the Completeness of Physics.

These points are important because as a result of the long hegemony of scientific reductionism it appears that many scientists and philosophers *implicitly assume* the Simple view, or associated theses such as the Completeness of Physics, without appreciating they are doing so. Thus such writers apparently use the Simple view, or the Completeness of Physics, as a *suppressed premise* when considering the Argument from Composition – whilst explicitly focusing simply upon premises endorsing the existence of compositional relations. Though *implicitly* endorsing reasoning based upon *both* the nature of compositional concepts *and* the Simple view (or the Completeness of Physics), which may provide a valid argument that we should not accept composed entities, these researchers consequently end up *explicitly* endorsing something like the invalid Argument from Composition.[18]

We thus have a diagnosis of how so many good thinkers have endorsed an invalid argument that the applicability of compositional concepts alone suffices for compositional reduction. And we can also see why these writers have also drawn a range of mistaken implications about the import of universal composition.

Driven by flawed reasoning like the Argument from Composition, the members of the Unholy Alliance and many others are left insisting that we have a *pair* of independently plausible but incompatible claims in the composed status of some entity and in this entity being determinative. However, my work has now shown that these two claims can be true together when we have the Conditioned view and the machresis it allows. My findings instead suggest that it is only plausibly a *trio* of theses that is incompatible and I explore the nature of the relevant theses in more detail in Part IV. But for our present purposes we can see that the following cannot plausibly be true together: the composed status of some entity, this entity being determinative *and* the reductionist's Simple view of aggregation (and/or the Completeness of Physics) holding.

---

[18] I explore the validity of the more complex parsimony argument in Chapter 8.

In response to a trio of incompatible theses we have at least *three*, rather than *two*, options in response to this situation corresponding to the pairs of claims we decide to retain. And, rather than a pair of viable forms of emergence corresponding to the viable options, in Weak and Ontological emergence, we have a corresponding third form of emergence going along with the overlooked option. The Unholy Alliance has thus endorsed a *false dichotomy* about emergence as limited to Weak or Ontological forms. The missed kind of emergence is the Strong variety and the corresponding missed position is a Mutualist view that retains compositional relations, and the determinativity of composed entities, whilst abandoning the Simple for the Conditioned view of aggregation with its new possibilities.

Just as Whitehead predicts, the influence of the epochal ontological assumptions of scientific reductionism has thus plausibly left many contemporary philosophers and scientists incapable of seeing the Mutualist option and thus without a clear view of all the live positions and the deeper issues in recent debates, especially the role of different views of aggregation and associated claims.

6.5 Scientific Emergentism, and the Utility of Theoretical Work, Vindicated

Far from being strange in the manner of a logically impossible square-circle, my work shows that Strong emergence is actually more akin to the historical situation of the duck-billed platypus – a creature that is odd only in conflicting with our prevailing presumptions about the way nature *must* be. Just as in the case of the platypus, and so many other times in the past, reflecting on the results of our empirical inquiries allows us to throw aside mistaken ontological assumptions and the theoretical frameworks and arguments based upon them.

Scientific emergentists have trenchantly pressed the point that we have been led astray about emergence by unsupported ontological assumptions of scientific reductionism, or what they term "myths," at odds with our empirical findings. Thus Laughlin tells us that:

The world we actually inhabit, as opposed to the happy world of modern scientific mythology, is filled with wonderful and important things we have not yet seen because we have not looked ... The great power of science is its ability, through brutal objectivity, to reveal to us truth we did not anticipate. (Laughlin (2005), p. xvi)

And Laughlin, and other scientific emergentists such as Ilya Prigogine, maintain we have now established these epochal claims as mere "myths" using our better empirical understanding in various examples.[19]

My examination of the kind of Strong emergence that Mutualists claim we find in such cases has served us well in our critical assessment of scientific reductionism. I have used a Mutualist scenario to highlight the invalidity of the Argument from Composition widely used by philosophical and scientific reductionists to wrongly conclude we should never accept Strong emergence. And I have subsequently illuminated how the epochal ontological assumptions of scientific reductionism have indeed led such writers to endorse this flawed reasoning and associated false dichotomies about emergence.

However, Laughlin, Prigogine, Freeman, and others contend the importance of Mutualism reaches far beyond its utility as a tool for illuminating flaws scientific reductionism. These scientific emergentists press the Mutualist picture of nature, incorporating Strong emergence and machretic determination, as well as the Conditioned view of aggregation, as the best account of specific scientific cases involving compositional explanation. And there is a very real danger that our faulty theoretical frameworks are solely responsible for our rejecting such Mutualist hypotheses. Carefully reconsidering what our empirical evidence actually shows once we put aside faulty arguments thus becomes a pressing undertaking. But before I can turn to this project in Part IV, we need to have a better grip on the detailed commitments of scientific emergentism and its Mutualism. In the next chapter, I therefore reconstruct scientific emergentism and its novel Mutualist position.

---

[19] See Prigogine and Stengers (1984), and Prigogine (1997), for Prigogine's own interesting discussion of such ontological "myths."

7 | *Understanding Scientific Emergentism: A Mutualist Nature and its Interdependent Levels, Laws, and Sciences*

The natural world is regulated both by the essentials and by powerful principles of organization that flow out of them ... Our conflicted view of nature reflects a conflict in nature itself, which consists simultaneously of primitive elements and stable, complex organizational structures that form from them.

– Robert Laughlin[1]

... there are two challenges for the friend of emergence. The first is to show that emergent properties do not succumb to the threat of epiphenomenalism ... This must be done without violating the causal/explanatory closure of the physical domain – or, if the physical causal closure is to be given up, a credible rationale must be offered. The second challenge is to give a positive characterization of emergence that goes beyond ... supervenience and irreducibility. Unless this is done, the thesis ... remains uninteresting and without much content ...

– Jaegwon Kim[2]

A Strongly emergent nature is Mutualist and hence has *two* sources of determinativity, as Laughlin intimates, with both the productive determination of components within their own lower level and the "upward" determination of their compositional relations, but also the productive determination of Strongly emergent composed entities at their higher level and their "downward" machretic determination. This is indeed, as Laughlin puts it in the title of his book, a "different universe" (Laughlin (2005). The goal of this chapter is to illuminate what a Mutualist scientific emergentism implies about ontology, laws, sciences and explanations, and methodology. As I work through the issues, I also show that Laughlin plausibly defends the kind of Mutualist position I reconstruct.

---

[1] Laughlin (2005), pp. ix–x.
[2] Kim (2006), p. 201.

As Kim makes clear in the second of our opening passages, defenders of such a novel position face at least three critical challenges. First, there is the task, which I began to address in the last chapter, of giving an ontological reason why we ought to accept that Strongly emergent properties really are determinative and exist. Second, Mutualism needs a positive and substantive characterization of any robust notion of "emergence." And, lastly, there must be a rationale for why scientific emergentism abandons both the Simple view of aggregation and hence the Completeness of Physics. As I move thorough the chapter, I use my findings to provide responses to each of these important challenges. However, I should also mark one caveat about my work at the outset. Although I can outline an initial theoretical framework for Mutualism, a number of issues can only be fully resolved by examining concrete scientific examples in detail. At various points, I note some of the questions that require further attention in this way.

My findings highlight why the Mutualist universe of scientific emergentism is not *too* different. I confirm that Mutualism is a coherent position with fresh insights about a nature that is a compositional hierarchy, as well as about the sciences, explanations, laws, and methodology needed to understand it. Crucially, I outline how we can accept "walls" of composition, and levels of determinative composed and components entities, rising upward in the house of nature when we have a "roof" of machretic determination, and Strongly emergent entities, pressing downward.

7.1 The Ontology of an Emergent Nature: Implications of Strong Emergence (I)

Given its centrality, I begin by exploring the mutual determination and interdependence we find between Strongly emergent entities and their components. In 7.1.1, I explore the complex web of determinative relations under Mutualism. I then fill out the ontological reasons why we should accept that Strongly emergent entities are determinative, in 7.1.2, to provide a more detailed answer to Kim's first challenge. Using these conclusions, in 7.1.3, I show that Strong emergence not only partially vindicates common sense and the Scientifically Manifest Image, but is plausibly necessary for an ontologically non-reductive view of a nature that is a compositional hierarchy with levels of determinative entities. I also use machresis to provide a positive characterization of Strong emergence in answer to Kim's second challenge.

7.1.1 *Mutually Determinative and Interdependent* *"Parts" and "Wholes"*

As I outlined in the last chapter, Strongly emergent composed entities and their components are both *mutually determinative* and *interdependent*. Given the complexity of the determinative relations that come along with Strongly emergent entities we need to start by carefully laying out the various relations in more detail before examining their implications.[3] (To keep my discussion simple, I again focus on property instances, though similar points hold for other categories of entity such as individuals, powers or processes, which can also be Strongly emergent under the framework I outline.)[4]

When we have Strong emergence we have two species of "vertical" non-productive determination relations. A Strongly emergent property is composed – and hence "upwardly" non-productively determined – by certain lower-level entities. But these components have some of their powers only as a result of the "downward" non-productive machretic determination of the Strongly emergent composed property instance. At a time, due to this pair of non-productive determination relations, we therefore have mutually determinative entities and, as Laughlin implies, *two* determinative vectors in nature. But, although the Strongly emergent entities and altered components are distinct determinative entities, we also need to keep in mind that Strongly emergent entities, and their altered components, are also in a sense a "unit" for a couple of reasons. First, these entities bear a compositional relation, so the higher-level entities are "nothing more than" the components, though we have seen that the reductionist is mistaken that this means we should never accept the composed entities are determinative or exist. Second, due to their mutually determinative natures, these entities only come along *together* and we have interdependent entities. Each of these points has important ramifications for all kinds of issues.

We also need to mark the distinct "horizontal" determinative relations of the relevant entities at higher and lower levels when we have

---

[3] This is a point that has often been emphasized by writers on complex systems and the "emergence" we find with them. See, for example, Scott (2007) and Mitchell (2013).

[4] Strongly emergent entities may therefore also be compositionally determinative. But to simplify my discussion, in this and later chapters, I leave aside the compositional determination of Strongly emergent entities unless it is directly relevant to the issue under discussion.

Strong emergence. Recall the difference between productive and causal relations, outlined in Part I, because causation is a much broader concept than production. Productive relations are manifestations of powers and are often examples of causation, but not all causal relations are productive ones. This distinction is particularly important in the case at hand because within the relevant compositional hierarchy the Strongly emergent property instance produces certain higher-level effects at its level, but its realizers do not produce effects at this level. Thus, in my scenario, the Strongly emergent instance G produces H, but the realizer instances of P1 do not because lower-level properties like P1 do not contribute powers that produce effects like H. Similarly, the instances of P1 produce instances of Pz at the lower level, the effect of their differential powers, but the Strongly emergent property G does not produce Pz because H contributes no power to produce a lower-level effect such as Pz.[5]

These conclusions fit well with our scientific accounts, since I noted in Chapter 2 that the productive relations posited with compositional relations in scientific explanations are *intra*-level relations. Within compositional hierarchies, and hence in compositional explanations, we thus typically find powers to produce effects at the *same* level. For example, we saw that higher-level properties are routinely taken to contribute powers to produce distinctive higher-level effects – thus the property of hardness in diamonds contributes powers to scratch glass, but not to keep carbon atoms in close spatial relation. Similarly, lower-level properties contribute powers to produce distinctive lower-level effects. As a result, the productive relations when we have Strong emergence, within higher and lower levels of entities, fit comfortably with a natural reading of our scientific accounts.

Some care is consequently needed when we turn to the *causal* relations we have with Strong emergence. Despite their not producing effects at other levels, the mutual determinativity of the Strongly emergent composed entity, and its components, means that these entities are still *joint causes* of both higher- and lower-level causal effects (and hence "diagonal" causes of these inter-level effects). For instance, the Strongly emergent property G in my scenario, although not producing

---

[5] It is worth emphasizing again that I am not claiming all production is intra-level, but only that all production by the composed/component entities within a compositional hierarchy is intra-level.

the event, is nonetheless still plausibly a *joint cause*, together with the instances of P1, of effects involving Pz. Similarly, we should note that the realizer instances of P1 *are* joint causes, together with the Strongly emergent property G, of the higher-level event instantiating the property H, although these instances of P1 do not produce this higher-level effect. And it is unsurprising that these entities should be joint causes of a variety of higher- and lower-level events given their interdependence, even though their productive relations in these cases are always intra-level.

My discussion in the last chapter might have led one to assume that Strongly emergent properties are *only* machretically determinative, but we have now seen that this conclusion is incorrect.[6] As well as being machretic, Strongly emergent properties are also productive of effects at their own levels and joint causes of higher- and lower-level effects along with their realizers (and may also compositionally determine still higher-level entities). We can therefore see why Strongly emergent entities have been aptly termed "structured structuring structures" by the philosopher of biology Charles Dyke, since they are multi-faceted in the determinative relations in which they are involved.[7] Strongly emergent entities are composed (i.e. "structured"), but are still entities that are productive at their own levels (i.e. real "structures") whilst also determining the powers of their components through the role-shaping of machresis (i.e. "structuring"). Overall, we thus also have multi-faceted ontological reasons for positing Strongly emergent properties that reflect these points.

7.1.2 Revisiting Parsimony and Causal Overdetermination Arguments

At this point, critics and many neutral observers will likely still be concerned that just the kinds of parsimony or overdetermination argument I looked at in earlier chapters are applicable. Surely positing such Strongly emergent entities involves "double counting" of some kind,

---

[6] This appears to be the reason why O'Connor and Churchill (2010) attack my account of Strong emergence as leading to an objectionably passive form of higher-level property instance. But it should now be clearer that this is a misimpression about what Strongly emergent properties must be like, for they can be, and usually are, both machretically and productively determinative.

[7] Dyke (1988), e.g. pp. 24 and 67.

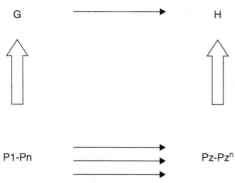

Figure 7A The vertical determinative structure of a case of compositional explanation assumed by scientific and philosophical reductionists such as Kim. At the earlier time, we only have an upward compositional relation between the realizers P1–Pn and the realized property G and at the later time only a compositional relation between Pz–Pzn and H.

to use David Lewis's phrase, or some other kind of ontological profligacy? It is worth briefly examining once again whether such arguments apply with Strongly emergent entities now that I have laid out their natures and determinative relations in fuller detail. Considering these kinds of objection at more length better illustrates the ontological reasons offered under Mutualist scientific emergentism for the existence of Strong emergence, so I briefly rehearse the challenge to Strongly emergent properties presented by the ontological parsimony, and also causal overdetermination, arguments.

The basic parsimony objection framed by the Argument from Composition is that in any situation where we have compositional relations, on reflection, we have reason to abandon the determinativity, and existence, of any entity beyond the broached components at that time. Crucially, the sub-conclusion of this reasoning is that when compositional concepts apply we can account for all the powers of individuals, at higher and lower levels, equally well by positing components alone. The relevant determinative relations posited by those who run such arguments are nicely framed by the kind of diagram, such as Figure 7A, made famous by Kim.[8] Given the relations outlined

---

[8] The diagram is adapted to reflect my finding in Chapter 2 that scientific composition is always a many–one relation, rather than being a one–one relation as Kim routinely assumes.

in 7A, it is argued that our compositional explanations show that the powers contributed by the realizer property instances P1–Pn account for all the powers of higher and lower individuals at that time. As Figure 7A illustrates, the powers of the realizers together produce the instances Pz–Pzn, and Pz–Pzn together realize the instance of H. So the sub-conclusion of the reasoning is that we need posit no further property instances that contribute powers than the realizer instances P1–Pn in order to account for the powers of *both* higher- *and* lower-level individuals. Given the latter, by applying the Parsimony Principle it is concluded that we should therefore accept the hypothesis that only takes P1–Pn to be determinative and hence to exist – thus abandoning the determinativity and existence of a composed property like G.

The same kind of concern has been framed by Kim (1993a) as a worry about causal or productive overdetermination, amongst other formulations. On this line of argument, the concern is that P1–Pn produce Pz–Pzn, but since Pz–Pzn realize H it is therefore plausible that P1–Pn at least cause H even if we grant that such properties do not produce this effect. But if we *also* assume that the Strongly emergent property G contributes powers that produce H, then it appears that the effect H is causally overdetermined – caused by both the realizers P1–Pn and the Strongly emergent property G. However, it is also maintained that causal overdetermination does not exist or is very rare. Consequently, the objection concludes, we ought to reject one of our starting premises and the premise most plausibly rejected is the assumption that the Strongly emergent property G contributes powers that produce the effect involving instantiation of H. We are thus again left with the conclusion that only P1–Pn contribute powers, are determinative, and hence exist.

In effect, these two objections, and related arguments, are each pressing us in different ways to give an ontological reason why we should accept that there is a determinative higher-level property instance when we have the compositional relations posited by successful compositional explanations. And the scientific emergentist's basic response is that Figure 7A, and the objections we have now seen are based upon it, *leave out* determinative relations that also hold in a Mutualist nature when we have Strong emergence.

The scientific emergentist contends that Figure 7B provides a better representation of all the relevant determination relations we find with Strong emergence, since it includes both machretic determination

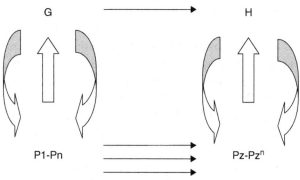

Figure 7B The structure of determinative relations that the scientific emergentist claims we find in some cases of compositional explanation. At the earlier time, the strongly emergent property instance G is realized (upward vertical arrow) by the properties P1–Pn, but G also machretically determines (curved downward arrows) the powers contributed by these realizers. Similar points hold at the later time.

as well as compositional and productive relations, and hence finally reflects the mutual determinativity and interdependence under Mutualism. And the scientific emergentist argues that these further relations are both compatible with the existence of compositional relations, as I established in the last chapter, and are crucial for understanding the ontological reasons why we should take a property instance like G to be determinative and hence to exist.

To appreciate these points, consider the Mutualist response to a simple parsimony argument like the Argument from Composition. Philosophers, such as John Dupré (1993), (2001), have pressed evidence from the higher sciences suggesting that realized higher-level properties like G contribute powers that produce higher-level effects such as that involving the instance of H. We therefore have a prima facie plausible reason to posit the property instance G in order to account for the effect H, since no lower-level realizer contributes powers to produce H. However, as I outlined in Chapter 3, the reductionist's parsimony arguments challenge this prima facie reason and this illustrates why the Mutualist does not offer this reason *by itself.*

Crucially, the scientific emergentist also contends that composed properties are machretically determinative of the powers contributed by their realizers *as well as* being productive through the powers the

Strongly emergent property instance contributes to the higher-level individual. As Figure 7B outlines, the Strongly emergent property G needs to be posited in order to account for the differential powers contributed to lower individuals by its realizers, such as P1 and others, as well as to account for the powers of the higher-level individuals. We are not "double counting" the determinative property instances, for positing G is required to account for the differential powers of components, as well as the powers of higher-level individuals, so contrary to the sub-conclusion of the Argument from Composition we simply cannot account for all the powers of individuals by solely positing P1–Pn despite having successful compositional explanations. Since in this case we cannot account for all the powers of individuals without positing the Strongly emergent property G in the broached example. Thus we see that it is not always profligate to posit G in examples where compositional concepts apply and that the Argument from Composition fails.

A similar response shows that we do not have a problem of causal or productive overdetermination when we have Strong emergence in a Mutualist nature. As I noted above, in the type of case illustrated by Figure 7B, it is not true that the lower-level realizer instances P1–Pn would produce the instance of Pz without the Strongly emergent property instance G – so G is plausibly a joint cause of the instance of Pz. Similarly, G would not produce H without the realizer instances. Thus the Pz's resulting in H, because they realize it, does not support the conclusion that G and P1–Pn are independent sufficient causes of the event involving H. How could these property instances be independent causes of this event when the altered instance of P1 is interdependent with the instance of G? G would not cause H without the realizers, at least on this occasion, or under these conditions, or universally if we lack the potential for multiple realization. And the instance of P1 would not produce the instance of Pz without the instance of G, universally or perhaps under these conditions or on this occasion. Thus G and P1–Pn are plausibly *joint* causes of both the higher-level effect involving H, and also the lower-level effects involving Pz–Pzn. So neither the higher-level effect involving H, nor the lower-level effects involving Pz–Pzn, are plausibly overdetermined.

At this point, a common complaint is that the latter responses of the scientific emergentist simply beg important questions. *If* we assume the situation is as Figure 7B lays out, the objector agrees, *then of course* the Argument from Composition, or causal overdetermination arguments, fail. But, the objector contends, the scientific emergentist simply begs the crucial questions by assuming that Figure 7B reflects the structure of nature, rather than Figure 7A. However, the objector is mistaken that the responses laid out in this section beg important questions because the objector has failed to see the differing responses that the scientific emergentist ought to make to the two distinct kinds of objection to the existence of Strong emergence – those that seek to show we should never accept Strong emergence and those that seek to show Strong emergence does not actually exist. My theoretical framework for Mutualism, and Strong emergence, was used in this section, and the last chapter, to illuminate reasons why we should reject the former sort of criticism. I have now shown that we may sometimes need to accept the existence of Strongly emergent entities to account for the differential powers of components and the production of higher-level events. In Part IV, I will return to objections that seek to show our empirical evidence establishes that there is no Strong emergence. Before turning to such concerns, however, I need to fill out our account of scientific emergentism and its Mutualist position.

7.1.3 Real Levels, the Scientifically Manifest Image, and a Positive Characterization of Strong Emergence

Once we accept the determinativity of Strongly emergent composed entities under Mutualism, then we can appreciate the biggest ontological implication of Strong emergence. Focusing on this news, Laughlin tells us:

In passing into the Age of Emergence we learn to accept commonsense and leave behind the practice of trivializing the organizational wonders of nature and accept that organization is important in and of itself – in some cases the *most* important thing. (Laughlin (2005), pp. 218–19. Original emphasis)

The Mutualist account of Strong emergence gives us an ontological reason to endorse a view of nature under which, as well as components entities and their collectives, there are also determinative composed entities, whether superconductors, populations of neurons, or even ourselves. Consequently, the Mutualist picture pressed by scientific emergentism partially vindicates the Scientific Manifest Image, as well as common sense, by showing there are real, rather than merely apparent, *levels* of compositionally related, but nonetheless determinative, entities as we commonly understand them.[9]

However, I should note that this vindication is only partial in an important sense. In the project of articulating the structure in nature underlying our successful explanations, we can understand scientific emergentism as defending a Mutualist view challenging the Fundamentalism of the scientific reductionist. But this Mutualist view is also different from that of common sense or the Scientifically Manifest Image, for neither of the latter endorses machretic relations. And I have now highlighted plausible reasons to think that without machresis the scientific reductionist may be correct that we only ever have the *appearance* of compositional relations and levels.

Without machresis, it appears that on their own compositional relations collapse in upon themselves and we are left with no good reason to accept that there is any composed entity or relation of composition. In contrast, when we have machretic and compositional relations alongside each other, then we have an ontological reason to accept the existence of composed entities that produce effects at their own levels as well as machretically determining the powers contributed by their components at lower levels. If this is correct, then machresis is necessary for composition – and composition and machresis are coadunative. Looking more widely, such a situation would also mean that to embrace an ontologically non-reductive physicalism, whether in a particular case or globally about all of nature, one must add machretic relations to the claims of the Scientifically Manifest Image. Consequently, Mutualism, or Mutualist physicalism, would be the only viable form of ontologically non-reductive physicalism. In the next chapter, I provide a formal argument that confirms these conclusions.

The centrality of machretic determination to Strong emergence should make it no surprise that we can answer Kim's first challenge using a positive characterization of "emergence" built around machresis. (Below I also present an alternative, but equivalent,

---

[9]   Elsewhere, borrowing from Cartwright (1999), I have also referred to this view as "Patchwork physicalism" (Gillett (2003b), (2006b)). However, I now think the centrality of machresis to this position makes the Mutualism tag more informative about the core of the position.

characterization using emergent laws.) Consider this theory schema for Strongly emergent property instances:

(Strong Emergence – Property and Machresis Version) A property instance G is Strongly emergent, in an individual $s*$ under condition $, *if and only if* (i) G is realized, in $s*$ under $, by properties/relation instances P1–Pn of the individuals $s1–sn$ (and $s*$ is constituted by individuals $s1–sn$); and (ii) G machretically determines, in $, some of the powers contributed by one or more of P1–Pn to $s1–sn$.

As I have emphasized, the empirical cases where scientists claim we have Strong emergence cover a far wider array of entities than just property instances as Strongly emergent, so let me also frame a general theory schema also covering powers, individuals, and processes.[10] I can give such a general theory schema as follows:

(Strong Emergence – General Machresis Version) An entity Y is Strongly emergent, under condition $, *if and only if* (i) Y is composed, under $, by entities X1–Xn; and (ii) Y machretically determines, in $, some of powers had by one or more of X1–Xn.

An entity that satisfies either schema passes the criterion for Strong emergence outlined in Chapter 5, since satisfaction of condition (i) of the schemata entails the entity is composed and satisfaction of (ii) means the entity is determinative. In the next chapter, I offer an argument supporting the necessity of machresis for Strong emergence, thus supporting my schemata as providing constitutive accounts of Strong emergence.

Overall, we can thus understand the scientific emergentist's Mutualist account of Strong emergence as illuminating the deeper nature of Strong emergence beyond the surface features outlined in the S-criterion offered in the last chapter. As a result, my schemata provide *positive* characterizations of Strong emergence in terms of machresis, and other features, and hence answer Kim's second challenge by utilizing the broader ontological account I have provided. The schemata also go beyond mere supervenience and irreducibility, and are framed in richer terms than a simple list of features or negative attributes. Rather than a laundry list of unconnected features, my theory schemata reflect a deeper ontological

---

[10] Notice also that neither schema is indexed to times, thus allowing diachronic cases such as those involving Strongly emergent processes.

account of Strong emergence. And below I show how my account of Strong emergence explains many of the features, such as downward causation, undeducibility, or unpredictability, often previously used in the lists putatively characterizing "robust" emergence.

7.2 More on Machresis

Machresis is at the heart of the Mutualism of scientific emergentism, so my goal in this section is to further explore its nature and its relations to other notions broached by emergentists to articulate "robust" emergence. I begin, in 7.2.1, by further outlining the features of machresis and clarifying its relations to the notion of "downward causation" that has often been used to explicate claims about emergence. I then outline some open questions about machresis, in 7.2.2, including the key question of whether, and in what ways, we can understand machretic relations and their productive role-shaping.

7.2.1 Machresis, Strong Emergence, and "Downward Causation"

I have already marked that machretic determination is non-productive and non-causal because it is a mass-energy neutral relation, is not identical to the triggering and manifestation of powers, is synchronous, involves entities that are in some sense the same, and does not involve the transfer of energy or mediation of force. With regard to the latter features, machresis is like composition and it is also alike in relating entities whose processes implement, and whose powers comprise, each other. However, in compositional relations, the entities that do the joint productive role-filling determine the entity whose role they fill. With machresis the situation is reversed. The entity whose productive role is filled determines the powers, and hence shapes the roles, of one or more of the entities that jointly fill its productive role.

To better appreciate machresis, and hence Strong emergence, let me briefly explore how they relate to the so-called "downward causation" that has routinely been included on the lists of features previously used to characterize "robust" emergence.[11] All too often, past defenders of

---

[11] Campbell (1974) was one of the scientists who popularized the use of "downward" causation to articulate emergence. See Andersen et al. (2000) for recent papers on this topic.

Strong emergence have claimed it involves a particular kind of downward causation where Strongly emergent entities are taken to *causally*, and probably productively, interact with their own components. Let me call this "direct" downward causation, since the higher-level property instance directly causally acts upon lower-level entities through the manifestation of its own powers without the mediation of the powers of any other entity. For example, the Nobel prize winning neurophysiologist, and pioneering emergentist, Roger Sperry often falls into endorsing such direct downward causation and we find many others who do so as well.[12]

However, the problem is that causal or productive relations only hold between *wholly distinct* entities. If the "emergent" property instance did directly downwardly causally or productively interact with its putative realizers, then this would give one reason to think these property instances are wholly distinct and hence *not* related by a compositional relation such as realization after all. One would thus have to conclude that the "emergent" property instance is really an empirically implausible Ontologically emergent property instance, thus precluding a compositional explanation of this emergent entity. If one endorses Strong emergence, we can thus see a key reason to take the "vertical" downward relation between an emergent property and its realizers to be machretic, rather than a causal or productive relation.

Direct downward causation is not the only species, since the work of the last section shows that Strong emergence actually results in a distinct form of downward causation. Although Strongly emergent properties do not productively or causally determine their own components, but only machretically determine them, Strongly emergent properties *do* causally (but not productively) interact with *other* lower-level entities in a compositional hierarchy. Under Mutualism, as I illustrated in the last section, Strongly emergent property instances are joint causes of the lower-level effects produced by the differential powers of their realizers. Let us therefore term this "mediated" downward causation, since the causal relations of the Strongly emergent property instance to such lower-level effects are always *mediated* by the productive relations, and powers, of its components. (In a similar

---

[12] Sperry (1986) and (1992). See Gillett (2002a) for critical discussion of Sperry's views.

fashion, upward causation of higher-level effects by realizers in cases of Strong emergence is also *mediated upward causation*, since it is mediated through the powers and production of the Strongly emergent realized property instance.)

Though downward causation has routinely been listed as one of the features individuating "robust" emergence, this has been dangerous for researchers focused on Strong emergence. It is all too easy to take such downward causation to be of the direct variety, thus inadvertently collapsing one's emergentism into an empirically implausible Anti-Physicalism. My deeper account of Strong emergence, following the Mutualism of scientific emergentists, highlights these dangers, but shows that the downward causation found with Mutualism is only the benign, mediated variety that avoids collapse into Ontological emergence. Using my wider ontological account, we can thus explain a feature included on earlier lists of the features of Strong emergence whilst avoiding a pitfall of such characterizations.

7.2.2 Explaining Machresis? Two Kinds of Explanation and the Need for Concrete Cases

A range of important questions remain about the nature of machretic relations. For example, if machresis is found at different levels does it have exactly the same features at each level or does machresis have sub-species in the different areas? Looking more widely, we also need to pursue the issue of the relata of machretic relations and hence the subjects of Strong emergence. Are property instances, individuals, powers, and processes ultimately all relata of machretic relations as I have left open? Or is one, or more, of these categories best understood as the primary relata of machresis? We also need to explore whether, or to what extent, instances of machresis are synchronic or diachronic in character. Engagements with real cases are plausibly necessary for helping us to fill out our account of the features of machresis in answering such questions.

Another looming issue is more general in character: What kind of account can we provide to understand the machretic relations in particular cases? Again, this is plausibly a topic on which we need guidance from concrete scientific examples, but I can begin to sketch the terrain over which such future work will range while noting some relevant existing research.

We can distinguish two varieties of account one might appropriately seek for a specific machretic relation where these are not exclusive of each other. The distinction is a simple one. A "genetic" account of a machretic relation provides a story of the productive or causal *genesis* of the specific complex collective where we find machresis and Strong emergence. The existence of machretic relations and Strongly emergent entities at various levels may be a precondition for the existence of machresis and Strongly emergent entities at still higher levels. Genetic accounts may therefore be complicated, but I should emphasize that by their nature such accounts still only tell us why the complex collective that is machretic comes to exist.

In contrast to genetic accounts, there are what I term "existential" accounts, which putatively explain why the composed and component entities in a certain complex collective bear machretic relations, rather than merely supplying a story about how this collective came to be. Given the centrality of productive role-shaping to machresis, a primary focus of any existential explanation is to allow us to understand why a certain Strongly emergent composed entity determines the contribution of powers by its components and hence shapes their productive roles. It appears that writers in earlier debates, such as Pattee (1970), (1973a), and also some interpreters of the sciences of complexity, such as Juarrero (1999) or Scott (2007), have sought to provide something like existential explanations of particular machretic relations.

To understand what is involved in providing an existential explanation, let me look one of the ideas found in these writers. For example, in Pattee's work in the 1970s, in debates in theoretical biology over the nature of complexity, we find Pattee presenting his ideas of "constraint" of components by composed entities and the resulting "reduction in degrees of freedom" of components.[13] More recently, Dyke has used these ideas to reconstruct a notion of "emergence" and the philosopher of science Alicia Juarrero has picked up these ideas in order to reconstruct the ideas of "emergence" found in the work of Ilya Prigogine and others working on dynamical systems.[14]

First, I should carefully note that simply endorsing *some* form of reduction in the degrees of freedom of components does not suffice

---

[13] See Pattee (1970) and (1973a).
[14] See Dyke (1988) and Juarrero (1999).

to provide a response to scientific reductionism. Recall that in Part II we saw that the scientific reductionist can happily acknowledge reductions in degrees of freedom of components when we have a collective – for example, bonded atoms have fewer degrees of freedom than isolated atoms – but still take other components, like the bonding relation or other component-level properties/relations, to fully determine these reductions. Reduction in degrees of freedom thus by itself poses no problem for scientific reductionism.

In contrast, if we interpret Pattee, Dyke, or Juarrero as endorsing machresis, then we have reductions in the degrees of freedom of the components resulting from their differential powers whose contribution by these components is machretically determined by the Strongly emergent composed entity. Consequently, we have a reduction in the component's degrees of freedom *not* solely determined by other components and hence a clear response to scientific reductionism about why we therefore ought to accept composed entities as determinative. Applying the Mutualist framework to these earlier emergentist claims about "constraint" and "reduction in degrees of freedom" provides a charitable, and highly fruitful, account of the central claims of these critics of scientific reductionism. And, in turn, the notions of constraint, and reductions in degrees of freedom, also provide a useful tool in characterizing, and understanding, examples where we have machresis.[15]

Second, however, we must now mark that the existence of reductions in degrees of freedom, or such constraint, is plausibly taken to be a feature of the machretic relation itself, rather than providing an existential explanation of why we have such machresis. We can thus see that supplying an existential explanation is a more challenging endeavor than many have supposed.

However, I should mark that various writers, including Pattee and Juarrero, have offered further ideas about how to supply an existential explanation. Exploring such ideas is important for the Mutualist, for if she can provide a better understanding of why we have a machretic relation in some case, then the existence of machresis is more plausible.

---

[15] Kistler (2010) defends an account of constraint and reduction in degrees of freedom explicitly abandoning Completeness which, like the earlier work of Dyke and Juarrero, can be comfortably defended with my Mutualist machinery.

However, to properly explore various ideas of existential explanations, and assess them, we plausibly need to engage concrete examples and their features in ways that go beyond the scope of my project here, so I put the exploration of whether we can provide an existential explanation of machresis to one side here.

7.3 The Laws of an Emergent Nature: Implications of Strong Emergence (II)

Working scientists, such as Anderson, Laughlin, and others, place a special importance on laws, so I now want to finally provide a more detailed treatment of the laws we find with Strong emergence. I use the scenario of my thought experiment, in 7.3.1, to outline some of the basic, common features of "organizational" or "emergent" laws. And I also provide an alternative positive characterization of Strong emergence using such laws.[16] Building on this work, in 7.3.2, I lay out two variations for the nature of emergent laws in what I term their "supercessional" and "supplemental" forms. I conclude, in 7.3.3, by detailing why scientific emergentists reject the scientific reductionist's version of the Completeness of Physics and, more importantly, I also answer Kim's final challenge by explaining the empirical rationale underlying this rejection.

7.3.1 The Basic Features of Emergent or Organizational Laws

In my scenario in the last chapter, the machretic determination of the Strongly emergent property instance, and the differential powers of the realizers that result, are plausibly the subject of laws and emergent laws to boot. For example, in the scenario the property P1 of blobs contributes the differential power C1 of attracting other blobs in a preferred direction only when machretically determined to do so by the Strongly emergent property G, thus producing lower-level events with the property Pz as effects. And remember that the instance of P1 would not contribute such a power *if* the laws governing its powers and behavior in this case were exhausted by the laws applying in simple collectives.

---

[16] For stylistic reasons, I follow Laughlin in using the terms "emergent law" and "organizational law" interchangeably.

Given these points, it appears that in the scenario the realizer properties are subject to an emergent law, call it "LE," that I can roughly
state as follows:

(LE) When instances of P1 are amongst the realizers of an instance of the
Strongly emergent property G, under conditions $, if *s* instantiates one of
these instances of P1, then *s* produces Pz.[17]

Let me draw out some of the features of an organizational law like LE.
First, LE embodies the fact that in my Mutualist scenario the contribution of one of P1's powers is *fundamentally* indexed to its realization of
a particular property and the machretic determination exerted by this
composed entity. The emergent law LE simply expresses the determinative role played by instances of the realized property G in such a case.
And the structure of LE reflects the role of the Strongly emergent entity
whose existence consequently figures as a condition on the law holding, hence echoing those like Michael Polanyi (1968) who take emergent entities to be some kind of boundary condition. Second, a law like
LE involves ineliminable reference to the Strongly emergent composed
entity, in this case the property instance G. Third, as I noted above,
the composed entity referred to by LE, and similar laws, machretically
determines the differential powers of its components that produce the
effect outlined in the law. And, fourth, emergent laws like LE are not
derivable from certain other laws and it is worth commenting further on
this last feature, since it is often the *only* feature ascribed to "emergent"
laws in the earlier, semantically oriented discussions in philosophy.

Emergent laws, like LE, are underivable not merely in an epistemic
sense, for example the inability of humans or their theories to allow
the derivation of laws. Rather, this underivability is the absolute failure of entailment, since emergent laws like LE *ultimately* fail to be
entailed by the relevant class of statements. There is a very rich range
of statements that may be involved in any attempted derivation of any
law. And it is important to note that even emergent laws can of course
be derived from all kinds of sets of statements – for most any statement can be derived from *some* set of statements. However, the mere

---

[17] Obviously, there may be more emergent laws that hold of P1 when it realizes
G and there may also be other microphysical realizers of G that are subject to
emergent laws.

derivability of a law from some set of statements is obviously not the question at issue. The key issue is what *kinds* of statements a law can be derived from, since only derivation from certain kinds of statement provides important clues about the ontological import of the law. And, given the features already noted, we can see that emergent laws like LE cannot be derived from any combination of the following: the laws governing the relevant entities in collectives simpler than those that the emergent law concerns; statements about lower-level background conditions; the compositional principles applying to simpler collectives than those that the emergent law concerns; any analytic statements or statements of other necessary truths; and any statements about the identity or composition of entities.[18] Consequently, the resulting law is plausibly a fundamental determinative law given its underivability from these kinds of laws and other statements.

Putting these features together, we can outline a basic theory schema for the "emergent" or "organizational" laws defended by the Mutualist scientific emergentist as follows:[19]

(Emergent Law) A law L is an emergent or organizational law *if and only if* (i) L directly and ineliminably refers to a composed entity X; (ii) X machretically determines some of the powers contributed to its components; (iii) L concerns some of the differential powers of the components machretically determined by X; and (iv) L is not ultimately entailed by any combination of: the laws governing entities in collectives simpler than those that L concerns; statements about lower-level background conditions; the compositional principles applying to simpler collectives than those that L concerns; any analytic statements or statements of other necessary truths; and any statements about the identity or composition of entities.

Once again, the machresis involved with Strong emergence takes center stage in this account of emergent laws. And I should mark that in my schema underivability does *not* underpin the claim that a law is emergent. Instead, the theory schema focuses on the law describing a Strongly emergent entity which is machretically determinative and this ontological phenomenon results in the status of the law as emergent

---

[18] Here I follow the insightful account of such underivability in McLaughlin (1992), adapted to reflect the laws in a Mutualist nature.

[19] The schema is only an account of "emergent law" involved with scientific emergentism.

and plausibly entails the underivability of the law. The machresis described by the emergent law, and framed in conditions (ii) and (iii), thus plausibly results in its underivability outlined in condition (iv), rather than vice versa.

As our earlier passages from Laughlin indicate, under Mutualism many scientists are attracted to characterizations of Strong emergence based around such emergent laws, so let me frame an alternative positive characterization of this kind. Consider this theory schema for property instances:

(Strong Emergence – Property and Law Version) A property instance G is Strongly emergent, in an individual s^* under condition $\$$, *if and only if* (i) G is realized, in s^* under $\$$, by properties/relation instances P1–Pn of the individuals $s1$–sn (and s^* is constituted by individuals $s1$–sn); and (ii) when realizing G, in s^* under $\$$, some of the powers contributed by P1–Pn to $s1$–sn are determined in accordance with one or more emergent laws ineliminably referring to G.

In addition, I can again provide an analogous general theory schema, also covering other ontological categories other than properties as Strongly emergent, as follows:

(Strong Emergence – General Law Version) An entity Y is Strongly emergent, under condition $\$$, *if and only if* (i) Y is composed, under $\$$, by entities X1–Xn; and (ii) when composing Y, in $\$$, some of the powers had by X1–Xn are determined in accordance with one or more emergent laws ineliminably referring to Y.

We thus have a second set of positive characterizations of Strong emergence. These theory schemata are equivalent to my earlier ones explicitly framed in terms of machresis, but are useful in bringing out the important connections between Strongly emergent entities and organizational laws.[20]

7.3.2 Supercessional versus Supplemental Emergent Laws

Appreciating their basic features now allows me to distinguish at least two importantly different forms that emergent laws may take. These

---

[20] The equivalence arises because the law-based schema for Strong emergence refers to emergent laws that are defined as involving machretic relations.

are what I dub the "supercessional" and "supplemental" options.[21] Under the former, fundamental laws governing the contributions of powers by component properties in simpler collectives are *superseded* in complex collectives by new fundamental laws when these component properties realize Strongly emergent properties. Thus, for example, if supercessional emergent laws were to apply to the fundamental microphysical entities, then some laws holding of microphysical properties in simple collectives would not hold in certain complex collectives and such laws would be superseded by emergent laws in such collectives. I term this kind of emergent law a "supercessional law."

Under the second option, all the laws holding of component properties in isolation or in the simplest collectives do hold universally, but in certain complex collectives these laws are *supplemented* by further fundamental laws modulating the contributions of differential powers of the realizer properties in ways consistent with, but not fully determined by, the laws holding in isolation or the simplest collectives. This type of emergent law is what I term a "supplemental" law. On the supplemental option, the laws holding of components in isolation or simpler collectives are not exhaustive of the determinative laws, although these laws might still hold universally since in more complex collectives these universal laws are only supplemented, rather than superseded, by emergent laws.

Potentially, Mutualist scientific emergentists may accept either supercessional or supplemental emergent laws, or a mix of the two across different cases. Either way, this leads to important conclusions that Laughlin apparently embraces. Thus we find Laughlin suggesting that:

From the reductionist standpoint, physical law is the motivating impulse of the universe. It does not come from anywhere and implies everything. From the emergentist perspective, physical law is a rule of collective behavior, it is a consequence of more primitive rules of behavior underneath (although it need not have been), and it gives one predictive power over a limited range of circumstances. Outside this range, it become irrelevant, supplanted by other rules that are either its children or its parent in the hierarchy of descent. (Laughlin (2005), p. 80)

Given the nature of the fundamental laws under Mutualism, Laughlin, and other scientific emergentists, thus reject the existence of a Final

---

[21] See Gillett (2003b).

Theory in a set of universal laws that hold in the simplest collectives and which *exhaust* the determinative laws holding everywhere. Instead, a Mutualist such as Laughlin contends that we have discovered *further* laws that are "children" of these laws. These emergent laws descend from the "parent" laws, but such emergent laws transcend the parent laws because they determine the further differential powers and behaviors that we find in complex collectives that would not exist if the laws holding in simple collectives were the sole laws holding in such complex collectives. Consequently, emergent laws thus often concern reductions in degrees of freedom for component entities, since they alter the powers and behaviors of these entities under certain conditions.

As a result, in cases involving Strong emergence we may get what Laughlin terms a "tower of truths."[22] Exploring such towers of emergent laws, or the relations of "descent" between such laws and other laws, is another important task. As I suggested earlier, such nested relations may be complex. For example, certain emergent laws may only hold at one level when other emergent laws hold at other levels. Given such complexities, and also the range of possibilities, examination of these complexities is best pursued by engagement with real examples, so let me move on to contrast the claims made about the fundamental laws under Mutualism and Fundamentalism.

7.3.3 The Abandonment of the Completeness of Physics and its Empirical Rationale

Scientific reductionism is commonly committed to a world where the landscape of fundamental laws is like a regular, flat salt pan, since these laws are always and everywhere exhausted by the set of laws holding in the simplest collectives.[23] In contrast, as I have now outlined, scientific emergentists such as Laughlin have an alternative and more complex picture of the fundamental laws. Under the supplemental and supercessional variants, we have different kinds of "towers" of fundamental laws, like a collection of step pagodas rising from the jungle. In this landscape, we thus have an array of different sets of

---

[22] Laughlin (2005), p. 208.
[23] In Chapter 8, I outline an overlooked variant of Fundamentalism that takes a different view of the structure of the fundamental laws.

fundamental laws, holding in particular conditions, and forming distinct, and potentially monumental, pyramids of laws.

Against this background, I can now return to the question of what Mutualism implies about the Completeness of Physics that Kim and others emphasize. Normally, the Completeness of Physics is bluntly stated as referring to the "laws of physics," but we can now see this is damagingly ambiguous and this explains why I earlier framed the thesis thus:

(Completeness of Physics) Microphysical entities are determined, in so far as they are determined, solely by other microphysical entities according to the laws covering their behavior in isolation or the simplest collectives.

The latter formulation disambiguates what the Fundamentalist means by the "laws of physics." And this is important, since both sides agree that by definition "physics" will supply a universal dynamics and that the "laws of physics" determine microphysical events. Crucially, what we can now see is that the two sides disagree over the nature of a final "physics" and hence over the character of the "laws of physics."

The scientific reductionist claims that the "laws of physics" are exhausted by the laws operative in systems of microphysical entities in isolation or simple collectives. The scientific emergentist claims further emergent laws, holding only in certain complex collectives, are also fundamental "laws of physics." Mutualism consequently *rejects* the Completeness of Physics that frames a key tenet of scientific reductionism and its philosophical allies. I have thus framed the Completeness of Physics to reflect this point.

Given this point, some philosophers will question whether scientific emergentism or Mutualism, taken as global views, can really be the fresh, and strikingly different, version of physicalism that I have claimed. For the reigning philosophical orthodoxy is that the truth of physicalism *entails* the truth of the Completeness of Physics.[24] However, when one stops to look, explicit arguments supporting such a necessary connection between physicalism and the Completeness of Physics are surprisingly hard to find. In fact, Kim gives lucid expression

---

[24] See for examples, Horgan (1993b), Melnyk (1995) and (2003), and Kim (1993a), amongst others.

to what appears to be the only substantive type of argument in defense
of such a connection when he tells us:

> I doubt ... that contemporary non-reductive physicalists can afford to be
> so cavalier about the problem of causal closure [i.e. the Completeness of
> Physics]: to give up this principle is to acknowledge that there can in prin-
> ciple be no complete physical theory of physical phenomena, that theoreti-
> cal physics, insofar as it aspires to be a complete theory, must cease to be
> pure physics and invoke irreducibly non-physical causal powers – vital
> principles, entelechies, psychic energies, élan vital, or whatnot. If that is
> what you are willing to embrace, why call yourself a "physicalist"? Your
> basic theory will have to be a mixed one, a combined physical-mental
> theory, just as it would be under Cartesian interactionism. And all this may
> put the layered view of the world itself into jeopardy ... (Kim (1993a),
> pp. 209–10)

Notice that Kim refers to the case of the biological, as well as the
mental, and his points generalize in the dilemma he offers us. Either
accept the Completeness of Physics and that all powers are comprised, hence endorsing universal composition but thus also accepting
Fundamentalism and an ontologically reductive physicalism. Or reject
the Completeness of Physics and accept uncomprised, non-physical
powers, and presumably embrace non-physical fundamental forces,
and hence reject universal composition and physicalism.

Other variants of this argument are found in both philosophy and
also in the sciences (Weinberg (2001), pp. 115–16). Unfortunately,
we can now see such reasoning is built upon the false dichotomies of
the Unholy Alliance. My work has shown that in its global form the
Mutualist position of scientific emergentism presents an option that
retains comprehensive composition in nature, accepting that all powers are comprised and that the physical forces are the only fundamental
ones, but which still coherently rejects the Simple view of aggregation
and hence the Completeness of Physics. Kim and others who press
such arguments are therefore mistaken in claiming that non-physical
fundamental powers or forces, or any other uncomposed entity, *must*
result from abandoning the Completeness of Physics.

This brings us to Kim's last challenge for proponents of emergence.
Even if we accept Mutualism is a coherent form of physicalism that
rejects the Completeness of Physics, Kim will press, *why* should we
adopt this position about the *actual* world? By now I hope the rationale

of scientific emergentism is clear. Laughlin lays out the motivation for the view as follows:

Ironically, the very success of reductionism has helped pave the way for its eclipse. Over time, careful quantitative study of microscopic parts has revealed that at a primitive level at least, collective principles of organization are not just a quaint side-show but *everything* – the true source of physical laws, including perhaps the most fundamental laws we know. The precision of our measurements enables us to confidently declare the search for a single ultimate truth to have ended – but at the same time to have failed, since nature is now revealed to be an enormous tower of truths, each descending from its parent, and then transcending that parent ... (Laughlin (2005), p. 208)

The scientific emergentist's claims are not based upon ignorance and these researchers are not positing an "emergence of the gaps" analogous to "god of the gaps" hypotheses. Thus, for example, Laughlin accepts that quantum mechanics, and related areas, have illuminated fundamental laws, but claims that empirical evidence about phenomena, such as high-energy superconductivity, shows these laws are *not exhaustive* of the determinative laws holding of the components found in complex collectives. Scientific emergentists thus press a thoroughly empirical rationale for abandoning the Completeness of Physics and adopting their alternative Mutualist picture of nature and its fundamental laws.

7.4 Higher Sciences and their Explanations in a Mutualist Nature: Implications of Strong Emergence (III)

Turning to the implications of Strong emergence for the nature of higher sciences, and their explanations, a number of points are worth emphasizing. To start, the existence of Strong emergence provides a distinctive reason for the indispensability of higher science predicates different in kind from the reasons I outlined in Part II for the scientific reductionist. Strongly emergent entities are composed entities that are determinative. Predicates of higher sciences refer to these determinative entities, so sentences of higher sciences will often express the truths about nature that we literally take them to express and which cannot be captured by lower-level predicates alone. Contrary to the claims of reductionist writers such as Weinberg or Bickle, Mutualism therefore implies that the complete determinative explanation of various natural phenomena cannot be given solely in terms of components and the predicates referring to them.

The latter point also highlights that Mutualism entails the existence of determinative laws in the higher sciences concerning Strongly emergent entities. For example, as we saw in 7.1.1, the Strongly emergent properties of higher sciences contribute powers that produce certain higher-level effects and many laws of higher sciences describe these productive relations. Such higher-level scientific laws are therefore determinative laws because they concern determinative entities. Unsurprisingly, the higher scientific laws we find with Strong emergence are quite complex, so I am going to put the task of further understanding such laws to one side here.

I can offer more insight about important kinds of explanation, prediction, or derivation that are always potentially available when we have Strong emergence. The shocking point for many critics of the existence of "robust" emergence in nature is that *all* Strongly emergent entities are, at least in principle, explicable through compositional explanations. The point is again relatively simple once we appreciate the character of Strong emergence and clearly distinguish it from Ontological emergence. Strongly emergent entities are composed entities and their compositional relations thus always present the potential for compositional explanations. For example, we can always potentially compositionally explain Strongly emergent properties using the realizer properties we find with them. The powers of the realizers comprise all the powers of the Strongly emergent property instance and the processes based by the manifestation of the powers of the realizers implement all the processes which are the manifestation of the powers of the Strongly emergent property. We thus have a ready basis for compositional explanations of Strongly emergent properties using these relations, though it bears emphasis that such explanations use the differential powers of the altered realizers that we find in the *special condition of composing the Strongly emergent, and machretically determinative, entity.*

In many cases, we may even be able to *predict* how a Strongly emergent entity behaves in certain previously unobserved situations *if* we are lucky enough to have a detailed account of the components, and their powers and behaviors, found with such a Strongly emergent entity. However, it once more bears emphasis that we only have such predictions as a result of our having an account of the components and their differential powers found in the *special condition of composing the Strongly emergent, and machretically determinative, entity.*

The latter point highlights the danger of using accounts of emergence from earlier debates, such as that of C. D. Broad (1925), that solely use bare unpredictability as a criterion of emergence unsupplemented by a deeper ontological account.[25] On one side, we saw earlier that scientific reductionism can accept such unpredictability, so the unpredictability of an entity does not suffice for ontological irreducibility. But, on the other side, I have now shown that scientific emergentism embraces one kind of predictability from the features of components that we find in certain collectives, and the laws about them, whilst *still* rejecting predictability from the powers of components, their principles of composition, and laws holding in simpler collectives. Rather than bare predictability or unpredictability, we thus need a deeper ontological account of Strong emergence in order to properly understand the claims of scientific emergentism.

If we are really fortunate in having sufficiently well advanced investigations, then we may even be able to *derive* the higher scientific laws holding of Strongly emergent entities from the emergent and other laws holding of the lower-level component entities found along with the Strong emergent entity, at least in combination with the principles of composition, statements about background conditions, and other statements. But this derivation of such higher-level laws, like the predictions just noted, is only a derivation from laws, including the emergent laws, that hold of the altered components and their differential powers found under the *special condition of composing the Strongly emergent, and machretically determinative, entity.* Once again, given the nuanced commitments of scientific reductionists and emergentists on derivability, we see the dangers of using earlier accounts of emergence relying solely upon underivability without a supporting ontological account.

---

[25] One recent account that makes similar points to Broad's about the kinds of predictability and unpredictability that we can expect with what looks like Strong emergence is Boogerd et al. (2005). As I outline in Chapter 10, Boogerd et al. unfortunately follow C. D. Broad a little too closely, and without a supporting ontological account, in ways that I show generate problems.

To avoid common misunderstandings, I have belabored the point that the explanations, predictions, and derivations we get of Strongly emergent entities all utilize the evidence, and laws, about the altered components found when composing the Strongly emergent entity. Although Strong emergence always in principle allows compositional explanations, as well as associated kinds of prediction or derivation, we thus have a contrast between the types of explanations, predictions, and derivations entailed by the Mutualism of scientific emergentism and the Fundamentalist framework of scientific reductionism. I explore their differences further in Chapter 9.

Against this background, I now want to return to what I termed in the last chapter the "Objection from Compositional/Mechanistic Explanation" against "robust" emergence pressed by Delehanty (2005). Delehanty takes on writers seeking to show that higher properties in certain conditions must have uncomprised powers of their own because in these concrete cases components behave differently in the relevant conditions.[26] For instance, a number of writers have claimed slime mold is an example of "emergence" in part because we cannot explain the behavior and/or powers of certain molecules except by reference to the conditions in which we find them in slime mold. In response, Delehanty argues convincingly that by looking at the compositional explanations using both the relevant component entities and the other *surrounding* entities *also at the component level* (in what she terms "extensions" to such compositional explanations), then we can establish that all the powers of composed entities, whether of the emergent higher-level entity, or the higher-level entities making up its surrounding conditions, are all comprised by powers of entities at the component level.

Delehanty's Objection from Compositional/Mechanistic Explanation thus makes a compelling case that the existence of compositional explanations in the relevant higher scientific areas precludes, or poses grave problems for, the existence of uncomprised powers and hence Ontological emergence. This is an important result that again confirms my earlier appraisal of the main problems facing Ontological emergentism. But the crucial point for my present purposes is that we cannot use the Objection to reach a similar conclusion about *Strong* emergence,

---

[26] This is what Delehanty terms the "Context Objection" in defense of "robust" emergence.

since Strongly emergent properties are all realized and all their powers are comprised. The charitable interpretation of Delehanty's critique is therefore that it is intended only to show that Ontological emergence does not exist, for the Objection from Compositional/Mechanistic Explanation does not resolve the distinct question of whether we have Strong emergence.

7.5 Scientific Methodology and the New Frontiers in an Emergent Nature: Implications of Strong Emergence (IV)

Inevitably, the ontological differences between Mutualism and Fundamentalism lead to methodological divergences.[27] I can most usefully provide an example to highlight the existence of such differences by returning to debates about the frontiers of science and the very possibility of fundamental scientific progress.

As I outlined in Part II, reductionist writers such as Weinberg place the frontier of fundamental research solely on research focused on the fundamental components and hence in particle physics and related areas. But the experimental apparatus needed to explore the nature of the fundamental particles consists of gigantic, and gigantically expensive, supercolliders leading to talk of the "end of science."[28] Laughlin summarizes this situation, and the resulting pessimism about fundamental research, as follows:

... eighty years after the discovery of the ultimate theory we find ourselves in difficulty. The repeated, detailed experimental confirmation of these relationships has now officially closed the frontier of reductionism at the level of everyday things ... There is even a best-selling book exploring the premise that science is at an end and that meaningful fundamental discovery is no longer possible. At the same time, the list of even very simple things found "too difficult" to describe with these equations continues to lengthen alarmingly. (Laughlin (2005), p. 5)

---

[27] For discussion of other methodological concerns see Anderson (1990), Laughlin (2005), and especially Mitchell (2009), which provides the most detailed discussions of methodological issues on which scientific reductionism and emergentism part ways.

[28] Horgan (1997).

Here we see Laughlin noting that we cannot completely account for many, many natural phenomena using the laws the scientific reductionist takes to exhaust the fundamental laws. The reductionist tries to explain away these deficiencies as resulting from our own cognitive difficulties, for example in making the necessary calculations, but Mutualism suggests a different diagnosis of these problems. Laughlin tells us:

> It turns out our mastery of the universe is largely a bluff – all hat and no cattle. The argument that all the important laws of nature are known is simply part of this bluff. The frontier is still with us and still wild.
>
> The logical conflict between an open frontier on the one hand and a set of master rules on the other is resolved by the phenomenon of emergence. (Laughlin (2005), p. 6)

The key point here is that its Mutualism means that scientific emergentism rejects the presumption that underlies worries about the "end of science." Namely, that investigation of the behavior of the most fundamental known particles in simple collectives is the *only* way to illuminate fundamental laws of nature.[29] As Laughlin points out, if Mutualism is true and Strongly emergent entities exist, then the "master rules" holding in the simplest collectives of microphysical entities will not exhaust the fundamental laws, since further emergent laws hold of such entities in complex collectives and are, as yet, undiscovered fundamental laws.

Far from the frontier of science being closed, scientific emergentists such as Laughlin thus contend that in a Mutualist nature there are many frontiers of fundamental research left to explore at many levels of the natural world and using many different scientific disciplines – from particle physics to biochemistry, neurobiology, and beyond. Given its strikingly different view of the frontiers of fundamental research, and the fundamental laws, Mutualism thus underpins a challenge to the reductionist's arguments about funding allocation.

Recall that Weinberg provided his Methodological Priority Argument to show that there was *still* one reason for prioritizing funding for particle physics that higher sciences always lack, since particle physics is the only discipline that studies the ultimate determinative entities or

---

[29] This is the commitment that we saw Papineau claiming in the last chapter it would be "absurd" to challenge.

the fundamental laws of nature. In response, we can now see that, for argument's sake, the scientific emergentist may grant Weinberg's premise that, all else being equal, investigating the nature of the determinative entities involved with some phenomenon of interest (and/or the determinative laws associated with this phenomenon) usually yields greater insight than investigating the nature of other entities. However, the scientific emergentist rejects Weinberg's further premise that the component entities studied by the relevant lower science(s) are the *sole* determinative entities in any case of successful compositional explanation. For in a Mutualist nature, Strongly emergent composed entities are also determinative and the fundamental laws do not solely involve component-level entities. Under Mutualism, particle physics is thus not the only science investigating the determinative entities, or even the fundamental laws, and we get a very different picture of which sciences ought to be prioritized for funding even if we focus solely on which disciplines investigate determinative entities or fundamental laws of nature.

Contrary to the claims of some philosophers of science, scientific emergentists have consequently not merely pressed empty rhetorical points about "emergence" for instrumental gains with regard to funding. One may be concerned that the wider claims of scientific emergentism have not been justified, an issue I examine in Part IV, but that is a far cry from using empty theoretical jargon to press nakedly self-interested funding claims. For I have now shown that scientific emergentism provides principled reasons for its key methodological conclusions that are firmly rooted in its wider ontological claims.

7.6 Are We All Reductionists Now? "Yes" (On the Everyday Understanding) and "No!" (On a Substantive Reading)

My work on scientific reductionism and emergentism confirms the truth of Weinberg's rallying cry that "We are all reductionists now." I have shown that both Fundamentalists such as Weinberg, and Mutualists such as Laughlin, are all everyday reductionists, i.e. they accept that we can and should provide compositional explanations. However, my work in this chapter also establishes at some length that there is no simple route from everyday reductionism, and the core empirical evidence it supplies, to the truth of the more substantive Fundamentalist claims of scientific reductionism.

Contrary to the contentions of scientific and philosophical reductionists, such as Weinberg, Kim, Heil, and others, I have further confirmed that the universal applicability of compositional concepts does not suffice for such an argument, nor does the provision of successful compositional explanations. By reconstructing the Mutualist framework of scientific emergentism, I have shown that this position is consistent with both universal composition and the ubiquity of compositional explanation, thus establishing that scientific and philosophical reductionists have drawn ontological conclusions that are far too strong from our core scientific evidence.

Mutualism is therefore not merely an interesting theoretical option, but also a substantive position in its own right, just as writers such as Anderson, Prigogine, Laughlin, Freeman, and other scientific emergentists have long contended. Although we are indeed all everyday reductionists now, my work shows that there are *two* coherent alternatives for everyday reductionists, and hence physicalists, to choose between in the Fundamentalism of scientific reductionists and the contrasting Mutualist account of nature articulated by such scientific emergentism. We have therefore found that E. O. Wilson was plausibly correct in the quote that started this book, for we are apparently very much in the thick of exciting, ongoing scientific debates, often metaphysical in nature, about the structure of successful cases of compositional explanation and their wider implications for laws, the species of determination, the status of the various sciences, methodology, and more.

PART IV

New Landscapes, New Horizons

8 | Our Competing Visions of Nature and the Sciences: Illuminating the Deeper Debates and Viable Positions

My work in earlier chapters shows our new debates about collectives, and their components, do indeed focus on a more complex, and even messy, array of positions than philosophers and scientists have often realized. The intellectual terrain is consequently rather different and this chapter does the necessary, albeit highly abstract, work of drawing together my conclusions to clarify the new landscape of arguments, theses, and viable positions that we actually inhabit. In the next chapter, I build on this abstract survey by illuminating the empirically resolvable differences between the viable positions. And, in the final chapter, I then use this work as a platform to examine whether evidence from scientific examples has, as yet, been successfully used to resolve the relevant disputes. Part IV will thus cumulatively provide an overview of our new debates, where they now stand, and how we can move them forward.

I start this chapter by explicitly laying out important arguments in the service of Fundamentalism and Mutualism respectively. For Fundamentalism, I provide a valid argument supporting compositional reduction, but I mark its high evidential requirements. Whilst for Mutualism, I outline an argument that shows that machresis is necessary for determinative composed entities – and hence composition itself. Building on these arguments, and my earlier findings, I then provide an overview of the theses actually driving our present problematic and systematically work through various positions to illuminate which views are live and those that are dead.

Overall, the chapter confirms that the most popular positions in the sciences and philosophy, in the views endorsing Scientifically Manifest Image and Naked non-reductive physicalism, are not even candidates for live positions. In contrast, I show that two kinds of the Fundamentalism pressed by scientific reductionists and philosophers are candidates for live positions. But I also show that the Mutualism defended by scientific emergentists and allies in philosophy is also one of the live views. Along the way, I begin to highlight the core disputes that are the focal points of our new discussions.

8.1 Revisiting Scientific Reductionism and its Arguments

My work in earlier chapters pointed to an alternative, valid argument supporting compositional reduction and I now want to explore whether such reasoning actually exists. I earlier noted that the Simple view of aggregation, or the Completeness of Physics for which it is a precondition, plays a large implicit role in underpinning the most widespread form of scientific reductionism. Still more importantly, as I noted in Chapters 6 and 7, once we make such epochal assumptions explicit, then we see how the truth of such theses shuts down the overlooked Mutualist avenue suggested by scientific emergentism. This is important, for it apparently points us towards an alternative, valid defense of compositional reduction.

To begin, for reasons that will become clear shortly, I want to provide an explicit definition of the *localized* claim that we have seen the scientific reductionist implicitly accepts about the components in complex collectives in particular cases. To this end, let us assume we have entities $X1$–Xm, at level n, which are taken in a successful compositional explanation to compose an entity Y at level $n+1$. In this type of case, we have seen that the reductionist routinely assumes that the powers of the components $X1$–Xm, under conditions \$ at a time t, are exhaustively, and hence solely, determined at t by *other components* at level n (or at still lower levels), in so far as their powers are determined at all. In such cases, I take what I term "determinative completeness" to hold for $X1$–Xn at time t under conditions \$ and hence take these components to be *determinatively complete*.[1]

---

[1] Three points bear emphasis. First, that certain components are determinatively complete does not mean that they are completely determined, rather the claim is that in so far as they are determined at all, which may be far less than completely, these components are determined by other components. Second, note that we have allowed that components at level n or *still lower levels* may often be the determinative entities, so we can have determinative completeness for the components at level n if they are solely determined at time t, in so far as they are determined, by components at level n, $n-1$, and so on. And, third, it is important to remember that by "other components" we mean other entities at the lower, component levels.

Although the truth of the Completeness of Physics, or the Simple view of aggregation, each entails that determinative completeness holds of components, below I outline how components can be determinatively complete even when both the Simple view of aggregation and the Completeness of Physics are *false*. Focusing on the more general notion of determinative completeness in framing the key reasoning supporting compositional reduction, rather than using the Simple view or Completeness, thus opens up overlooked options for the ontological reductionist – or so I argue in later sections.

With the latter notion in hand, I now want to reconstruct a more complex parsimony argument that adds determinative completeness as a premise alongside the applicability of compositional concepts found with a successful compositional explanation. To provide a detailed exemplar of the reductionist's alternative reasoning I again focus on property instances and their realization relations, though the same type of argument plausibly applies to individuals, processes, and powers.[2]

Our new argument has to be a little elaborate not only due to its extra premise, but also because we obviously now need to cover a number of further options for a property instance to be determinative at a time beyond its own contribution of powers to an individual. So consider once more a case where we have provided a successful compositional explanation of a higher-level property instance G, at time t under conditions \$, using instances of lower-level properties P1–Pn that realize G, and where we also know that P1–Pn are determinatively complete. What determinative property instances should we accept at t, under \$, once we reflect upon the features of this case and their implications?

The reductionist now accepts that a property instance can only make a difference to the powers of individuals at a time by either (i) solely contributing powers to an individual itself; or (ii) solely determining powers contributed by some other property instance to an individual; or (iii) by both contributing powers to an individual itself and also by determining the powers contributed to some individual by some other

---

[2] And given the coadunative character of these entities a compositional reduction in our commitments about the entities in one of these categories again probably requires compositional reductions in our commitments to the entities in each of these categories.

property instance. But, the scientific reductionist claims, in this case we know that we have determinative completeness with regard to P1–Pn at t under $. Thus, concludes the scientific reductionist, the realized property instance G does not make a difference to the powers of individuals through options (ii) or (iii), since both require that G makes a difference to the contributions of powers of its realizers P1–Pn and the determinative completeness known to hold of P1–Pn excludes this.

Furthermore, continues the scientific reductionist, our successful compositional explanation allows us to account for the powers of G, and/or higher-level effects of these powers, solely in terms of the effects of the powers contributed by P1–Pn to lower-level individuals. And our earlier sub-conclusion was that components, and not G, determine all of the contributions of the powers of its realizers P1–Pn. The contributions of powers by P1–Pn also accounts for all the lower-level effects of the relevant lower-level individuals. Therefore, concludes the reductionist, we can account for *all* of the powers and effects of the relevant higher- and lower-level individuals simply using the contributions of powers by the realizer property instances P1–Pn. Applying the Parsimony Principle the reductionist therefore concludes that we should *only* accept that the realizer property instances P1–Pn contribute powers to individuals at t under $, rather than taking both G and P1–Pn to contribute powers to individuals. Consequently, concludes the reductionist, we should not accept that option (i) holds of G either.

Bringing the various sub-conclusions together, argues the reductionist, we can now turn to the two hypotheses about the determinative property instances we should take to exist at a time in this case of compositional explanation. The hypothesis associated with the Scientifically Manifest Image (or Mutualism) is that we should accept that both G and P1–Pn are determinative instances at t; whilst the scientific reductionist's favored Fundamentalist hypothesis is that we should only accept the component properties P1–Pn are determinative. But, notes the reductionist, we have shown that we should not accept that G is determinative through options (i), (ii), or (iii). Since options (i), (ii), and (iii) are the only ways for a property instance to make a difference to the powers of individuals, the reductionist concludes we should not take G to be determinative and we should only accept that P1–Pn make a difference to the powers of individuals at t. Since it is also true that the only property instances that exist make a difference to the powers of individuals, then we should only accept that the realizer property instances P1–Pn exist at t. We therefore conclude that, really, we should accept that there are "nothing but" the component properties P1–Pn in this case.[3]

The latter kind of reasoning is what I am calling the "Argument from Composition and Completeness," whether adapted to property instances, individuals, powers, or processes, since it relies on determinative completeness at the level of components in addition to the applicability of compositional concepts in a successful compositional explanation. Scientific reductionists such as Weinberg, or reductionist philosophers such as Kim or Heil, have often used the invalid Argument from Composition, or related reasoning, but the Argument from Composition and Completeness is valid and has the potential to deliver successful compositional reductions.

It is important to note the connections between the Argument from Composition and Completeness and some earlier reasoning in philosophy. Although many of Kim's arguments focused on supervenience, causal overdetermination or exclusion are close in nature to the Argument from Composition, Kim's arguments deploying the Completeness of Physics are similar to the Argument from Composition and Completeness.[4] The same holds of related reasoning based around the Completeness of Physics independently offered in defense of ontological reductionism by Barry Loewer.[5] One key difference from even these philosophical arguments, to which I return below, is that the Argument from Composition and Completeness can utilize a number of kinds of evidence to support the determinative completeness of components, rather than simply using the Completeness of Physics as Kim and Loewer do in their narrower formulations.

It is important to emphasize that the Argument from Composition and Completeness still has the same conclusion as the Argument from Composition – that we should only accept that there are component entities or at least that they are the only determinative entities. Consequently, the Argument from Composition and

---

[3] The permissive scientific reductionist will again settle for the sub-conclusion that we should accept P1–Pn as the only determinative property instances.
[4] Kim (1998).
[5] Loewer (2008).

Completeness suffices to drive compositional reductions supporting the Fundamentalist commitments of scientific reductionism laid out in Chapters 3 and 4.

However, I do need to carefully mark the increased epistemic burdens placed on reductionists by the move to the Argument from Composition and Completeness given the richer set of premises used in this new reasoning. We consequently get these revised requirements for a successful compositional reduction:

(Compositional Reduction*) If we know entity Y is composed by entities X1–Xn, at time t under conditions $, and also know that the entities X1–Xn are determinatively complete, at t under $, then we ought to reduce our commitment to both Y and X1–Xn to a commitment solely to X1–Xn.

This thesis highlights how simply having a successful compositional explanation, and knowing compositional concepts apply in some case, does not *alone* suffice for a compositional reduction, since we *also* need to establish that determinative completeness holds of the relevant components to drive such a reduction. The looming question is whether in various scientific examples, let alone globally, we are justified in accepting determinative completeness of components. I explore this key question further in the remaining chapters.

8.2 The Argument from Exhaustion and the Necessity of Machresis for Strong Emergence

Having clarified a key argument of the Fundamentalist, let me now turn to reasoning that establishes a central claim of the Mutualist. I have already shown that the existence of machresis in a certain kind of situation can be sufficient for us to accept that Strong emergence exists, but I now want to confirm that machresis is *necessary* for the existence of composed entities that are also determinative. To support this important conclusion, I outline what I term the "Argument from Exhaustion," which works through the different species of determination to assess whether they alone suffice, without machresis, for a composed property to make a difference to the powers of individuals and hence be Strongly emergent. The Argument shows that none of the other kinds of determination alone suffice for Strong emergence, either singly or in combination, hence establishing that machretic determination is indeed necessary for a composed property instance to be determinative and Strongly emergent.

Once again, I frame my exemplar of the Argument from Exhaustion using property instances, but the reasoning can plausibly be extended to other categories of entity and we need to provide such extensions given the coadunative character of the entities posited in the sciences. With regard to the Argument, it is important to remember that it is solely focused upon cases of successful compositional explanation where compositional concepts have been shown to apply comprehensively, so the only relevant entities are either composed or component entities.

Consider a property instance, G, instantiated in an individual s^* where we have true compositional explanations illuminating, amongst other compositional relations, how lower-level property instances P1–Pn, instantiated in $s1$–sn, realize G, in s^*, at time t under conditions $. And let us also assume we have true compositional explanations of all the property instances realized by G in combination with other instances at its level such as G^*–G^*m. There are then three routes by which G might determine the powers of individuals at a time. (a) G can *itself* contribute powers to individuals. Or G can determine the powers had by other individuals where, to be comprehensive, I want to look at two ways this can occur;[6] (b) G can *compositionally* determine, through its joint role-filling, that some *other* property instance exists and contributes powers to individuals;[7] or (c) G can *machretically* determine, through its role-shaping, that some *other* property

---

[6] Following my earlier procedure, I did not consider a realized property instance's compositional determination of still higher-level properties/relations in the Argument from Composition and Completeness for an obvious reason. If that reasoning suffices to show that a composed instance that composes no other instances is not existent or determinative, then any composed instance added on top of the first composed instance can still be shown, by an application of the Argument from Composition and Completeness, not to be determinative or existent. And so the original composed instance should not be taken to determine the powers of such a composed instance. I therefore left such cases aside in my presentation of the Argument from Composition and Completeness, but one may wonder whether compositional determination could underpin Strong emergence.

[7] Given the nature of composition, various other implications follow. Thus for example s^* is also constituted, under $, by $s1$–sn along with other individuals at the same level that realize s^*'s other properties/relations.

instance contributes powers to individuals. Furthermore, I should note that G might be determinative through combinations of (a)–(c).

The Argument from Exhaustion focuses upon whether G can be determinative, and Strongly emergent, solely through (a), or solely through (b), or solely through (a) and (b) in tandem. To begin, the Argument examines whether G can be determinative, and Strongly emergent, *solely* through option (a) and G's own contributions of powers to s^* at t. I am assuming we have a compositional explanation of G, so all of G's powers are comprised by the powers that its realizers P1–Pn contribute to $s1$–sn. We then confront a dilemma: either determinative completeness holds of P1–Pn or it does not. On the first horn, where P1–Pn are determinatively complete, we can apply the Argument from Composition and Completeness to establish that we should not accept that G contributes powers to s^* and hence G should not be taken to be determinative or Strongly emergent. On the other horn, P1–Pn are not determinatively complete at t, so an entity other than another component determines the contributions of powers by P1–Pn at t. But the only entities are either components or composed entities. Thus a composed entity must determine the powers of P1–Pn and hence a composed property instance like G is not determinative *solely* by its own contribution of powers to individuals. Consequently, a realized property instance like G cannot be determinative and Strongly emergent *solely* by (a) contributing powers to an individual itself.

The Argument from Exhaustion then turns to option (b) and considers whether a property instance like G can be determinative, and Strongly emergent, *solely* by its compositional relations to other property instances. For example, G in combination with G^*–G^*m might be argued to make a difference to the powers of individuals by bringing into existence a realized property instance, M, of an individual s^{**}, along with M's contribution of powers to s^{**}. By assumption, we again have successful compositional explanations of G, G^*–G^*m, and M, in terms of instances of the lower-level properties P1–Pn, and the lower-level property instances involved in the realization of G^*–G^*m. Once again, at t, either P1–Pn and the realizers of G^*–G^*m are determinatively complete or they are not.

On the first horn, P1–Pn, and the other realizers, are determinatively complete, so given the transitivity of compositional relations the realizers of G and G^*–G^*m also realize M in s^{**}. But the Argument from Composition and Completeness then applies to show that we should only accept that P1–Pn and the other realizers are determinative and that we should not accept that either M or G are determinative or Strongly emergent. On the other horn, P1–Pn and the realizers of G^*–G^*m are not determinatively complete. But, again, the only entities are either components or composed entities. So a composed entity must determine the powers contributed by P1–Pn, or the other realizers, but this determination cannot be through a compositional relation that is based upon joint role-filling, since the composed properties do not have powers that jointly fill the roles of, and hence comprise, the powers of P1–Pn or the other realizers. Consequently, we can conclude that G, or some other composed property instance, cannot be determinative, and Strongly emergent, *solely* through option (b) and the compositional determination of this property instance.

Lastly, the Argument from Exhaustion turns to the relevant mixed case to consider whether G can be determinative, and Strongly emergent, as a result of *both* option (a) in G's contribution of powers to s^*, *and* option (b) in the compositional relations of G. Once again, at t, we face the same dilemma: Either the realizers P1–Pn and the realizers of G^*–G^*m are determinatively complete or they are not. On the first horn, where the realizers are determinatively complete, then the Argument from Composition and Completeness establishes that we should not accept that anything other than P1–Pn, and the realizers of G^*-G^*m, determine the powers of individuals. So the realized property instance M and G should not be taken to be determinative or Strongly emergent. Whilst on the other horn, where P1–Pn, and the other realizers, are not determinatively complete, then the composed entity G (or M) must determine some of the powers contributed by P1–Pn or the other realizers. Consequently, a composed property like G cannot be determinative, and Strongly emergent, *solely* by (a) contributing powers itself *and* (b) composing other properties.

Bringing the various sub-conclusions of the Argument together, we have shown that a composed property instance like G cannot be determinative, and hence Strongly emergent, solely through option (a), solely through (b), or solely through (a) and (b) together. But the only options by which G can be determinative are options (a), (b), and (c), or some combination of them. Consequently, the Argument from Exhaustion concludes that option (c), and machresis, is required in order for a composed property like G to be determinative and Strongly emergent.

Overall, the conclusion of the Argument from Exhaustion fits nicely with a number of my earlier findings. First, it is supportive of my independent conclusion from Chapters 7 and 8, built around my scientific thought experiment, that if a composed property instance G machretically determines the powers of its lower-level realizers, then we do have a reason to take G to determine the powers of individuals that is not open to a valid parsimony argument.[8] Second, I should carefully note that the conclusion of the Argument from Exhaustion is only that machresis is necessary for Strong emergence and this is compatible with my earlier conclusion that compositional and machretic relations are coadunative, i.e. that they are always found together.

The Argument from Exhaustion plausibly shows, against the background of my other findings, that the existence of machresis is necessary for the acceptance of an ontologically non-reductive physicalism and I further confirm this point in my survey in coming sections. One can thus have walls of composition, and levels of determinative composed and components entities, rising up in the house of nature *only if* we have a roof of machretic, Strongly emergent entities pressing downward. My earlier pair of theory schemata for Strong emergence constructed to require machresis have therefore been vindicated. Scientific emergentists have illuminated the deeper character of Strong emergence, rather than just one way to underpin its existence.

8.3 Surveying the Actual Landscape of Conflicting Theses, Possible Options, and Viable Positions

The key arguments for the Fundamentalist and Mutualist I have made explicit allow me to highlight the deeper theses driving the contemporary problematic and to lay out the available options built around combinations of the relevant theses. However, it is useful to start by reminding ourselves of a common, and highly influential, account of the problematic before contrasting it with what I have highlighted about its actual character.

---

[8] The Argument from Composition is not valid and the Argument from Composition and Completeness does not apply in cases of machresis, since we lack determinative completeness in such examples.

As I outlined in earlier chapters, driven by their acceptance of the Argument from Composition philosophers such as Kim, and other dichotomists in the Unholy Philosophical Alliance, often explicitly assume we have an incompatibility between *two* theses: composition (or at least the applicability of compositional concepts) and the existence of determinative higher-level entities. The resulting taxonomy of options and possible positions is framed in the table in Figure 8A.

My work establishes that the Argument from Composition is invalid and that the applicability of compositional concepts, and the determinativity of composed entities, are compatible in the Mutualist

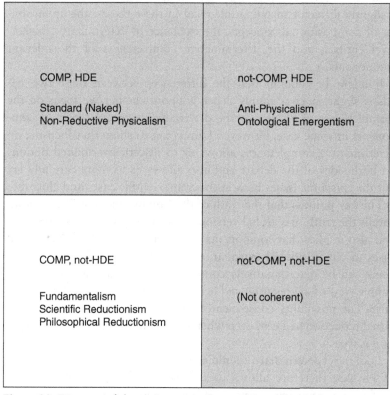

COMP, HDE	not-COMP, HDE
Standard (Naked) Non-Reductive Physicalism	Anti-Physicalism Ontological Emergentism
COMP, not-HDE	not-COMP, not-HDE
Fundamentalism Scientific Reductionism Philosophical Reductionism	(Not coherent)

Figure 8A Diagram of the options as understood by philosophical critics such as Jaegwon Kim and Barry Loewer. Boxes in grey indicate combinations of theses taken by these reductionists to be untenable. (COMP = composition; HDE = the existence of determinative higher-level entities.)

scenarios broached by scientific emergentists. So the received picture of the problematic driving the debates is mistaken. Articulating the valid Argument from Composition and Completeness has further confirmed these earlier conclusions. Rather than a *pair* of incompatible theses, the latter reasoning shows we actually have an incompatibility between *three* theses forming an incompatible *triad*.

As I noted above, in one strand of reasoning philosophers such as Kim and Loewer have come close to framing something like the deeper problematic by highlighting an incompatible triad in these theses: the applicability of compositional concepts, the existence of determinative higher-level entities, and the Completeness of Physics.[9] My focus on determinative completeness allowed me to show that we also have a slightly different incompatible triad in these theses: the applicability of compositional concepts, the existence of determinative higher-level entities, and the determinative completeness of the relevant components.

It might be thought that the differences between these two triads is slight, since it simply depends upon whether we focus on the Completeness of Physics or the determinative completeness of components in some case. However, I am going to show that focusing on determinative completeness allows us to discern overlooked options on both sides of the debate and also allows us to more carefully lay out the epistemic issues both in concrete scientific cases and globally.

The key point is that the truth of the Completeness of Physics both entails the truth of a global version of the Simple view of aggregation and also implies that components are determinatively complete in all cases of compositional explanation. But, as I outline below, components can be determinatively complete in some case when *both* the Simple view of aggregation *and* hence the Completeness of Physics are false. The possibility consequently opens up for pressing a compositional reduction in a case even when the Conditioned view of aggregation is true.

Looking at determinative completeness, rather than the Completeness of Physics, therefore allows us to more easily appreciate the role of the two opposing accounts of aggregation I have highlighted in earlier chapters and which have been given little explicit attention.

---

[9] Again, I loosely talk of "composition," but to be fair to the reductionist this is strictly evidence for the applicability of compositional concepts.

Philosophers have plausibly missed the key role that the Simple view of aggregation plays in the arguments of reductionists and the possibility, let alone the implications, of the alternative Conditioned view of aggregation that is central to the Mutualism of scientific emergentism.

To reflect the importance of the opposing views of aggregation, I therefore want to quite literally overlay acceptance of the Simple or Conditioned views of aggregation *on top* of the theses philosophers recognize, outlined in the table Figure 8A, to produce the more complex table of positions, and underlying theses, in Figure 8B. What should immediately be apparent is that the table in 8B has a far wider array of positions than are normally recognized or discussed. Most

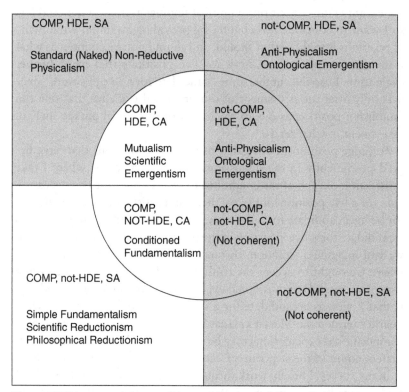

Figure 8B Diagram of the possible combinations of theses factoring in views of aggregation. Boxes in grey indicate non-live combinations of theses. (COMP = composition; HDE = the existence of determinative higher-level entities; SA = Simple view of aggregation; CA = Conditioned view of aggregation.)

obviously, in each of the cells we now have *two* options, rather than one, given the choice between whether we combine the relevant theses with the Simple or Conditioned views of aggregation. Furthermore, at least one of these variants has novel implications. The Simple view of aggregation entails determinative completeness of components, but when the alternative Conditioned view holds of some example then this is compatible with *either* determinative completeness of the components *or* its failure.

In the final two chapters, I am going to directly engage arguments about how our empirical evidence, primarily from cases of successful compositional explanation, does, or does not, support the various positions laid out in this new landscape of positions. However, surveying extant claims about our empirical findings is a large undertaking, so I want to conclude this chapter by providing an initial winnowing of positions into those that should, and should not, be counted as what I am going to term *candidates* for live positions that merit further evaluation. I use the cumbersome "candidate for a live position," since it is only after an assessment of our empirical evidence that one can establish a position as one of the live positions, and I pursue such an assessment in Chapter 10.

To judge positions, I am going to use two conditions that have figured prominently in the discussions of earlier chapters, where I take satisfaction of each of the conditions to be necessary for being a candidate for a live position and I take their joint satisfaction to be sufficient for being a candidate for a live view. My first condition is that a view is a candidate for a live view only if the position is coherent and such that we can imagine a situation, through a scientific thought experiment, where we ought to accept the truth of all of the position's theses in a case of successful compositional explanation. And my second condition is that a view is a candidate for a live position only if it passes what I earlier termed the Meta-Explanatory Adequacy Test of showing how compositional explanations may be successful under the position's ultimate account of the structure of nature, i.e. under its favored ontology.

In my survey, I briefly work through *all* of the cells in the table in 8B aside from the bottom right cell, whose combination of theses again does not appear to be coherent.[10] Although it might seem excessive

---

[10]   Putting the point in terms of my two conditions, the two views in the bottom right cell of table 8.2 are combinations of theses that fail the first condition.

to work through all of these options there are two good reasons for this approach. First, we have a number of novel views that need to be articulated and evaluated. And, second, I want to avoid being influenced by earlier conclusions about positions that we have found are all too often based upon flawed reasoning like the Argument from Composition or related false dichotomies about reduction or emergence. My brief survey will thus give us a clearer picture of which positions potentially offer viable hypotheses about cases of compositional explanation that I can further examine in the final chapters.

8.4 Assessing Two Minority Views and Their Challenges: The Top Right Cell and Going "Off the Grid"

To start my survey, in 8.4.1, I am going to revisit the Anti-Physicalism, and associated forms of Ontological emergence, that we find in the top right cell of Figure 8B in order to gauge whether our new resources make this kind of view any more viable in a case of compositional explanation. Perhaps unsurprisingly, I show that Anti-Physicalist views are still not candidates for live views. In addition, in 8.4.2, I then look at an exemplar of a view that is "off the grid" in having commitments not easily fitted into the table in Figure 8B to illustrate some of the challenges that face such positions rejecting assumptions central to the debates I have been engaging.

8.4.1 *The Novel Options and Unresolved Difficulties of Anti-Physicalism*

My work now reveals that we may endorse two distinct forms of Anti-Physicalism depending upon whether one embraces the Simple or Conditioned forms of aggregation about the collectives formed by the entities of physics or other components.[11] The outermost part of

---

For we have yet to provide any scientific thought experiment that establishes the coherence of either view by showing the relevant combination of theses can all be true together, let alone when we also have a true compositional explanation.

[11] One can multiply options still further if Anti-Physicalism endorses non-physical individuals which then also themselves aggregate, since we can then have Simple or Conditioned aggregation with both the physical *and* non-physical collectives – thus giving us four options.

the top right cell is the position discussed in earlier sections, which I term "Simple Anti-Physicalism," which we can take to implicitly endorse the Simple view of aggregation about the components in the natural world. In addition, I now also need to mark the novel position in the innermost part of the top right cell, which I call "Conditioned Anti-Physicalism," which endorses the Conditioned view of aggregation for components.

Given its novel commitments, this latter form of Anti-Physicalism may thus endorse differential powers of components. Consequently, a novel option for the Conditioned Anti-Physicalist is to claim that such contributions of differential powers by component entities is *productively determined* by uncomposed Ontologically emergent entities.[12] However, notice that the Conditioned Anti-Physicalist still cannot accept machretic determination of the powers of component entities by the uncomposed higher-level entities they endorse, since machresis requires a compositional relation between the two entities that is obviously lacking with the Ontologically emergent entities endorsed under Anti-Physicalism.

Turning to evaluation of whether either Simple or Conditioned Anti-Physicalism is a candidate for a live view in a case of compositional explanation, we find that the novel commitments provide no help with the difficulties laid out in Chapter 5. Most importantly, Ontologically emergent property instances are not realized and thus still do not fit within a case of compositional explanation. Both forms of Anti-Physicalism thus fail my first condition, since we cannot imagine a case of successful compositional explanation, with the comprehensive applicability of compositional concepts that it involves, in which the commitments of either Simple or Conditioned Anti-Physicalism are also true. Consequently, Simple and Conditioned Anti-Physicalism are plausibly not amongst the candidates for live positions about cases of successful compositional explanation.

8.4.2 Going "Off the Grid": The Unresolved MEAT of the Problem for Schafferian Monism

It will be helpful at this point to examine a position not easily framed by the theses in table in Figure 8B, since this is instructive in illustrating

---

[12] This type of view may be suggested in O'Connor and Churchill (2010).

the rather different nature, and scale, of the challenges for positions stepping outside the common assumptions of the debates I have examined. As my example of such a view, I want to look at Jonathan Schaffer's "Monism" (Schaffer (2007), (2010)), whose central claim is that the whole universe is the only determinative entity or the only entity. Schaffer's Monism comes in a strict variant under which the universe is the only entity, or a permissive form that allows there are also non-determinative "parts" of the universe that depend upon it.

However, I put scare quotes around the term "part," since the sense of "whole" and "part" here cannot be the scientific notions outlined in Chapter 2 because the "whole" under Monism is not related to its "parts" in the manner required for scientific composition. For example, "parts" are not determinative and do not fill the roles of "wholes" under Monism. And, given the nature of the universe, it is hard to see how the "whole" even has a role.[13] However, as compositional explanations are commonly understood, at the lower level we find determinative individuals that have roles that allow them together to fill the role of some higher-level individual, so at the higher level we have an individual with a role. Consequently, Schaffer's Monism does not accept the entities commonly understood as posited at *either* the higher *or even* the lower levels in our compositional explanations.

Schaffer's Monism thus falls outside the shared commitments of the positions examined in my discussions in earlier chapters, framed in the table in Figure 8A, which all accept some of the entities posited in compositional explanations in the sciences. I now want to assess whether Monism is a candidate for a live view. Fortunately, it is easy to understand Schaffer's defense of Monism since it partially overlaps with the arguments I have examined so far. Schaffer uses a two-step defense of Monism (Schaffer (2007)). First, Schaffer endorses the success of ontological parsimony arguments of the type I have examined at length and which I have now established must take the form of the Argument from Composition and Completeness if they are to be successful. Second, Schaffer then uses further arguments to show that Monism should be favored over versions of the Fundamentalism favored by

---

[13] Under Monism, the "whole" is the entire universe, but there appears nothing else for this whole to productively interact with – hence this "whole" is not a working entity.

scientific and philosophical reductionists. Most importantly, Schaffer argues that Monism provides the only adequate account of quantum entanglement.

In the next two chapters, I examine whether anyone has, as yet, shown we can run the Argument from Composition and Completeness in an actual example. And, given the nature of Schaffer's defense of Monism, whether we are justified in accepting this position depends upon our showing the Argument from Composition and Completeness can be successfully applied to a range of cases and that Mutualism or Anti-Physicalism have not been established about any case of compositional explanation. So Schaffer's view has large epistemic obligations overlapping with those of Fundamentalism and my work in the next two chapters shows that such obligations have yet to be satisfied. However, the difficulties facing Schaffer's Monism are deeper than whether applications of its key supporting argument have been established in various examples of compositional explanation. For we can now see that Monism has difficulties even in passing my two conditions for being a candidate for being a live position about a case of compositional explanation.

In this case, let me start with the second condition focused on whether a view has been shown to satisfy the Meta-Explanatory Adequacy Test. Remember that the Monist uses parsimony reasoning focused on compositional explanations to drive her claims, so Monism plausibly needs to provide an account of the success and/or truthmakers of such explanations. But obvious problems now arise because Monism accepts no part of the ontology of compositional explanations as commonly understood at either the *higher or lower levels* of such explanations. Monism thus faces a monumental task with two elements. First, Monism needs to provide an ontology for compositional explanations and this will be a completely new ontology, since the Monist's ultimate account of the structure of nature does not involve any of the entities (whether powers, properties, individuals, and processes) at either the higher or lower levels in a compositional explanation given in any science. In itself providing such a completely new ontology is a large task and one that the Monist has yet to complete.

However, I should emphasize that there must also be a second element of any account that a Monist gives in order to satisfy MEAT. For a Monist also needs to show that the new ontology she provides as an interpretation of the structure of a compositional explanation can, and does, underlie the success and/or truth of our compositional explanations. This is a daunting prospect, since philosophers of science now take the best test of the utility of some practice that it has successfully used. But the novel ontology of Monism has obviously never been posited in actual practice, so the Monist faces real challenges. So far as I know, the Monist has therefore yet to show how compositional explanations can be successful, and/or true, under the Monist's ultimate account of the structure of nature. I therefore conclude that Schaffer's Monism has yet to provide an adequate account that satisfies the Meta-Explanatory Adequacy Test and hence my second condition for being a candidate for a live view.

Related concerns also arise when we turn to my first condition for a view to be a candidate for a live view. Without the ontology Monism offers as an interpretation of the structure of a case of compositional explanations, we cannot presently imagine a situation that serves as the basis for a scientific thought experiment in which we have a true compositional explanation and the commitments of Monism are also true. So, as yet, Monism also fails to satisfy the first condition as well. (And I should also note that both sets of difficulties for Monism, with the first and second conditions, are not just confined to the compositional explanations that have been my focus here. For Monism also faces similar difficulties with all the successful and/or true intra-level causal or productive explanations, and laws, in the higher or lower sciences.)

My conclusion is therefore that Monism is not one of the candidates for a live position about a case of compositional explanation. And similar problems plausibly confront other positions that do not grow from a direct engagement with compositional explanations in the sciences. For all of these positions consequently face the difficult task of providing a completely new ontology to underpin their interpretation of the structure of such explanations no part of which has ever been posited in such an explanation in actual practice. Consequently, these "off the grid" views must address the still further task of establishing that this novel ontology plausibly underwrites the success and/or truth of such explanations even though this ontology has never been posited in practice.

8.5 The Varieties of Fundamentalism and Reductionism Redux: Exploring the Bottom Left Cell

Returning to positions that do take the ontology posited in scientific explanations as a starting point, I now want to look at the bottom left cell of Figure 8B, which I started the book by noting has been widely abandoned as dead by mainstream philosophers. The cell is built around the acceptance of the applicability of compositional concepts and the denial of determinative higher-level entities, but we now have two options for the resulting view depending upon whether the Simple or Conditioned views of aggregation are added to these theses. In 8.5.1, I look at the position standardly accepted by scientific and philosophical reductionists that accepts the Simple view of aggregation. In 8.5.2, I then outline a novel form of Fundamentalism that endorses the Conditioned view. Contrary to the received philosophical wisdom, I show that Fundamentalism is amongst the candidates for a live view. However, contrary to assumptions of most contemporary reductionists in the sciences and philosophy, I show that Fundamentalism actually has *two* options for understanding compositional explanations and *two* candidates for a live view.

8.5.1 Scientific Reductionism, Reduction in Philosophy, and Simple Fundamentalism: Our First Successful Candidate for a Live View

My discussion of scientific reductionism so far has focused almost exclusively on the position in the *outermost* part of the bottom left cell. Let me call this "Simple Fundamentalism," since it endorses evidence for the applicability of compositional concepts, accepts the Simple view of aggregation, and rejects the existence of determinative composed entities. It appears a scientific reductionist such as Weinberg endorses a global form of Simple Fundamentalism, for it plausibly underpins his conception of a Final Theory and conforms to his stated claim that the only determinative laws are those that hold of fundamental components in isolation or the simplest systems. And Simple Fundamentalism is plausibly the view endorsed by philosophers such as Kim, Loewer, or Rosenberg, given their commitment to the Completeness of Physics that entails the truth of the global version of the Simple view of aggregation.

To start, let me briefly reprise how my findings have supported philosophical and scientific reductionists who have all recently defended the viability of something like Simple Fundamentalism against the received wisdom in philosophy. Although philosophers take reductionism to be a dead position, I have shown that this only applies to semantic reductionism and that the Simple Fundamentalism pressed by scientific reductionists is untouched by extant philosophical objections to "reduction." Many reductionists have mistakenly thought that the Argument from Composition, or related reasoning, can drive such a view, but I have established that this reasoning is invalid. However, I have now illuminated valid reasoning, in the Argument from Composition and Completeness, that can establish compositional reductions using the applicability of compositional concepts and determinative completeness.

Let me therefore turn to assessing Simple Fundamentalism as an account of a compositional explanation. With regard to the first condition for being a candidate for a live position, it is easy to imagine a situation underpinning a scientific thought experiment in which we have a compositional explanation, the components are determinatively complete and hence the commitments of Simple Fundamentalism are all true. In such a case we can run the Argument from Composition and Completeness to the conclusion that either we only have components or that they are the only determinative entities. So Simple Fundamentalism passes the first condition.

In addition, in Chapters 3 and 4, I have already outlined how Simple Fundamentalism passes the Meta-Explanatory Adequacy Task. Unlike Monism, Simple Fundamentalism accepts the entities posited at the lower level in the compositional explanations deployed so successfully in actual practice in the sciences. Building on this basis through its Collectivist ontology, I detailed how Simple Fundamentalism can consequently account for the truth and success of compositional explanations. Thus MEAT, and hence the second condition, is plausibly satisfied by Simple Fundamentalism.

The Simple Fundamentalism pressed by scientific reductionists such as Weinberg, as well as philosophical reductionists such as Kim, Loewer, or Rosenberg, amongst others, is therefore a candidate for a live position about compositional explanations in the sciences. Contrary to the received wisdom in philosophy, as well as much of the higher sciences, at least one substantive species of reductionism therefore needs to be carefully considered in a key type of scientific example.

8.5.2 Conditioned Fundamentalism as an Overlooked Option for Reductionism: Our Second Successful Candidate for a Live View

My focus on determinative completeness in setting out the dialectic, rather than the Completeness of Physics, pays dividends by allowing us to appreciate that there is another option for the ontological reductionist that conforms to the *innermost* part of the bottom left cell. In Chapter 6, I termed this position "Conditioned Fundamentalism," since it endorses the applicability of compositional concepts, denies the existence of determinative composed entities, but also embraces the *Conditioned* view of aggregation.

Conditioned Fundamentalism goes along part of the way with the claims of the Mutualist by accepting the Conditioned view of aggregation and hence endorsing the existence of the differential powers.[14] However, the key claim of Conditioned Fundamentalism is that although we find discontinuous aggregation, which fits the Conditioned view of aggregation, the novel differential powers of components in the relevant complex collective are solely determined by *other components*. Conditioned Fundamentalism thus also accepts that there are further determinative laws holding of components in complex collective that either supersede or supplement the laws holding in simpler collectives, but the position claims that these are *not* emergent laws because such laws *only involve component entities*. Notice that the Completeness of Physics therefore fails to be true under Conditioned Fundamentalism, since the laws in the simplest collectives do *not* exhaust the determinative laws.[15] Nonetheless, in a particular case

---

[14] This appears to the position that Sydney Shoemaker (2003b) highlights and which he labels an "emergentist" view. Given its contrast to Simple Fundamentalism, we can now see why one might take such a stance – especially when one explicitly articulates the position's endorsement of the Conditioned view of aggregation. However, once we focus on determinative completeness, then we see how the position can run the Argument from Composition and Completeness to establish compositional reductions, so it plausibly is an ontological form of reductionism.

[15] Conditioned Fundamentalism consequently endorses this alternative thesis referencing supplemental/supercessional laws:

(Completeness of Physics*) The entities of physics are determined, in so far as they are determined, solely by other microphysical entities according to the laws covering their behavior in the simplest collectives and any relevant non-emergent supplemental/supercessional laws.

of compositional explanation components are still the only entities determining the components in the complex collective, so determinative completeness still holds of these components. Consequently, the Conditioned Fundamentalist can run the Argument from Composition and Completeness to secure compositional reductions.

Appreciating Conditioned Fundamentalism shows that although the Simple view of aggregation, and/or the Completeness of Physics, have been the most common ways recently used in the sciences and philosophy to justify claims of determinative completeness for components, neither thesis is necessary for determinative completeness to hold, or to be known to hold, in some concrete scientific case. Consequently, neither the Simple view of aggregation, and/or the Completeness of Physics, is necessary to drive the Argument from Composition and Completeness, or a compositional reduction, or scientific reductionism, or a Fundamentalist account of some case of compositional explanation.

Conditioned Fundamentalism is thus an overlooked option for ontological reductionists and when I turn to concrete scientific cases in coming chapters I show that this position presents a substantive, and important, hypothesis about such examples. However, this is to get a little ahead of myself, since I now need to consider how Conditioned Fundamentalism fares with the two conditions for being a candidate for a live view about such cases.

In Chapter 6, I already made a plausible case that we can imagine situations where we have a compositional explanation and we ought to accept the truth of Conditioned Fundamentalism. I went on to argue that it is a mistake to think that Conditioned Fundamentalism should be accepted about *any* example where the Conditioned view holds, regardless of the other details of the case, but that point does not undercut the existence of *some* imaginable examples where we ought to plausibly accept Conditioned Fundamentalism. All we need to do is imagine a case where the Conditioned view of aggregation holds, but we have determinative completeness of components in the relevant complex collective so that any differential powers of components are determined by other components. Consequently, Conditioned Fundamentalism passes my first condition for being a candidate for a live position.

Conditioned Fundamentalism also plausibly has the resources to pass the Meta-Explanatory Adequacy Test in cases of successful compositional explanation, since this position once again accepts the entities posited at the lower level in such explanations. So Conditioned Fundamentalism can thus also use a Collectivist ontology to provide an ultimate understanding of the predicates deployed at the higher level in compositional explanations and hence illuminate both the success and the truth of such explanations.

Conditioned Fundamentalism takes the natural world to be much less tidy and a lot more nomologically complex than Simple Fundamentalism. But I have now shown that Conditioned Fundamentalism is also a candidate for a live position about cases of compositional explanation. Scientific reductionists, as well as reductionists in philosophy, consequently have another substantive option to use in understanding scientific examples. And I should also mark that Conditioned Fundamentalism may well provide an attractive alternative for some scientific emergentists who have opposed the dominant Simple Fundamentalist framework pressed by scientific reductionists.

8.6 The Varieties of Non-Reductivism and Their Differing Fortunes: The Two Sides of the Top Right Cell

The last cell we need to examine, in the top left of Figure 8B, is where we find so-called "non-reductive" positions. I start, in 8.6.1, by assessing the most popular positions in the sciences and philosophy falling in the outer part of this cell, based around the Scientifically Manifest Image and Naked non-reductive physicalism. I show these popular views are not candidates for live positions. In contrast, in 8.6.2, I look at the overlooked option in the inner part of the cell that corresponds to the Mutualism defended by the scientific emergentist and her allies in philosophy. With regard to Mutualism, I use my earlier work to establish that it is a candidate for a live view – and is the only form of ontologically non-reductive physicalism that has this status.

8.6.1 Why the Most Popular Scientific and Philosophical Positions are Not Live: On the Inadequacy of Naked Non-Reductive Physicalism

The most widely held view in the sciences in positions endorsing the Scientifically Manifest Image, and also the Naked non-reductive physicalism so popular in philosophy, each derive from a natural interpretation of the combined implications of our intra- and inter-level scientific explanations. Following compositional explanations, these positions consequently accept, first, that there is a layered compositional hierarchy in nature, involving levels of composed and composing entities; and, second, following intra-level explanations, the views accept that the composed entities at higher levels in this hierarchy are nonetheless still determinative as well as their components.

We have begun to see at various points in earlier chapters that, although popular, endorsing the Scientifically Manifest Image, or Naked non-reductive physicalism, may well be unsustainable. Using the work of this chapter, I can now more crisply confirm the difficulties facing the proponents of these positions, given their analogous commitments, who are caught on the horns of what I term the "Completeness Dilemma."

The proponents of the Scientifically Manifest Image, or Naked non-reductive physicalism, each accept we have a true compositional explanation and the qualitative account of components that it supplies. And the defenders of these positions either take components to be determinatively complete, at a time t under conditions $, or they do not. On the first horn, the proponent of the Scientifically Manifest Image, or Naked non-reductive physicalism, takes components to be determinatively complete at t. And this is the horn that most philosophers who are Naked non-reductive physicalists find themselves facing, since acceptance of the Completeness of Physics is widespread amongst such philosophers and it entails that components are determinatively complete in all cases.[16] On this horn, the defenders of the popular views endorse both composition and the determinative completeness of components. But the Argument from Composition and Completeness consequently establishes that we ought to accept nothing but the components. The first horn therefore entails *subtracting* key elements of the Scientifically Manifest Image, or Naked non-reductive physicalism, and these positions thus collapse into Fundamentalism of one kind or the other.

On the other horn of the dilemma, if components are *not* taken to be determinatively complete at t under $, then one holds that

---

[16] Remember that I am not taking a commitment to the Completeness of Physics to suffice for a view to be an Enriched position, since this claim can now be seen to only make the situation worse for the existence of Strong emergence and the truth of ontologically non-reductive physicalism.

something other than a component determines the powers contributed by components at t under $. But the Scientifically Manifest Image, and Naked non-reductive physicalism, each take everything to be either a composed entity or a component, so this horn involves taking composed entities to determine components at t. But the Argument from Exhaustion shows that this commitment involves *adding* machretic determination, from composed to component entities, to the Scientifically Manifest Image or Naked non-reductive physicalism. Thus, on this second horn, the Scientifically Manifest Image, and Naked non-reductive physicalism, consequently collapse into Mutualism.

Using the Completeness Dilemma, we can consequently see that the Scientifically Manifest Image, and Naked non-reductive physicalism, cannot be sustained on their own terms and collapse into either Fundamentalism or Mutualism. These popular positions therefore fail my first condition for being a candidate for a live view, since we cannot imagine a situation where we have compositional explanations and these popular positions hold true, rather than some form of Fundamentalism or Mutualism. And similar points plausibly hold for any other view trying to hold the combination of theses in the outermost part of the top left cell.

8.6.2 Scientific Emergentism and Mutualism: Our Third, Ontologically Non-Reductive, Candidate for a Live View

Given the problems of the popular views that are the only widely recognized examples of "non-reductive" views, anyone sympathetic to ontologically non-reductive positions might be depressed. However, my alternative framework again shows its strengths by illuminating another option for non-reductivists in the *innermost* part of the cell, which encompasses what I am terming the "Mutualism" that has been pressed by scientific emergentists and various philosophers. Mainstream philosophers have overlooked this view, which combines acceptance of the Conditioned view of aggregation with composition and the existence of determinative composed entities. Let me briefly rehearse my findings about this philosophically neglected position.

Crucially, using my thought experiment, in Chapter 6, I highlighted how the Conditioned view of aggregation allows the existence of determinative, and Strongly emergent, composed entities that machretically determine the differential powers of their components. Mutualism thus denies that components are determinatively complete and hence also rejects the Completeness of Physics. The Argument from Composition, and related reasoning, has routinely been used to reject this type of position, but my thought experiment shows that such arguments are invalid. Furthermore, when Mutualism is true of an example of compositional explanation then we cannot run the Argument from Composition and Completeness, since we do not have determinative completeness of components.

My scientific thought experiment thus establishes that Mutualism is a coherent position that we ought sometimes to accept about natural phenomena when we have compositional explanations. So I have already established, through the latter work in Chapter 6, that Mutualism passes my first condition for being a candidate for a live view. Furthermore, in the last chapter, I detailed at some length how Mutualism is built around the existence of composition, albeit alongside machresis, and entails the in principle availability of compositional explanations. Crucially, Mutualism provides ontological reasons why we should accept determinative composed entities alongside their lower-level components. Mutualism thus accepts the ontology posited by working scientists in compositional explanations at *both* lower *and* higher levels. Consequently, Mutualism passes MEAT because its accepts an ontology that comfortably illuminates the truth and success of compositional explanations because this ontology overlaps with that posited so successfully by all working scientists in actual practice – although with an *added* commitment to machresis and Strong emergence on top of those commitments.

Mutualism thus passes my two conditions and offers us a third candidate for a live position. The Argument from Exhaustion, and the Completeness Dilemma, also confirm that Mutualism is plausibly the only version of an ontologically non-reductive physicalism with this distinction. So ontological "non-reductivists" need not be downcast, since Mutualism offers them a candidate for a live view about cases of compositional explanation in the sciences. And Mutualism also offers a charitable interpretation of the various philosophers and scientists, noted in earlier chapters, who have defended something like Strong emergence using various theoretical notions whether of "downward causation," "boundaries conditions," "self-organization" etc. However, the arguments of this chapter establish that if writers, whether in science or philosophy, intend to defend ontologically non-reductive accounts of compositional explanations, or the structure of nature, then they must endorse the Conditioned view of aggregation, the existence of machresis, and the truth of Mutualism.

8.7 The Candidates for Live Views, the Deeper Issues, and the Price of Physicalism

... what is becoming increasingly clear from the continuing debate over the mind–body problem is that currently popular middle-of-the-road positions, like property dualism, anomalous monism, and nonreductive physicalism, are not easily tolerated by robust physicalism ... Physicalism cannot be had on the cheap. (Kim (1998), p. 120)

My work in this chapter has highlighted why when confronted with complex scientific examples we need to recognize positions of a more complex and nuanced kind than theoreticians such as Kim have supposed when thinking purely in the abstract. Nonetheless, I have also shown that Kim is exactly right in his broader contention in the passage above that physicalism, i.e. a world where compositional concepts apply comprehensively, cannot be "had on the cheap" and that an ontological price must be paid to accommodate the evidence of successful compositional explanations.

However, I have also established that, contrary to Kim's own diagnosis, the contemporary problematic is driven by an incompatibility between *three*, rather than two, theses in the existence of determinative higher-level entities, the applicability of compositional concepts, and the determinative completeness of components. Widening the number of theses that are in conflict allowed me to highlight the central role of opposing accounts of the nature of aggregation in our new debates and to illuminate a wider range of positions than philosophers have previously engaged, including a number of novel, overlooked options.

In assessing which of these positions are even candidates for live views of a compositional explanation, I reached some surprising negative and positive conclusions. On the negative side, I have confirmed in detail Kim's contention that the most popular view in philosophy, in forms of Naked non-reductive physicalism, but also the most popular view in the sciences in the Scientifically Manifest Image, are not in fact live positions. On the positive side, I have shown that there are *three* candidates for live positions in Simple Fundamentalism, Conditioned Fundamentalism, and Mutualism. All of the three live views are physicalist, because they accept compositional concepts are comprehensively applicable, but each also comes, as Kim warns us, with an ontological price of commitments that entail either *subtracting* from, or *adding* to, the Scientifically Manifest Image.

As the abstract survey of this chapter begins to illuminate, the disputes between the three candidates for live views about cases of compositional explanation are complex in having *two* broad axes that cross-cut the positions. On one axis of the disputes, the central issue focuses on whether composed or just component entities are determinative and the distinct price one must pay with the two options in this dispute. With one option, the two Fundamentalist positions line up in requiring that we pay the price of *subtracting* composed entities from the Scientifically Manifest Image. In contrast, with the option on the other side of this axis of the debate, Mutualism requires that we pay the price of *adding* machretic determination to the Scientifically Manifest Image – hence preserving composed entities, compositional relations, and compositional levels.

Cross-cutting this first dispute, my survey highlights how we have a second axis to the new debates focused upon whether nature is continuous or discontinuous in some case of compositional explanation and specifically about the character of aggregation in the example. On one side of this axis, Simple Fundamentalism accepts the Simple view of aggregation, defends continuity in nature, rejects differential powers, and endorses the laws holding of components in simple collectives as exhausting the fundamental laws. On the other side, Mutualism and Conditioned Fundamentalism line up in endorsing the Conditioned view of aggregation, discontinuities in nature, the existence of differential powers, and the existence of further fundamental laws concerning complex collectives of components.

We consequently have two broad axes in our new debates between three candidates for live positions. Though the complexity of the issues and positions revealed by my survey may be exasperating to some, it is unsurprising when we remember that we continue to struggle to understand the complex character of the entities in all manner of phenomena from superconductors, to the biochemical networks in cells, to populations of neurons, and beyond. In the next chapter, I therefore turn to the important question of what my findings imply for the live accounts about such scientific examples. And, over the final pair of chapters, I use such cases of compositional explanation to show that, however potentially exasperating, our empirical evidence forces us to accept such complexity in our ongoing disputes about concrete scientific examples.

9 | Making the Issues Concrete: The Scientific Hypotheses and Their Empirical Differences

My response to reductionists and anti-reductionists alike is "If it is doable, then *do* it." We have had enough hand-waving on both sides ... Stop gesturing toward possible reductions. Actual accomplishments are always more convincing than in-principle arguments. If anti-reductionist holists, in their turn, want to be taken seriously, they best produce holist accomplishments ...

– David Hull[1]

David Hull's Nike-esque injunction to "Just do it" in concrete scientific cases sensibly reminds us that debates over reduction and emergence are supposed to be empirical in character. Before we take seriously the candidates for live views about examples of compositional explanation we can therefore reasonably ask whether these hypotheses can be confirmed, or disconfirmed, by empirical evidence. My goals in this chapter are therefore to articulate the detailed commitments of the viable positions about a case of compositional explanation, and then to examine whether the views have any empirical implications after all and, if they do, what evidence suffices to confirm or disconfirm these positions.

The brunt of the chapter is therefore focused on sketching what I term the "basic" hypothesis offered by each of our three candidates for live views about a case of compositional explanation and outlining their competing accounts of a complex collective and its components along with the kinds of laws and explanations accompanying it. Building on this work, I highlight how our present debates are very unlike those of the last generation, since they do not focus on disputes over the applicability of compositional concepts or explanations, nor on the implications of non-linear frameworks, multiple realization, etc. Instead, I summarize the more fine-grained issues over which the

---

[1] Hull (2002), p. 171, original emphasis.

live views now differ. More importantly, I outline how empirical evidence plausibly suffices to resolve these disputes and hence resolve whether we should endorse the Simple Fundamentalist, Mutualist, or Conditioned Fundamentalist hypothesis about a case of compositional explanation. By the end of the chapter, I therefore lay out what it empirically takes to move beyond the de facto stalemate that reigns in scientific discussions and have Mutualism, or the versions of Fundamentalism, "just do it" to provide the best account of a concrete case of compositional explanation in the sciences.

9.1 The Scientifically Manifest Image as a Working Hypothesis Alone

Applying the Scientifically Manifest Image leaves us with a hypothesis about the structure of a case of compositional explanation that posits both determinative higher- and lower-level entities with compositional relations holding between them. However, the last chapter framed, through the Completeness Dilemma, why we simply cannot accept such an account as finally reflecting the structure of a case of successful compositional explanation.

Even given this conclusion, I need to emphasize, as I noted in Part II, that scientists are comfortable, adept, and highly successful pursuing research based around what I earlier termed the common understanding of compositional concepts central to the Scientifically Manifest Image. The latter approach is plausibly the practice that has produced most of our successful explanations. Using a hypothesis based around the Scientifically Manifest Image in our research thus continues a spectacularly successful practice of discovery. Although not viable as a final account of the structure of cases of compositional explanation, an account based around the Scientifically Manifest Image consequently has appeal as a *working hypothesis* in everyday research, while it is yet to be resolved which position provides the best explanation of the structure of a case.

At what point should we cease to use the Scientifically Manifest Image as our working hypothesis in a case? Answering this question leads us back to the nature of the viable hypotheses about the case and the kind of evidence that makes discussion of such accounts productive. Let me therefore put the latter question to one side for the moment and examine in turn the detailed implications of our three candidates for live views.

9.2 Constructing a Basic Hypothesis for Simple Fundamentalism

To begin, let me look at the basic hypothesis offered by Simple Fundamentalism about a case of successful compositional explanation. Such a hypothesis claims that, ultimately, nature is continuous across simpler and more complex cases of aggregation. Crucially, the Simple Fundamentalist hypothesis claims that the powers contributed by components across the relevant simpler and complex collectives are continuous in character. If laws are involved, the Simple Fundamentalist hypothesis thus claims the laws holding of components in simpler collectives exhaust the determinative laws holding of components in the relevant complex collective. Consequently, the Simple Fundamentalist hypothesis takes the Simple view of aggregation to hold of the case and hence determinative completeness holds of the components in the complex collective.

In combination with the application of compositional concepts established by our successful compositional explanation, the Argument from Composition and Completeness can therefore be applied in the case under this account, so the Simple Fundamentalist hypothesis claims we can establish a compositional reduction and conclude that either only components exist or at least that components are the only determinative entities. Figure 9A illustrates the resulting Simple Fundamentalist hypothesis about the ontological structure of the case.

Although the Simple Fundamentalist hypothesis takes nature to be continuous in its ontological structure across the simpler and complex collectives relevant to the case, the hypothesis allows that theories and explanations across the collectives may be radically discontinuous in various ways. As a compromising view, the Simple Fundamentalist hypothesis happily endorses the indispensability of higher sciences for expressing true explanations and laws about the collectives, and other collective phenomena, we find at various levels in nature. Simple Fundamentalism allows that we often cannot derive/compute/predict the laws or theories holding of the complex collective, or its components, from the laws holding of such components in

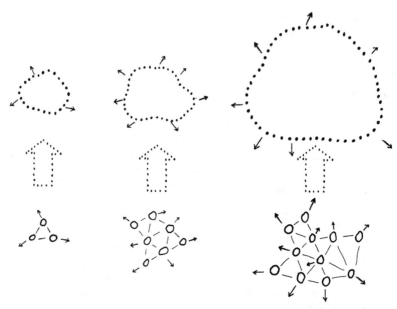

Figure 9A An illustration of what Simple fundamentalism says about the structure of a case involving successful compositional explanation of a complex collective, on the left, and its relations to simpler collectives, on the right. (The large dotted circles at the top are collectives; the larger dotted arrows are collective relations; smaller circles at the base are components; straight solid lines are bonds; small solid arrows are forces where the size of arrow equals the magnitude of the force.)

simpler collectives. And it accepts that we often have multiple composition through collective propensities of complex collectives to which higher science predicates refer that may cross-cut the kinds of components giving rise to such propensities in other cases. Lastly, we saw in Part II that Simple Fundamentalism can allow the use of non-linear frameworks or the existence of feedback loops. Simply finding indispensable higher-level predicates or explanations, or theories utilizing non-linear dynamics, or discovering feedback loops in the example, does not therefore preclude the Simple Fundamentalist hypothesis from holding of a case.

In contrast to the complexity of our higher- and lower-level explanations, theories etc., the Simple Fundamentalist hypothesis still takes the case to be simple in its ontological structure, claiming that we ultimately only have components, and collectives of them, that the only determinative laws are the laws holding of components in simpler collectives, and that the only determinative relations hold between components. And I now also want to briefly outline the kinds of explanatory structures that go along with the Simple Fundamentalist hypothesis. This takes the form of what I roughly term the "series" of compositional and other explanations going along with the basic hypothesis of each position. There are as many variations in such series as there are variations in adequate explanations – some may use laws, others may simply offer mechanisms; some may be highly detailed, others may be schematic; some may offer richer accounts of the simpler collectives, some of the complex collectives; and so on. Given these myriad variations, I cannot provide a comprehensive story about such series, but I can now provide a broad brushstrokes account.

In the case of Simple Fundamentalism, its hypothesis implies we can ultimately secure what I term a "Simple Series" to explain the complex collective and its components. Such a series has at least two stages. The initial stage is focused on the explanations of the behaviors and powers of the relevant component entities in simpler collectives. These may or may not involve laws, but let me assume that they do since it makes my presentation easier.

The other stage of a Simple Series has two parts concerning the explanation of the composed and component entities found in the complex collective. In one part, Simple Fundamentalism claims we can provide fully adequate explanations of the *components* we find in the complex collectives using the laws and/or other machinery offered for the components in the first stage – that is, using the laws illuminated for components in the *simpler* collectives, along with information about conditions in the complex collectives etc. In the second part, Simple Fundamentalism claims that we have fully adequate explanations of the *composed* entities using compositional explanations based around compositional relations to the components found in these complex collectives. Notice that in the first part we explain the components in the complex collectives using the *same* laws, and/or principles of composition, that we used in simpler collectives. Concerning the complex collective, under Simple Fundamentalism we are thus only seeking new explanatory resources about the *composed* entities in the shape of compositional explanations.

Summarizing, the basic hypothesis of Simple Fundamentalism about a case of successful compositional explanation has these features:

(1) We have successful and true compositional explanations of the complex collective and concepts of composition apply comprehensively.

(2) We may have higher-level predicates, laws, explanations, and theories that are indispensable for expressing truths about the case; and/or higher-level predicates, laws, explanations, and theories that are uncomputable/underivable/unpredictable from laws, explanations, and theories concerning components; and/or the case involves non-linear dynamics; and/or the case requires explanation by simulation; and/or we have relations of multiple composition in the examples; and/or the case involves feedback loops; amongst other features.

(3) The same laws and/or principles of composition hold about components in this complex collective as hold with components in simpler collectives, i.e. the Simple view of aggregation holds of this case, and components contribute no differential powers.

(4) The laws holding of components in simpler collectives exhaust the determinative laws holding of the complex collective.

(5) Ultimately, components are the only determinative entities, the components are determinatively complete, determination amongst components is the only species of determination, and machresis and Strongly emergent entities do not exist in the case.

(6) After sufficient investigation, at least in principle, we can supply one or other version of a Simple Series of explanations for the simpler and complex collectives relevant to this case.

9.3 Constructing a Basic Hypothesis for Mutualism

Mutualism presents a basic hypothesis about a case of successful compositional explanation that posits *both* ontological *and* epistemic/semantic discontinuities alongside each other. Most importantly, in a case of successful compositional explanation, the basic Mutualist hypothesis claims that we have a discontinuity between some of the powers contributed by components in the complex collective and simpler collectives. Thus the Conditioned view of aggregation is taken to hold in this case. Furthermore, the Mutualist hypothesis takes components in the complex collective to contribute differential powers as a result of the machretic determination of the Strongly emergent entities they compose. In explaining complex collectives, and their components, Mutualism thus posits an underlying ontological structure whose complexities, and untidiness, to an extent mirrors many of the complexities we find in our theories, explanations, and findings.

Once again, for ease of exposition, I assume we can identify the laws in the case. The Mutualist hypothesis may take the laws holding of components in simpler collectives to hold in the complex collective if it endorses supplemental emergent laws, although it can also endorse supercessional emergent laws. But Mutualism denies that the principles of composition and laws holding of components in simpler collectives exhaust the determinative laws holding of components in the complex collective. The basic Mutualist hypothesis thus takes emergent laws, referring to composed entities, to hold of components

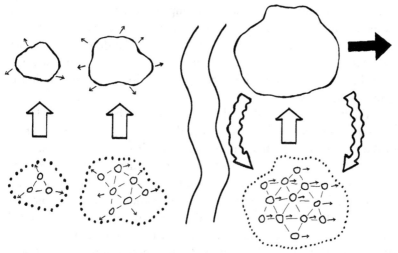

Figure 9B An illustration of what Mutualism says about the structure of a case involving successful compositional explanation of a complex collective. (Again, the large circles at the top are composed entities, small circles at bottom are components; the enclosed upward arrows are relations of composition; the enclosed downward wavy arrows are machretic relations; straight solid lines are bonds; solid arrows are forces where the size of arrow equals the magnitude of the force.)

in the complex collective and to be amongst the determinative laws holding in the case.

What the basic Mutualist hypothesis says about the structure of the case, and the relation of the complex collective to simpler collectives, is represented in Figure 9B. In addition to production at higher and lower levels, the varieties of determination include two species of "vertical" determination in upward compositional relations and machretic relations downward. Determinative completeness thus fails for the components in the complex collective. Consequently, the Argument from Composition and Completeness is unavailable to drive a compositional reduction and there are levels of mutually determinative, interdependent composed and components entities that are productive at their own levels. As well as compositional concepts being applicable, we thus also have actual relations of composition, and hence composed entities, as they are commonly understood.

Mutualism entails that we can ultimately secure what I term a "Conditioned Series" of explanations for the complex collective and its components. This series again has at least two stages. The first stage once more concerns components in simpler collectives where we may assume that we have explanations of the relevant component entities based upon certain laws. The other stage of a Conditioned series has two parts, one focused on the explanation of the components, and the other on explanation of the composed entities, that we find in the complex collective.

In the first part of the second stage, given that the Conditioned view of aggregation holds, we *cannot* provide a fully adequate explanation of the *components* we find in the complex collectives simply using the laws and/or other machinery offered for simpler collectives. For the components have differential powers that would not be contributed if the laws holding in simpler collectives were the only laws in complex collectives. In the second part, the composed entities are explained using compositional explanations, but this time based around compositional relations to components with differential powers *only found in the complex collective.*

The Mutualist hypotheses thus implies that, if laws are used, we must explain the components in the complex collectives using emergent laws, whether supplemental or supercessional, positing a role for the Strongly emergent higher-level entities that machretically determine some of the powers of the components. Putting the point without laws, we explain the components as contributing powers that they would not contribute except for the presence, and machretic determination, of a composed entity. The Mutualist hypothesis thus deploys new explanatory resources to understand the *components*, as well as the composed entities, we find with the complex collective.

The basic hypothesis of Mutualism about a successful example of compositional explanation therefore has these features:

(1) We have successful compositional explanations of the complex collective and concepts of composition apply comprehensively.

(2) We may have higher-level predicates, laws, explanations, and theories that are indispensable for expressing truths about the case; and/or higher-level predicates, laws, explanations, and theories that are uncomputable/underivable/unpredictable from laws, explanations, and theories concerning components; and/or the case involves non-linear dynamics; and/or the case requires explanation by simulation; and/or we have relations of multiple composition in the examples; and/or the cases involve feedback loops, amongst other features.

(3*) Further distinct determinative laws (whether supplemental or supercessional) and/or principles of composition hold about components in this complex collective than hold with components in simpler collectives, i.e. the Conditioned view of aggregation holds of this example, and/ the components in the complex collective contribute differential powers.

(4*) The determinative laws in this example include emergent laws, either supplemental, supercessional, or both, governing the components in the complex collective.

(5*) Both component and composed entities are determinative in this example, the species of determination include production at both higher and lower levels, as well as compositional and machretic relations between the entities at higher and lower levels, the components are not determinatively complete in the complex collective, and machresis and Strongly emergent entities exist in this example.

(6*) After sufficient investigation, at least in principle, we can supply one or other version of a Conditioned Series of explanations for the simpler and complex collectives relevant to this case.

9.4 Constructing a Basic Hypothesis for Conditioned Fundamentalism

As with Mutualism, the basic Conditioned Fundamentalist hypothesis about a case of compositional explanation also embraces the Conditioned view of aggregation and hence accepts ontic discontinuities across simpler and more complex collectives. Like Mutualism, the Conditioned Fundamentalist account also endorses *both* epistemic/semantic *and* ontological discontinuities. And Conditioned Fundamentalism, again like Mutualism, also accepts that there are differential powers (and/or new laws) of components in the complex collectives. However, Conditioned Fundamentalism posits a very different underlying ontological situation to explain the structure of the relevant complex collectives, and crucially the differential powers of their components, from that posited in the Mutualist hypothesis.

The Conditioned Fundamentalist account takes the differential powers of components, and hence the discontinuities, to result *solely* from the determinative influence of component entities, i.e. entities at the same (or lower) levels as the components of the complex collective. Consequently, if it posits laws, the Conditioned Fundamentalist account still only embraces further *non-emergent* supplemental or supercessional laws and/or principles of composition – that is, laws and principles making no reference to composed entities – to account for the differential powers of the components in the complex collective.

Although under the Conditioned Fundamentalist hypothesis we do have differential powers we still have determinative completeness of components. As a result, Conditioned Fundamentalism uses the Argument from Composition and Completeness to drive compositional reductions. The basic hypothesis of Conditioned Fundamentalism, like that of Simple Fundamentalism, only posits components and collectives of them and the only determination is found at the level of components. However, the Conditioned Fundamentalist hypothesis also claims components in collectives do have differential powers and non-emergent supplemental or supercessional laws are amongst the relevant determinative laws. Figure 9C sketches the structure of the example under the Conditioned Fundamentalist hypothesis.

If the basic hypothesis of Conditioned Fundamentalism is correct, then we can ultimately secure what I term a "Semi-Conditioned Series" of explanations. Once more, the first stage concerns the components

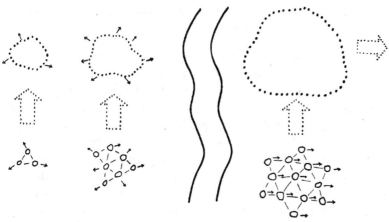

Figure 9C An illustration of what Conditioned fundamentalism says about the structure of a case involving successful compositional explanation of a complex collective. (The large dotted circles at the top are collectives; the upward dotted straight arrows are collective relations; the horizontal dotted arrow at the top is the collective force of the components together; smaller circles at the base are components; straight solid lines are bonds; solid arrows are forces where the size of arrow equals the magnitude of the force.)

we find in simpler collectives where, again assuming we find laws, the laws illuminated in the simpler collectives are taken to explain these components and their behaviors. The second stage again has two parts focused on the explanations of composed and component entities in the complex collective.

Like the Conditioned Series that is available if Mutualism is true, under a Semi-Conditioned Series the relevant composed entities are explained using compositional explanations based around compositional relations to components with differential powers *only found in the complex collective*. However, the latter explanations are only taken to ultimately concern components and collectives of them. And, in the second part, although a Semi-Conditioned Series again uses new laws to explain the components, the crucial point is that *non*-emergent supplemental or supercessional laws are deployed which refer solely to components. In the second stage, the Conditioned Fundamentalist hypothesis thus deploys new explanatory resources to understand *both* the components *and* composed entities, but the entities used in these further explanations are ultimately taken to draw solely on determinative entities at the component level.

Summarizing these points, the basic Conditioned Fundamentalist hypothesis has these features:

(1)   We have successful compositional explanations of the complex collective and concepts of composition apply comprehensively.

(2)   We may have higher-level predicates, laws, explanations, and theories that are indispensable for expressing truths about the case; and/or higher-level predicates, laws, explanations, and theories that are uncomputable/underivable/unpredictable from laws, explanations, and theories concerning components; and/or the case involves non-linear dynamics; and/or the case requires explanation by simulation; and/or we have relations of multiple composition in the examples; and/or the cases involve feedback loops, amongst other features.

(3*) Further distinct determinative laws (whether supplemental or supercessional) and/or principles of composition also hold about components in this complex collective than hold with components in simpler collectives, i.e. the Conditioned view of aggregation holds of this example, and/or the components in the complex collective contribute differential powers.

(4**) The determinative laws in this example include *non*-emergent supplemental or supercessional laws governing the components in addition to the laws holding of these components in simpler collectives.

(5)   Ultimately, components are the only determinative entities, determination amongst components is the only species of determination, the components are determinatively complete, and machresis and Strongly emergent entities do not exist in the case.

(6**) After sufficient investigation, at least in principle, we can supply one or other version of a Semi-Conditioned Series of explanations for the simpler and complex collectives relevant to this case.

9.5 Not Your Grandma's Reductionism and Emergentism: What the Live Hypotheses Do, and *Do Not*, Differ Over

My survey of the commitments of the basic hypotheses of Simple Fundamentalism, Mutualism, and Conditioned Fundamentalism, building on the work of earlier chapters, frames the points of agreement and disagreement between the positions. Although my treatment of the hypotheses, just like my abstract survey in the last chapter, might be thought to be too detailed or complex, it shows its worth by highlighting where the competing accounts do, and do not, diverge. This work consequently confirms that the viable positions, and the issues between them, are not the "emergentism" or "reductionism" of the last century, as all too many writers in the sciences and philosophy often presume about their opponents.

Those leaning toward reductionism all too often take the debate to be between positions differing over the applicability of compositional concepts or compositional explanation. However, as Part III clarified at length, Mutualist scientific emergentism accepts both the applicability of compositional concepts and also implies that, at least in principle, we can supply compositional explanations for all higher-level entities. The latter points are enshrined in the commitment of the basic Mutualist hypothesis to thesis (1) and so the common construal of the debates as focused on the applicability of compositional concepts is widely off the mark.

On the other side, those sympathetic to emergentism or anti-reductionism have commonly taken the opposing views to be battling over the indispensability of higher sciences, or the existence of multiple composition, or the necessity of using non-linear dynamics, or the existence of Qualitative emergence, etc. Once again, however, as my findings in Part II detailed, the Fundamentalist framework of scientific and other reductionists is compatible with each of the latter phenomena and this point is framed in thesis (2) of the basic hypotheses of both Simple and Conditioned Fundamentalism.

As my treatment of the basic hypotheses highlights, all sides accept everyday reductionism, the search for compositional explanations, and illumination of Qualitative emergence that it entails. And the various parties also all endorse the indispensability of higher sciences and non-linear dynamics or related approaches. All three basic hypotheses therefore accept theses (1) and (2), highlighting how under the impetus of our scientific research, both in supplying successful compositional explanations and in providing quantitative accounts of components, our new debates have simply moved on from the battles of earlier generations.

Despite this agreement, Simple Fundamentalism, Mutualism, and Conditioned Fundamentalism do differ. Broadly put, the hypothesis of Simple Fundamentalism is that, underneath all the semantic and epistemic complexity, aggregation really is continuous, the Simple view of aggregation applies to the case, and we only have component entities and the determination between them. In contrast, the hypotheses of Mutualism and Conditioned Fundamentalism both assert that aggregation is discontinuous, that components have differential powers in the relevant complex collective, and hence that the Conditioned view of aggregation applies in the case. However, Mutualism and Conditioned Fundamentalism differ over the entities that determine the differential powers of components, the nature of the determinative entities, and the species of determination.

Narrowing in on how these broad differences manifest themselves in particular issues, my work on the basic hypothesis of each view allows us to frame the theses in dispute. The hypotheses differ over theses (3) and (3*) and hence the character of aggregation, determinative laws, and the powers we find in the components in simpler and complex collectives; the hypotheses diverge over theses (4), (4*), and (4**) and their claims about the character of the determinative laws holding of components in collectives; the hypotheses also differ over (5) and (5*) about the nature of the determinative entities and the species of determination; and finally the hypotheses endorse the distinct series of explanations outlined in theses (6), (6*), and (6**).

9.6 The Empirically Resolvable Differences of Our Competing Scientific Hypotheses

The new focus of debates upon collectives, their components, and the nature of aggregation should be clear from the differences between the competing hypotheses outlined in the last section. But what we now need to consider is whether, and how, such disputes can be resolved using scientific evidence. This is especially important given the de facto stalemate in scientific discussions. I therefore look at one way to confirm/disconfirm the various hypotheses using empirical evidence, looking at the Simple Fundamentalist hypothesis in 9.6.1 and the Mutualist or Conditioned Fundamentalist hypotheses in 9.6.2. Given my conclusions, I argue in 9.6.3 that the three opposing hypotheses are all substantive scientific hypotheses and I also frame the threshold of evidence we need in order to productively engage the new debates in a particular example.

9.6.1 Confirming/Disconfirming the Simple Fundamentalist Hypothesis

In any case of compositional explanation a key difference between the Simple Fundamentalist hypothesis, and its Mutualist or Conditioned Fundamentalist rivals, concerns the existence of differential powers. I can thus frame the relevant issues using a pair of questions and our answers to them. Consider this question:

(Q1) Do the components in the complex collective contribute differential powers?

If in some case of compositional explanation we answer this question affirmatively, and components are found to have differential powers, then we must *reject* Simple Fundamentalism in this example because the Simple view of aggregation is mistaken. On the other hand, if we answer negatively to Q1, and there are no differential powers of components in the example, then we must reject the Mutualist and Conditioned Fundamentalist accounts about this example of compositional explanation. For those who prefer to frame the issues in terms of laws we can put the same question in this way:

(Q1*) Are there other determinative laws holding of the components in the complex collective aside from the determinative laws holding of such components in simpler collectives?

If we answer affirmatively, and the laws holding of the components in simpler collectives do not exhaust the determinative laws holding of components in the complex collective, then the Simple view is again plausibly false and we must abandon the Simple Fundamentalist hypothesis. On the other hand, if we answer Q1* negatively, and the laws in simpler collectives exhaust those in the complex collective, then we must plausibly reject the Mutualist and Conditioned Fundamentalist hypotheses about the case.

Notice that empirical evidence can plausibly suffice to answer either question Q1 or Q1*. The issues simply concern the powers of components in simpler and more complex collectives. And scientific inquiry routinely allows us to discover what powers are had by component entities and/or what laws hold of such entities in different situations, thus allowing us to answer Q1 and/or Q1*. However, one point that does bear emphasis is that it appears we can only answer questions Q1 and Q1* when we have a sufficiently detailed, quantitative account of the nature and effects of components in both simpler collectives and the relevant complex collective, since only then can we determine if some power is a differential power or resolve whether further laws hold of the complex collective.

It thus appears that we can empirically confirm or disconfirm Simple Fundamentalist hypotheses about cases of compositional explanation. And, given these points, I can also diagnose another reason why the scientific reductionists and philosophers have wrongly thought that a compositional explanation *alone* resolves the issues, thus falling into something like what Sandra Mitchell aptly terms the "snap-shot fallacy" of considering only what goes in a complex collective for which we have a successful compositional explanation (Mitchell (2012)).

Proponents of Simple Fundamentalism implicitly assume the Simple view of aggregation or its proxy in the Completeness of Physics. Consequently, it is simply assumed that the laws, and principles of composition, holding of components in simpler collectives suffice to explain the components in the complex collective. Beyond discovering the laws in simpler collectives, proponents of Simple Fundamentalism thus assume we never need to worry about accounting for the powers of lower-level *components* in complex collectives – hence concluding that the only open issue is whether we can explain *higher-level* entities by supplying compositional explanations.

But one of the disputes between Simple Fundamentalism on one side, and Mutualism and Conditioned Fundamentalism on the other, concerns whether the Simple or Conditioned views of aggregation hold in some example – and hence whether components can be explained solely using the laws holding in simpler collectives. And we have now found that one can only resolve the latter issue by considering the powers and behaviors of components in *simpler collectives as well as the complex collective*. Seeking to resolve the ongoing debates by relying upon the existence of the compositional explanation about the complex collective is therefore aptly termed a "snap-shot fallacy," since it only looks at one element of the evidence required to resolve the issue.

9.6.2 Confirming/Disconfirming the Mutualist or the Conditioned Fundamentalist Hypothesis

Moving on to whether we can empirically resolve the dispute between the Mutualist and Conditioned Fundamentalist hypotheses about some case, for argument's sake let me assume we have established that the Conditioned view of aggregation holds in the relevant example and that we know there are differential powers of components in this case. In such a situation, how one answers the following pair of questions can then resolve the choice between the Mutualist and Conditioned Fundamentalist hypotheses. Consider this question:

(Q2) Are the contributions of differential powers by the components in the complex collective machretically determined by composed entities?

If we answer affirmatively, then we must plausibly accept the Mutualist over the Conditioned Fundamentalist hypothesis in the example. If we answer Q2 negatively, then we must endorse the Conditioned Fundamentalist over the Mutualist account. Again, we can frame the same question in terms of laws as follows:

(Q2*) Are the supplemental/supercessional laws holding of the components in the complex collective emergent laws referring to at least one strongly emergent composed entity and its machretic determination?

Here one must accept the Mutualist account if we answer affirmatively and find only emergent supplemental/supercessional laws; and one must endorse the Conditional Fundamentalist hypothesis if we answer Q2* negatively because we only discover non-emergent supplemental/supercessional laws.

With regard to these questions, it again appears that we can answer them if our empirical evidence is sufficiently detailed. As I outlined in Chapter 6, in addressing what I termed the Objection from Conditioned Fundamentalism, the key point is that the Mutualist and Conditioned Fundamentalist hypotheses each make opposing claims about the determination relations between various entities and hence entail diverging predictions about the existence of differential powers when certain entities are present or absent. The Conditioned Fundamentalist hypothesis claims the contribution of differential powers by a specific component is determined by certain other components, call them $X1$–Xn. Crudely put, the Conditioned Fundamentalist hypothesis predicts that the differential powers of the component will be absent if $X1$–Xn is absent, and present when those components are present, all else being equal. In contrast, the Mutualist hypothesis contends that the contribution of differential powers by a component is determined machretically by a certain Strongly emergent composed entity, call it

H. Thus, again putting things roughly and ignoring a range of complications, the Mutualist account predicts that the differential powers will be absent if H is absent, and that the differential powers will be contributed if H is present, all else being equal.

If we can secure quantitative accounts of the nature of components across simpler and complex collectives, then it appears plausible that we can potentially empirically resolve whether we have differential powers when X1–Xn, and H, are present or absent. Such quantitative accounts describe the powers of components in various situations, so it is plausible that the same methods can allow us to explore whether we have differential or other powers across such collectives with permutations of the relevant components and composed entities being absent or present. Using this richer array of evidence, which is often supplied by scientific inquiry, we can consequently resolve whether the predictions of the Mutualist or Conditioned Fundamentalist hypotheses are true or false.[2]

I should note that the latter conclusion flies in the face of many people's initial judgment that the choice between Mutualist and Conditioned Fundamentalist hypotheses is empirically unresolvable.[3] However, the motivating idea behind such intuitions appears once again to be a version of the "snap-shot fallacy" and the associated assumption that if we can resolve disputes between the opposing hypotheses at all, then these disputes will be resolvable solely using the evidence of a compositional explanation alone. But, claims the objector, the Mutualist and Conditioned Fundamentalist hypotheses each account for the features of the compositional explanation equally well. So, the objector concludes, we cannot empirically resolve whether we should endorse Mutualism or Conditioned Fundamentalism when we have differential powers.

But this intuitive judgment is badly mistaken. Crucially, we have now seen the flaws with the assumption underlying the "snapshot fallacy," namely that only evidence from the compositional explanation, and hence about the complex collective with all its components, is

---

[2] For example, in Chapter 6, I highlighted a situation where it appears plausible that a Conditioned Fundamentalist hypothesis is undermined and a Mutualist one is supported. But we can just as easily sketch cases where the nature of the evidence is such that we should reject the Mutualist hypothesis in favor of the Conditioned Fundamentalist hypothesis.

[3] This concern was raised by a referee and has often been raised by graduate students in classes where I have described the choice between Mutualism and Conditioned Fundamentalism.

relevant to assessing the Mutualist and Conditioned Fundamentalist hypotheses. In fact, evidence about components, and their powers, is required from across both simpler and complex collectives, for only then can we resolve whether some power is a differential power or not. The intuition driving many people's initial judgment is therefore mistaken and we can potentially resolve whether to accept the Mutualist or Conditioned Fundamentalist hypothesis if we have sufficiently detailed empirical evidence of the right kind.[4]

9.6.3 Substantive Scientific Hypotheses and the Empirical Threshold for Moving Beyond the Scientifically Manifest Image

Far from the scientific debates over reduction and emergence being merely rhetorical posturing to secure funding, my work provides good reasons to take the hypotheses of Simple Fundamentalism, Mutualism, and Conditioned Fundamentalism to be *substantive scientific hypotheses* about cases of compositional explanation. First, being empirically confirmable or disconfirmable is one of the distinctive features of a scientific hypothesis and my work in the previous sections has shown the hypotheses have this feature. Second, these accounts press the general kinds of claims about powers, properties, laws, and determinative relations that are routinely broached by working scientists in ongoing inquiry in the sciences. And, third, in the final chapter I provide actual examples where just such opposing hypotheses have been defended by working scientists in ongoing debates about actual cases.

Putting these points together, we therefore have good reasons to treat the local hypotheses of the candidates for live positions as *substantive scientific hypotheses* – albeit competing ones. And my work has also clarified when, and why, it is productive to put the Scientifically Manifest Image aside as a working hypothesis and explore which of the three rivals is the best account of some case of compositional explanation in what I term "Threshold":

(Threshold) It is fruitful to put aside the Scientifically Manifest Image as a working hypothesis about an example and engage which of the viable

---

[4] Thanks to Louis Gularte, amongst others, for pressing me to address the issues of this section.

hypotheses we should endorse *only if* (a) we have a successful compositional explanation in the example and a plausible qualitative account of all the relevant components; (b) we have plausible quantitative accounts of the relevant components in the complex collective in the case; and (c) we have plausible quantitative accounts of the relevant components in simpler collectives.

Threshold reflects the fact that we actually need at least three kinds of evidence before we can put the Scientifically Manifest Image to one side in a case. So the question of whether to endorse Simple Fundamentalist, Mutualist, or Conditioned Fundamentalist hypotheses about some example is actually *preceded* by the important prior question about whether in a specific case we possess evidence satisfying Threshold and can productively engage such questions.

The latter points are also significant because they show that *both* Mutualism *and* Fundamentalism each plausibly depend upon the new wave of quantitative evidence about components in collectives to prevail in their claims about specific cases. Only if we have detailed quantitative accounts about the behavior and powers of the components can we confirm or disconfirm whether the Simple or Conditioned view of aggregation holds in the case, whether there are differential powers and, if there are, whether composed or component entities determine the contribution of such powers.

9.7 Appreciating the Differences, and Relations, between Local and Global Hypotheses

The last issue I want to briefly explore concerns the relations of "local" and "global" versions of the candidates' for live positions. Throughout this book, I have focused "locally" upon particular cases of compositional explanation, since such examples plausibly supply our core evidence in debates over reduction and emergence. And, so far in this chapter, I have therefore considered what we can dub "local" forms of Simple Fundamentalism, Mutualism, and Conditioned Fundamentalism – that is, hypotheses based on these views about specific cases of compositional explanation. However, earlier discussions, especially in philosophy, have focused on what I term "global" forms of these positions, which make claims about the *whole* of nature, rather than the structure of just one example. Given the widespread discussion of global views, I want to round out my discussion by briefly examining what it takes to confirm/disconfirm the *global* variants of each of the candidates for live views.

It appears plausible that conclusions about their local hypotheses in specific scientific examples are the primary vehicle through which the competing global positions, which are all versions of physicalism, are confirmed or disconfirmed. So let me work through the three positions in turn, highlighting what we need to show about local hypotheses in order to establish the global view. (In the final chapter, I examine an alternative route to defending a global form of Simple Fundamentalism, but I show this putative alternative is flawed and trades upon the false dichotomies outlined in earlier chapters.)

To begin, the global form of Simple Fundamentalism is committed to every single scientific example involving a compositional explanation being such that a local hypothesis of Simple Fundamentalism provides the best account of the case. Consequently, it is not just the case of the chemical and the quantum mechanical, but also the biochemical and chemical, the cellular and the biochemical, the cellular and the organ levels, and so on, that must be such that the local hypotheses of the Simple Fundamentalist are the best accounts of these cases. In order to justify the global form of Simple Fundamentalism it thus appears that we need, first, a range of examples of compositional explanation, across various levels, where we have established that the Simple Fundamentalist hypothesis about the case is the best account. And, second, we must *lack* any examples of where a local Anti-Physicalist account, or either a Mutualist or Conditioned Fundamentalist hypothesis, has been established as the best account of the case, since the latter hypotheses all undercut global Simple Fundamentalism.

As an aside, it also appears that a similar range of local cases is needed to establish a *global* version of the Simple view of aggregation, and hence the global claims of the Completeness of Physics for which it is a precondition. Both the global Simple view of aggregation, and the Completeness of Physics, each imply that every single example of compositional explanation, across all levels, has the same ontological structure – thus implying there are never differential powers of components and that the laws of components in simpler collectives always exhaust the determinative laws. Given these points, we therefore plausibly need a range of cases of compositional explanation where the Simple view of aggregation has been shown to be the best account and we must lack any examples where the Conditioned view of aggregation has been established as the best account of the case. And notice that establishing these global theses consequently means that in each of the range of cases we have the detailed evidence about components allowing us to answer Q1 or Q1* in these examples.

If we turn to the global form of Conditioned Fundamentalism, then the requirements are slightly different from those for global Simple Fundamentalism. To justify the global form of Conditioned Fundamentalism we need to establish a local variant of Conditioned Fundamentalism as the best account of at least *one* example, to show that local variants of Simple *or* Conditioned Fundamentalism apply in a range of cases, and we must *lack* any examples where a local Anti-Physicalist or Mutualist hypothesis is the best account of the case. In effect, the global Conditioned Fundamentalist must do everything the Simple Fundamentalist does, but can use either form of Fundamentalism in most cases with at least one example where Conditioned Fundamentalism is the best account.

When we turn to the global form of Mutualism, then we find different requirements as befits the challenger hypothesis in global debates. Basically, the global Mutualist is claiming that nature does not everywhere go as the Fundamentalist says. Thus all that the global Mutualist needs to do is to establish that we have a range of cases of successful compositional explanation across various levels and that just *one* of these successful compositional explanations is such that the local variant of Mutualism provides the best account of its nature, and we lack any case where Anti-Physicalism provides the best account. Strictly speaking, every other case can go the way that Simple and/or Conditioned Fundamentalism says, but if one case goes their way (and we lack successful Anti-Physicalism hypotheses as appears plausible), then Mutualism provides the correct global picture.

Though the global forms of the candidates for live views are established by rather different combinations of successful local hypotheses, I want to emphasize that global forms of Simple Fundamentalism, Mutualism, and Conditioned Fundamentalism are alike in depending upon successfully establishing *local* variants of these views as the best accounts of cases of compositional explanation. The global positions thus all depend upon establishing their local variants in order to be justified.

Given the latter point, I can finally return to what I termed Inclusivism in the Introduction – that is, the view that we need evidence from a range of sciences to resolve the scientific debates over reduction and emergence. I earlier assumed Inclusivism as a working hypothesis, but my findings now support the acceptance of Inclusivism as the correct epistemic position. For I have shown that to resolve debates about reduction and emergence globally, as well as to resolve which global theses are correct, we plausibly need evidence from a *range* of sciences and not just physics. So Inclusivism is plausibly true, rather than the Exclusivist position that assumes physics alone can resolve the debates.

9.8 Empirically Resolving Our Epochal Disputes over Reduction and Emergence?

My findings in this chapter confirm that the scientific reductionism and emergentism of today are strikingly different from the positions that went under those names in the last century. Unfortunately, it is still too often the case that arguments are offered by proponents of reduction and emergence against their opponents based upon reading their views as being those of the *last* century's "reductionists" or "emergentists." But I have highlighted how today's debates come down to a rather different, and more fine-grained, set of ontological disputes about cases of compositional explanation.

Two of the central issues are whether a case of compositional explanation involves components contributing differential powers and, if it does, whether contribution of such powers is machretically determined by composed entities or solely by other components. Given their underlying nature, I have shown that these key issues are empirically resolvable if we work case by case to acquire the relevant evidence – so the de facto stalemate presently reigning in scientific disputes can be broken once more adequate theoretical frameworks clarify the deeper issues and evidence relevant to resolving them.

Just as Hull rightly demands, whether we endorse the Mutualist or Fundamentalist hypotheses of today's reductionism or emergentism comes down to which of these positions can "just do it" in concrete scientific cases. Against this background, a question consequently looms very large: Now that we know what empirical evidence resolves their disputes, have either Fundamentalists or Mutualists successfully used empirical evidence to establish their views about some scientific case or even globally? In the final chapter, I use my frameworks to explore this important question.

10 | *The Age of Reduction versus the Age of Emergence: What Has Been Successfully Shown from the Sciences (So Far)*

Is this justifiably the Age of Reductionism? Are we instead warranted in accepting the Age of Emergentism? Or is it too early to make any decision? Scientific reductionists and emergentists, as well as various philosophers, have all argued that our empirical evidence has already allowed us to answer these questions, though they defend contradictory answers. Such defenses need to be treated with some care, since I have shown that the "established views" on both sides of the debate, in philosophy and the sciences, have failed to address crucial theoretical issues, endorsed bad arguments and false dichotomies, and overlooked candidates for live positions. Against this background, since theory appraisal in the sciences is comparative where one assesses a hypothesis against its strongest rivals, one may therefore be immediately suspicious about previous assessments of the best theories. In this final chapter I show that such concerns are indeed warranted, but I also highlight promising signs about the prospects for future work and its chances of making progress.

At the outset, I should note that I am limiting myself to assessing *extant* defenses – existing arguments that this, or that, empirical evidence establishes one or other of the candidates for live positions globally or locally. To provide balance to my discussion, I examine three defenses of Simple Fundamentalism and then three of Mutualism. I show that there are two kinds of reasons why, as yet, neither Fundamentalism nor Mutualism has been successfully established as the best hypothesis about even these favored scientific examples. One reason I highlight for such failures is that in some of the scientific cases there is presently no scientific consensus about the nature of the phenomena or there are reasons why we lack the detailed evidence meeting Threshold and hence sufficient to resolve the key issues between the competing hypotheses. However, even where we might have a settled consensus and suitably deep evidence, I also show a second difficulty still undercuts extant defenses. I detail how proponents of the various positions have not appreciated, and hence engaged, the strongest opposing hypotheses and the key issues in dispute with them. As a result, I show that extant defenses fail in their comparative assessments because they do not engage their strongest rivals.

However, as well as these negative points, my work also establishes a positive conclusion by confirming that our Battle of the Ages is indeed empirically resolvable in ways that the sciences are beginning to engage. However, my final conclusion is that we are still very much in the midst of these new debates about specific scientific cases, let alone about the global structure of nature, where the three live positions are Simple Fundamentalism, Mutualism, and Conditioned Fundamentalism. In concluding, I therefore provide a brief road map about where we should, and should not, focus in the next phase of research on reduction and emergence in nature.

10.1 Assessing Simple Fundamentalism Globally and in Particular Cases

I consider one of the few philosophical defenses of the Completeness of Physics, in 10.1.1, as a potential way to support *global* Simple Fundamentalism. I then consider John Bickle's defense of Fundamentalism in molecular neuroscience in 10.1.2. Finally, in 10.1.3, I look for support for Simple Fundamentalism based on the "approximate" derivations of chemical phenomena from quantum mechanics. My conclusion, in 10.1.4, is that in each of these cases a Fundamentalist hypothesis has yet to be successfully established.

10.1.1 Defending Simple Fundamentalism (I): Philosophical Defenses of the Completeness of Physics and Global Fundamentalism

One way some philosophers have sought to defend a global form of Simple Fundamentalism is to establish the Completeness of Physics, or a global form of the Simple view of aggregation, which can then be used with our ubiquitous compositional explanations to drive the Argument from Composition and Completeness to secure compositional reductions at a range of levels. Although the Completeness of Physics is very widely endorsed by naturalistic philosophers, shockingly few explicitly justify their acceptance of this thesis. However, a few philosophers have defended the Completeness of Physics and they use a far less onerous route to its justification than the one I laid out in the last chapter based around finding a range of cases from different levels. Such arguments are offered by David Papineau (2001), (2002) and David Spurrett (1999), but given the greater prominence of Papineau's work I focus on his defense, though the concerns I raise apply equally to Spurrett's justification.

Papineau's defense, as he himself admits, has evolved across a series of papers and we need to mark two important points to start. First, Papineau takes the situation to be a comparative one, for he takes "emergentism" to be the position challenging the Completeness of Physics and reductionism. But Papineau (2002) follows the received wisdom in philosophy in taking such "emergentism" to be committed to fundamental non-physical forces, which he terms "special" forces, in a species of Ontological emergentism. Second, as Papineau frankly outlines, he has changed his mind about whether conservation laws alone justify the Completeness of Physics. Papineau now accepts that Anti-Physicalists can accept conservation laws, for they simply need to endorse deterministic, fundamental non-physical forces that are conservative in nature. However, Papineau still presses the role of evidence for the conservation laws as "a boundary principle that limits the range of possible forces and helped rule out special animate forces" (Papineau (2002), p. 255, n. 15).

In his defense of the Completeness of Physics, Papineau basically argues that work on conservation laws, as well as other findings about the fundamental forces, in the nineteenth and early twentieth centuries, increasingly supported the existence of only four fundamental forces, which are all physical in character. Papineau calls this the "Argument from Fundamental Forces" (Papineau (2002), p. 253). But what further evidence sufficed to justify the Completeness of Physics against this background and "boundary principle"? The evidence Papineau points to comes from various types of physiological research that grounds what Papineau terms the "Argument from Physiology." Papineau tells us:

During the first half of the [twentieth] century the catalytic role and protein constitution of enzymes were recognized, basic biochemical cycles were identified, and the structure of proteins analysed, culminating in the discovery of DNA. In the same period, neurophysiological research mapped the body's neuronal network and analysed the electrical mechanisms responsible for neuronal activity. Together, these developments made it difficult to go on maintaining that special forces operate inside living bodies ...

In this way, the argument from physiology can be viewed as clinching the case for the completeness of physics against the background provided by the argument from fundamental forces. (Papineau (2002), pp. 253–4)

Papineau therefore points to work providing compositional explanations of key bodily and cellular features in terms of lower-level components. One might wonder whether we should not also include the vast array of such explanations from the full range of higher sciences as well? Or work, again in the early part of the twentieth century, in quantum mechanics that provided "approximate" derivations of the features of atoms?

Fortunately, I need not resolve this issue for our purposes, since the same problems arise whether we take Papineau's argument to be based on compositional explanations solely from physiology and biochemistry or from other sciences as well. The underlying difficulties can clearly be seen when we frame Papineau's justification as a deductive argument. Obviously, Papineau's defense can be formulated in different ways, but the deductive version nicely highlights the underlying problems relevant to any of these formulations. So consider this reconstruction of Papineau's reasoning:

(1) Either robust "emergentism" is true or the Completeness of Physics is true;
(2) If robust "emergentism" is true, then there are fundamental non-physical forces;
(3) The sciences establish that all higher-level entities are plausibly fully composed and that there are no fundamental non-physical forces (Argument from Fundamental Forces plus the Argument from Physiology);

Therefore from (2) and (3) we conclude that:

(4) Robust "emergentism" is not true.

And from (1) and (4) we may further conclude that:

(5) The Completeness of Physics is true.

Here we have an argument moving from very plausible empirical evidence framed in (3), however wide we take the base of evidence to be (i.e. just from physiology or also including biochemistry, chemistry, quantum mechanics, and other sciences), the received wisdom about "emergentism" in (2) and a connected summary of the options in (1), to the conclusions that "emergentism" is false and the Completeness of Physics is true.

Such reasoning is indeed close to objections that have been commonly used against robust "emergentism." As I noted in Chapter 5, Brian McLaughlin's (1992) widely accepted critique of "emergentism" treats the view as a coherent form of Anti-Physicalism based around Ontological emergence and the existence of fundamental, non-physical forces. But McLaughlin argues that "emergentism" so understood has consequently been empirically falsified by our panoply of compositional explanations, and by work in quantum mechanics, which provide plausible evidence against the existence of uncomposed Ontologically emergent entities, and hence fundamental non-physical forces, in nature.[1]

We can agree with Papineau, as well as McLaughlin, about the difficulties raised for the existence of *Ontological* emergence by our now ubiquitous compositional explanations and findings in quantum mechanics. However, my work in earlier chapters shows that a premise like (2), equating robust "emergentism" with Ontological emergentism, is simply false. Papineau, and arguably McLaughlin, fail to appreciate a Mutualist scientific emergentism that is a robust "emergentism," but which does *not* imply the existence of anything more than the four (possibly three) fundamental physical forces. Premise (2) is therefore mistaken and both Papineau and McLaughlin are plausibly enmeshed in the false dichotomies of the Unholy Alliance. More importantly for my present concerns, Papineau's, and similar, justifications of the Completeness of Physics consequently fail because the scientific evidence that underpins this justification is equally compatible with a Mutualism that rejects the Completeness of Physics.

---

[1] However, see the work of the historian Peter Bowler (2001) for a contrasting view of the historical factors that undermined the "emergentism" that was so prominent in many areas of intellectual life between the two world wars. Bowler suggests that the two world wars destroyed the plausibility of the central claim of this earlier "emergentism" that within the world there is moral progress over time and this, rather than problems with the general ontological picture, undercut the previously widespread interest in "emergentism."

Putting aside flawed views of the viable positions, we thus see there is no quick route to justifying the Completeness of Physics that avoids the hard work, laid out in the last chapter, of establishing the Simple view of aggregation in a range of local cases across various levels, hence excluding any downward machretic determination. As I suggested in the last chapter, this reliance upon a range of concrete cases in defending global Simple Fundamentalism, or its global claims like the Completeness of Physics, should surely be no surprise.[2] So let me now examine whether its proponents have succeeded in the more modest, but necessary, task of establishing Simple Fundamentalism in a particular example of compositional explanation in either the higher or lower sciences.

10.1.2 Defending Simple Fundamentalism (II): Simple Fundamentalist Hypotheses in Molecular Neuroscience

Defending Simple Fundamentalist hypotheses about particular cases of compositional explanation, as I noted in Chapter 1, appears to be the approach of philosophers of science such as Bickle (2003), (2006), (2008), Rosenberg (2006), and Schaffner (2006). Each of these writers has sought, to use Schaffner's apt term, "creeping" local reductions, rather than the "sweeping" conclusions of a global form of reductionism. Each of these writers also looks to a science dealing with molecular components – Rosenberg looks at molecular biology, Schaffner considers molecular behavioral genetics, and Bickle takes molecular neuroscience as his material. I am going to focus in my discussion on Bickle, who proclaims his local case supports a "ruthless

---

[2] There is a similar problem with Papineau's approach to the justification of physicalism, as well as related defenses such as Loewer (1995), using the Completeness of Physics where implicit commitment to the reductionist's epochal assumptions presumably explains this focus. But being justified in accepting physicalism is plausibly a *precondition* for justifying the Completeness of Physics, since it is hard to see how one rules out fundamental non-physical forces without establishing universal composition. Thus using the Completeness of Physics to justify belief in physicalism appears unlikely to succeed. (See Foster (1991) and Bishop (2006a), who make what I take to be related points.) As I noted in Chapter 1, a more plausible approach to justifying physicalism is the abductive approach building on local cases of compositional explanations in the manner laid out in Melnyk (2003).

reductionism," since my conclusions about Bickle are applicable to Rosenberg's and Schaffner's claims as well.

The case that Bickle considers is a nascent molecular explanation of the psychological property of spatially remembering a location. As Bickle masterfully lays out, we have a budding compositional explanation of such psychological properties in terms of molecular components. Furthermore, Bickle also emphasizes that our advanced experimental techniques in this area now allow us to intervene at the molecular level to alter psychological behaviors. This is what Bickle terms the "intervene cellularly/molecularly and track behaviorally" approach based on ingenious, and astonishingly successful, genetic and other methods of intervention. I refer the reader to Bickle's accessible treatment of the case (Bickle (2003), (2006)), since it is one where we basically have a budding compositional explanation of certain cognitive phenomena.

However, the more pressing questions concern the substantive conclusions Bickle wants to draw from this example and his argument, or arguments, for them. It is clear that Bickle has some very substantive claims that he thinks follow in this kind of case, since he tells us that:

Heuristically, higher-level investigations and explanations are essential to neuroscience's development. But once they have isolated the relevant neuroanatomy and the candidate cellular and molecular mechanisms, the explanatory investigation shifts to the "intervene cellularly/molecularly and track behaviorally" approach. Once these heuristic tasks are complete, then there is nothing left for the higher-level investigations to *explain*. (Bickle (2006), p. 428. Original emphasis)

Bickle tells us that the cognitive level is "fully explained by the dynamics of the interactions at the lowest level at which we can intervene directly (Bickle (2006), p. 429). And he also claims that, ultimately, such practices justify moving from accepting psychological entities to merely embracing "psychological *descriptions*" (Bickle (2006), p. 430, emphasis added). Although not a reductionist who holds a compromising stance, since he claims that higher-level sciences can ultimately be dispensed with, Bickle thus looks very much like a Simple Fundamentalist.

This brings me to the key question of the argument that Bickle uses to support a Simple Fundamentalist hypothesis or whatever his final position may be. Unfortunately, as I noted in Chapter 1, Bickle does not explicitly outline the needed reasoning. However, using my taxonomy of various arguments, we can identify what appears to be Bickle's reasoning in the last quote. For Bickle contends that the successful compositional explanations offered in this case establish there are no higher-level entities and hence Bickle apparently endorses something like the Argument from Composition. But the problem for Bickle is that, as I have shown in detail, such reasoning is invalid.

The question that immediately arises is whether Bickle could instead run the Argument from Composition and Completeness. In order to do that, one must plausibly be justified in the relevant case in accepting either that the Simple view of aggregation is true or have some other route to establish determinative completeness for the molecular components in the case. However, in this regard Bickle himself highlights difficulties when discussing whether we presently get laws from the research in molecular neuroscience. Bickle tells us that:

In current cellular and molecular neuroscience, as in cell and molecular biology generally, few explanations are framed in terms of laws or generalizations. Many interactions are known to occur and have both theoretical and experimental backing; but biochemistry has not even provided molecular biology with a general (and hence generalization-governed) account of how proteins assume their tertiary configurations. Molecular biologists know a lot about how specific molecules interact within restricted contexts, but few explanatory generalizations are found in molecular biology publications. In addition, what generalizations there are do not by themselves yield extensive predictions or explanations of lower-level interactions. (Bickle (2006), p. 429)

These points about the state of molecular biology generally raise real concerns about whether we can presently answer questions like Q1 or Q1* and hence whether our evidence in this discipline passes Threshold. Remember that we need a precise quantitative account of the behavior of components across *both* the complex collective *and* in simpler collectives to pass Threshold and productively engage Q1 or Q1*. But Bickle points to features of contemporary molecular biology that suggest we presently lack such evidence.[3] We can therefore see

---

[3] Schaffner also makes comments that confirm this state of affairs in the case he examines (Schaffner (2006), pp. 389 and 397).

that, at the very least, it has yet to be shown that our present evidence in molecular biology, and related areas, is sufficient to successfully run the Argument from Composition or Completeness in Bickle's example. And similar points plausibly hold for the claims of Rosenberg and Schaffner in the scientific areas they focus upon.[4]

Contrary to Bickle's, Rosenberg's, or Schaffner's claims, Simple Fundamentalist accounts have yet to be established as the best hypotheses about the relevant examples. It is not just that Bickle, Schaffner, and Rosenberg do not engage the strongest rival views, in Mutualist hypotheses, although this is in fact the case.[5] In the relevant examples, the more important problem is that our understanding of the relevant components is yet to be shown to pass Threshold.

However, I should explicitly note that my brief discussion of Bickle only shows that extant work does not successfully establish a local Simple Fundamentalist hypothesis. For all I have argued it may therefore be the case that using my theoretical frameworks existing scientific evidence in molecular neuroscience, or in other areas of molecular biology, can be marshaled to successfully support such a hypothesis. But whether such defenses can be successfully provided is important work for the future. My conclusion here is that, so far as I know, no one has yet provided a successful defense of a Simple Fundamentalist hypothesis in molecular neuroscience, molecular behavioral genetics, or molecular biology more generally.

10.1.3 Defending Simple Fundamentalism (III): Using Quantum Mechanical Derivations of Atomic/Molecular Features

Many proponents of Fundamentalism may not be concerned by my findings so far, since they primarily point to the derivations of atomic or molecular features using quantum mechanics when they seek to defend their views. Most famously, we can derive features of the

---

[4] Below I examine Mutualists who contend our evidence about the molecular biology of certain cases passes Threshold, but such writers contend the evidence establishes a Mutualist hypothesis about these examples.

[5] Bickle and Rosenberg do not consider Mutualist rivals at all. Schaffner at least discusses the possibility of "emergence" in the case he considers, but Schaffner does not include Strong emergence and hence falls into the flawed Weak/Ontological emergence dichotomy (Schaffner (2006), pp. 382–3).

hydrogen atom using quantum mechanics and we can give what are termed "approximate" derivations of the features of other atoms and some molecules.

One might therefore expect this case to be a promising one for Simple Fundamentalism, but a couple of points must immediately be confronted now that I have more carefully articulated the nature of Fundamentalism and its required support. First, it is far from clear how, even if one is successful in defending a Simple Fundamentalist hypothesis about the case, this *alone* could justify the Completeness of Physics, a global version of the Simple view of aggregation, or global Simple Fundamentalism. Global Simple Fundamentalism, and its associated global theses, need to be established across a far, far wider range of phenomena and levels, and the collectives they involve, in order to be accepted. However, let me set these concerns to one side and simply consider whether extant defenses have successfully shown that a Simple Fundamentalist hypothesis is the best account even just in the specific case of atoms or molecules and quantum mechanical phenomena.

This brings me to the second set of issues, since we actually lack a detailed defense of Simple Fundamentalism in this case that I can even consider. And that is no accident. When one tries to reconstruct both the putative Simple Fundamentalist hypothesis, or the basis of it in compositional explanations and their compositional concepts, then amongst other issues one confronts the problems about quantum mechanics I noted in earlier chapters. To start, one faces the well-known, and long-standing, disputes over the ontological interpretation of quantum mechanics. In addition, we also need to be alive to the equally substantive, and connected, questions about whether the "composition," "components," and "composites" posited in interpretations of quantum mechanical phenomena are the same as the concepts of composition, articulated in Chapter 2, that we find in compositional explanations from other sciences. Given these issues, there are consequently important questions about whether we are presently justified in accepting that there are compositional explanations in this case of the kind that can drive the compositional reductions underlying a Simple Fundamentalist hypothesis.

So there is substantial work to do even in *articulating* a plausible Simple Fundamentalist hypothesis and its justification in this example, let alone showing it is the best hypothesis about the case. But proponents of Simple Fundamentalism still may not be daunted, for they may suggest that given the existence of the *derivations* from quantum mechanics of atomic and molecular features, whether "approximate" or otherwise, then we can still use such bare derivations to establish some kind of local Simple Fundamentalism even if we lack compositional explanations.

In response, I need to reiterate the points I flagged in Chapter 7 about derivations and their import. Recall that Mutualism, unlike Ontological emergentism, does not reject the existence of derivations of statements about Strongly emergent entities from statements about their components. The mere existence of some derivation of the features of atoms or molecules therefore resolves very little. Instead, we need to be given a detailed justification of why the features of the quantum derivations of atomic and molecular features support a Simple Fundamentalist hypothesis, rather than its Mutualist rivals. But, so far as I know, such comparative arguments have yet to be given.

Furthermore, the need for the Fundamentalists to defend their claims about this case in more detail is only heightened by an important critique recently pressed by both scientists and philosophers of chemistry. For example, scientific researchers, such as the quantum chemist Rodger Woolley, have long complained about the fact that, beyond the hydrogen atom, when one looks at the "approximate" derivations of the features of simple molecules used in practice, then these involve putting in information "by hand" that appears to concern the composed entities in these cases.[6] And such criticisms have recently been defended in a sophisticated form by the philosopher of chemistry Robin Hendry, who focuses on the "approximate" derivations of atomic or molecular features.[7]

Hendry argues at length that these "approximations" all plausibly use information about the relevant composed entities, whether implicitly or not, in combination with the machinery of quantum mechanics to derive the features of these macro-entities. Consequently, we thus

---

[6] See Woolley and Sutcliffe (1977) and Woolley (1985). Along similar lines, we also find Laughlin and Pines stressing that "the schemes for approximating are not first-principles deductions but rather art keyed to experiment" (Laughlin and Pines (2000), p. 28).

[7] See Hendry (2006) and (2010). Related concerns about the nature of the derivations of chemical features from quantum mechanics are raised in a series of papers by Eric Scerri (Scerri (1994), (1997), (1998a), and (1998b)).

have real problems, putting the points in my terms, in taking such cases to clearly support a Simple Fundamentalist hypothesis. For it is puzzling why reference to, or other use of the composed entity, should be necessary in the relevant derivation if the Simple view of aggregation holds. As I outlined in the last chapter, the Simple Series of explanations found under Simple Fundamentalism lack such references to composed entities. At the very least, we thus need a compelling story about why information about the composed entities must be used in these derivations.

In fact, given points like Woolley's and Hendry's about the nature of the derivations of atomic and molecular features, it appears that a *Mutualist* hypothesis actually becomes plausible in these cases. An advocate of Mutualism can argue that our failure to produce such derivations *except* by feeding in "by hand" information about the composed entity shows that this composed entity is playing a machretic determinative role. And Hendry does indeed go on to argue that these features of our derivations plausibly ground a case for what he terms "emergentism," though deciding whether Hendry is defending the existence of Ontological or Strong emergence is a difficult exegetical matter. However, applying my frameworks, one interesting interpretation of Hendry is that he espouses Mutualism and defends the existence in chemical practice of something closer to a *Conditioned* series of explanations making reference to differential powers of components, and/or emergent laws containing ineliminable reference to composed entities. And, regardless of Hendry's own view, such a Mutualist hypothesis offers an important competing hypothesis about this case.

When I directly examine claims about Mutualism, in section 10.2 below, I assess Hendry's claims under the Mutualist interpretation of their nature. But the relevance of the Mutualist hypothesis to the assessment of extant defenses of Simple Fundamentalism is clear. Defenders of Simple Fundamentalism have not engaged this rival Mutualist hypothesis, nor the key issues in dispute between it and their own hypothesis. Given the face plausibility of the Mutualist hypothesis just noted, and the comparative nature of theory appraisal, until the Simple Fundamentalist hypothesis is shown to be superior to its rival, then we cannot accept the Simple Fundamentalist account as the best picture of the case.

To conclude, I have found that even in the talismanic case of quantum mechanics and chemistry a Simple Fundamentalist hypothesis is yet to be established for two reasons. First, we have yet to actually be provided with such a defense built around either compositional explanations and their compositional concepts, or derivations that are shown to have features plausibly supporting a local Simple Fundamentalist hypothesis. Second, even if we had such a defense in hand, I have now shown that the key disputes with rival Mutualist accounts have yet to be addressed. Someone might reasonably ask whether my better theoretical frameworks could sharpen a defense of Simple Fundamentalism into a successful form. That is indeed an interesting question, but one that must again be addressed in future research. My conclusion here is that a Simple Fundamentalist hypothesis is yet to be justified as the best account of the case of quantum mechanical derivations of atomic and molecular phenomena.

10.1.4 Simple (or Conditioned) Fundamentalism has Not (Yet) Been Established Either Locally or Globally

My work in this section shows that the Fundamentalist has work to do. Looking at a direct philosophical defense of the Completeness of Physics, and hence global Simple Fundamentalism, I highlighted how it was based upon the same flawed arguments about the implications of compositional explanations I outlined in earlier chapters. I reconfirmed that to establish the Completeness of Physics we plausibly require a range of cases of successful compositional explanation where Simple Fundamentalism has been established, rather than simply the universal applicability of compositional concepts.

But when I turned to defenses of Simple fundamentalist hypotheses in such scientific examples I also illuminated problems, albeit of a different character. With defenses of Simple Fundamentalist hypotheses in molecular neuroscience, or molecular biology more generally, our evidence is yet to be shown to be detailed enough to pass Threshold and differentiate between the Simple Fundamentalist hypothesis and its rivals. And, even if we take the evidence to be sufficient to pass Threshold in molecular biology, then I further noted how defenders of Simple Fundamentalist accounts have not engaged their relevant rivals, such as Mutualist scientific emergentist accounts, so their defenses of Simple Fundamentalism fail for another reason.

Retreating to examples at still lower levels, I then looked at the case of quantum mechanical derivations of chemical phenomena, but

I found that Simple Fundamentalist hypotheses have yet to be articulated in such cases. And, once more, I also highlighted how, even if we had such accounts, then the relevant Simple Fundamentalist hypotheses have still not been shown to be better than Mutualist rivals. I concluded we have a couple of reasons why Simple Fundamentalist accounts have yet to be successfully established even in the case of quantum mechanical derivations of chemical phenomena.

Given the background state of the debates I have detailed in earlier chapters, I contend that we can plausibly expect these two kinds of difficulty, in lack of sufficient evidence or a failure to engage relevant rivals, will also plausibly undermine other extant defenses of Simple Fundamentalism and also Conditioned Fundamentalism, if there are any. Overall, I thus conclude that we are yet to successfully establish Simple or Conditioned Fundamentalist hypotheses in particular scientific cases of successful compositional explanation.

We can now also extend these conclusions to global claims, since global Simple or Conditioned Fundamentalism must plausibly be established through showing Simple or Conditioned Fundamentalist hypotheses hold in a range of such local cases. I therefore also conclude that we are yet to successfully confirm global Simple or Conditioned Fundamentalism as well.

Focusing on global claims, it also bears emphasis that the problems highlighted in my survey extend to the justification of the disputed global theses associated with Simple Fundamentalism. Given my findings, for example, we are also plausibly yet to establish either the Completeness of Physics or the global form of the Simple view of aggregation. Remember that the Completeness of Physics implies that the fundamental microphysical components are such that they never have differential powers and hence that the laws holding of them in isolation or the simplest collectives exhaust the determinative laws holding of such components in *every* collective across *every* level. The truth of the global form of the Simple view of aggregation is a precondition for the truth of the Completeness of Physics. Thus for the Completeness of Physics to be true, then the Simple view of aggregation must hold of the fundamental microphysical components in *every* level of collective such as the chemical, biochemical, cellular, tissue, organ, organismic, and on, and on. To establish the Completeness of Physics or a global form of the Simple view of aggregation, we plausibly need to show the Simple view holds across a range of cases of successful compositional explanation. But given my findings about the failure to establish Simple Fundamentalist hypotheses over relevant rivals in any local case, it is plausible that we lack such a range of cases. Consequently, we are yet to establish either a global version of the Simple view or the truth of the Completeness of Physics.

Despite these negative conclusions, I want to reiterate that all I have shown is that *extant* defenses of Simple Fundamentalism presently fail. And, as I have noted throughout this section, the more adequate theoretical frameworks I have developed provide a platform for the Fundamentalist to pursue the better defenses she needs. My twin hopes are that my brief survey of the deficiencies of existing defenses will provoke Fundamentalists to pursue more adequate defenses; and that the better theoretical frameworks I have provided can facilitate the needed projects. Until we are given successful comparative defenses of Simple (or Conditioned) Fundamentalist hypotheses in concrete scientific examples, then we are not justified in endorsing scientific or other versions of ontological reductionism, or its key theses, either locally or globally.

10.2 Assessing Mutualism in Concrete Examples

The aspirations of defenders of Mutualism have been firmly focused on particular cases, rather than global claims. So, to keep some balance with my examination of Fundamentalism, I look in detail at three representative examples where Mutualist hypotheses have recently been defended. To begin, in 10.2.1, I examine a case from a higher science in the arguments about the emergence associated with the self-organization in eusocial insects. I then look at the new area of systems biology, in 10.2.2, and emergence in biochemical networks. Finally, in 10.2.3, as my third formal example of a defense of a Mutualist hypothesis I consider claims about high-energy superconductivity. Overall, I show that Mutualist defenses are unsuccessful for the same two problems that I highlighted with Fundamentalist justifications.

10.2.1 Defending Mutualism (I): Wilson and Holldobler (and Mitchell) on Eusocial Insects

I want to start by looking at the claims that E.O. Wilson and Bert Holldobler (1988) make about eusocial insects, since this also provides an example where a prominent scientific reductionist, in Wilson, switched sides to Mutualism on empirical grounds – and then eventually switched back again.[8] The example is also useful because it involves "self-organization," and related phenomena, of a type common across the sciences of complexity. Lastly, the case allows us to engage the claims about similar examples in other eusocial insects independently broached by the philosopher of science Sandra Mitchell (2009), (2012) in support of what appears to be Strong emergence.

Wilson and Holldobler are famous for their decades-long research on ants, which include so-called "eusocial" species that have some of the most complex social organizations and artifacts to be found in the natural world.[9] Focusing on the genesis of these singular phenomena, Wilson and Holldobler are struck by what they term the "self-organization" they find in colonies of eusocial ants, which leads them to defend the existence of Strong emergence. Wilson and Holldobler tell us that:

recent research has shown an ant colony is a special kind of hierarchy, which can be usefully be called a heterarchy. This means that the properties of the higher levels affect the lower levels to some degree, but induced activity in the lower units feeds back to influence the higher levels. (Wilson and Holldobler (1988), p. 65)

An ant colony is composed, claim Wilson and Holldobler, but we nonetheless have determination in both directions – from the properties of the colony to those of individual ants, as well as from component ants to the colony and its properties. Hence Wilson and Holldobler posit a "heterarchy," rather than the usual hierarchy. And Wilson and Holldobler thus conclude that an ant colony is "a superorganism with emergent properties" (Wilson and Holldobler (1988), p. 66). I leave aside the claim about "superorganisms" and focus on the reasons

---

[8] As well as Wilson and Holldobler (1988), Wilson lays out his commitment to something closer to Mutualism in his interview in Lewin (1992). In contrast, Wilson (1998), pp. 88–90, later ruefully explains why he takes his flirtation with Strong emergence to be mistaken and reaffirms his commitment to Simple Fundamentalism (see pp. 266–7 amongst others).

[9] Their own, and other work, on ants is outlined in a magisterial volume (Holldobler and Wilson (1990)).

Wilson and Holldobler give to defend the existence of Strong emergence in this case.

Wilson and Holldobler discuss a number of concrete examples of the type they have in mind, but let us simply consider one that they cite based upon the findings of Sorenson, Busch, and Vinson (1985) about foraging in fire ants. Wilson and Holldobler tell us:

An excellent example recently discovered is the regulation of food flow into a colony of the fire ant *Solensopsis invicta*. When the foragers are starved as a group, they collect disproportionately more honey; when they are well fed but the nurse workers or larvae are starved, the foragers collect more vegetable oils and egg yolk. Sugars are used mainly by adult ant workers, lipids by workers and some larvae, and proteins by larvae and egg-laying queens. Hence the fire ant foragers ... respond to the nutritional needs of the colony as a whole and not just their personal hunger. (Wilson and Holldobler (1988), p. 66)

The idea is apparently that a higher-level property of the colony, in something like the satiation of the colony and/or its levels of stored resources of a certain foodstuff, alters the powers and behaviors of the component ants. And Wilson and Holldobler note that we now have a compositional account of the basis of this phenomenon whereby individual ants respond to the higher-level property of the colony. The same basic process is found in a number of eusocial insects including both ants and bees. Wilson and Holldobler tell us this about how the fire ants regulate their foraging for various foodstuffs:

How do the foragers monitor this generalized demand?... When the demand encountered ... declines in any sector of the colony, they accept the corresponding food less readily from the foraging workers. The foragers are unable in turn to dispose of their loads as quickly as before, and they reduce their efforts to collect more of the same. As a result, they shift their emphasis as a group ... (Wilson and Holldobler (1988), p. 66)

The claim is thus that a property putatively of the colony, in its stores and satiation with certain foods, leads to longer hand-off times in unloading particular food stuffs, which in turn leads foragers to modify their foraging for such foodstuffs. Wilson and Holldobler note that there are numerous types of feedback loops involved here and imply that although the individual ants only use simple "rules of thumb" this still leads to highly complex behaviors.

We therefore have a case where composed properties putatively influence properties of the relevant components and where there are various feedback loops. Like other writers, these features are apparently the ones leading Wilson and Holldobler to defend a Mutualist hypothesis about this example and hence to conclude that we have Strongly emergent properties in the colony or "superorganism" machretically determining their components. However, as I noted above, Wilson apparently abandoned this Mutualist hypothesis, for by the time of his (1998) book Wilson is again espousing something very close to Simple Fundamentalism and offers a somewhat bitter critique of views in the sciences of complexity with scientific emergentist leanings.[10] What motivates this turnaround?

If we consider whether there is a viable Simple Fundamentalist hypothesis about the example, then we may be able to understand this reversal and also assess whether a Mutualist account has been established as the best hypothesis about the case. The key point is that the Simple Fundamentalist can seek to offer a hypothesis about the example where all the determination is at the component level of ants, and their properties and relations, and other properties, relations, and individuals with which they productively interact, such as specific food stores, which we plausibly may also count at the component level. Thus the Simple Fundamentalist hypothesis will take the individual ants, on either side of the interaction of passing off food, to be determinative, as well as the other ants and food storage units, and their quantities of particular foodstuffs. Crucially, the Simple Fundamentalist takes other ants and specific food stores to be at the lower level, since ants productively interact with other ants and particular storage units, filling them for example, which is a key test of a same-level productive relation within a compositional hierarchy. Notice that the interactions at the lower level, for example between particular storer ants and various food stores, suffice for the slow-down at the exchange site between storers and foragers. The ants responsible for storing go slower given their interactions with particular storage sites that are full and these interactions result from properties that count as component-level

---

[10] For Wilson's pessimistic musings on certain "emergentist" approaches, primarily the work of Stuart Kauffman, see Wilson (1998), pp. 88–90.

properties, rather than colony-level properties. And the slower turna-round time that results prompts the foragers to change the foodstuff they search for.

Under this Simple Fundamentalist hypothesis, we again have feed-back loops, but they are all at the lower level of components. As I noted in Part II, Fundamentalism can, and often does, accept feed-back loops at the level of components. Similarly, I should also mark that given the large number of interacting actors, and variables, the resulting dynamics describing the behavior of foragers (and storers) may well also be non-linear in nature and/or only be explainable by simulation. However, as I outlined in Parts II and III, both features are compatible with the Fundamentalism.

We thus have available an alternative, and seemingly plausible, Simple Fundamentalist hypothesis about the structure of the case accommodating most of its main features whilst avoiding Strong emergence and machresis. To establish the Mutualist account in the case of social insects and colonies we need to show it is better than this Fundamentalist rival. One way to establish the Mutualist over that of the Simple Fundamentalist hypothesis would be to answer Q1* and show that the laws holding of the components, i.e. ants, in simpler systems do not alone suffice to determine all of their powers in the complex collective, i.e. the colony. At the least, we need to answer Q1 and make it plausible that the ants have differential powers in the complex collectives. However, for all the empirical information that Wilson and Holldobler point to, they offer nothing that answers Q1 or Q1*. The crucial issues are thus left unresolved between the competing Simple Fundamentalist and Mutualist hypotheses. Similar points plausibly apply to Mitchell's (2009), (2012) claims about bees and Strong emergence.

My conclusion is that, as yet, Holldobler and Wilson have not established the Mutualist account of the case over even a Simple Fundamentalist hypothesis of the type outlined. And more complex Conditioned Fundamentalist accounts loom as rivals as well where such hypotheses endorse the existence of differential powers and new laws, but offer an alternative account of their natures. As yet, a Mutualist hypothesis has not been shown to be superior to such Conditioned Fundamentalist views. In fact, it bears emphasis that in this example it even remains to be shown that our present evi-dence satisfies Threshold, including whether we have the quantitative accounts of components in simpler and complex collectives that allow us to fruitfully engage the disputes between the Fundamentalist and Mutualist hypotheses.

Could a defense of Strong emergence and Mutualism of the kind offered by Holldobler and Wilson, or Mitchell, now be sharpened into a successful one? Once more that is an important question I must leave to the future. My conclusion here is that the extant defenses offered by Wilson and Holldobler, and Mitchell, have not (yet) successfully established a Mutualist hypothesis as the best account of the foraging behavior of eusocial insects.

10.2.2 Defending Mutualism (II): Boogerd, Bruggeman, Richardson, Stephan, and Westerhoff on Systems Biology and Biochemical Networks

I am drawing from the new area of systems biology for my second example because it often throws up claims about emergent phenomena. I should note that not all systems biologists are emergentists and a few are clearly scientific reductionists.[11] O'Malley and Dupré (2005) point out that the majority of systems biologists take "system" to be a vague, but convenient, term to use in pursuing everyday reductionism. However, O'Malley and Dupré also note there is a smaller group, which they term "systems theoretic biologists," who take an ontologically robust view of systems and emergence close in character to Mutualism. As my example of such systems theoretic biologists I want to examine the work of a team consisting of both philosophers of science, in Robert Richardson and Achim Stephan, as well as prominent systems theoretic biologists in Fred Boogerd, Frank Bruggeman, and Hans Westerhoff.

In a recent article, Boogerd et al. present a theoretical framework for emergence and then defend the existence of such emergence in biochemical networks in the eukaryotic cell.[12] This work is interesting for my purposes, since some of the features of the relevant notion of emergence, and the concrete case, are strikingly similar to features found under Mutualism. However, although Boogerd et al. lay out a

---

[11] See Sorger (2005) for an explicitly reductionist approach in systems biology.
[12] See Boogerd, Bruggeman, Richardson, Stephan, and Westerhoff (2005). In addition, see Richardson and Stephan (2007).

notion of emergence, and also term it "Strong" emergence, some care is needed with regard to their account. (To avoid terminological confusion, in this section I therefore revert to my practice in Chapter 5 of talking about "S-emergence" when I mean a composed property instance that is still determinative.) Boogerd et al. explicitly take their lead from the British emergentist C. D. Broad (1925) by defining an "emergent" property, putting it in my terms, as a realized property had by a constituted individual "if the properties of the parts within the system cannot be *deduced* from their properties in isolation or other wholes, even in principle" (Boogerd et al. (2005), p. 135, emphasis added). At various points, Boogerd et al. substitute being *predicted* for being deduced, since they assume that predictions take the form of deductions.[13] For ease of exposition, I follow them in their assumption. Boogerd et al. articulate other notions of emergence, but the latter conception based around failure of deduction/prediction is the one they argue is significant in scientific cases.

With this notion of emergence in hand, Boogerd et al. then look at examples of the biochemical networks we find in eukaryotic cells and in the simpler collectives forming naturally in bacteria or produced artificially using laboratory techniques such as knockout mutants. As Boogerd et al. highlight, the biochemical networks in eukaryotic cells are formidably complex, involve all manner of feedback loops, and are often only describable using non-linear dynamics. When we look at these complex collectives, and crucially their components, as well as such components in simpler collectives (whether natural or artificial), then Boogerd et al. contend that "there are cases of non-deducibility if we restrict the deduction base accordingly" (Boogerd et al. (2005), p. 142) and hence examples of their distinctive form of emergence based around non-deducibility.

This interdisciplinary team obviously defends claims using a notion of emergence not so distant from the Mutualist account of emergence because Boogerd et al. discuss behaviors and powers, and possibly laws, of components found in complex collectives, but not in relevant simpler collectives of these same components. So let me examine

---

[13] Boogerd et al. tell us that the relevant forms of emergence as failure of deducibility from laws an behaviors in simpler collectives "means the properties (and behaviors) of the parts within the whole cannot be predicted from the properties (and behaviors) in other systems" (Boogerd et al. (2005), p. 136).

whether Boogerd et al.'s work suffices to establish the existence of composed property instances that are nonetheless determinative, i.e. the S-emergence of Chapter 5.

To begin, I need to mark some problems with the theoretical framework used by Boogerd et al. since their focus on unpredictability in defining their key notion of emergence, largely inherited from Broad, is unfortunate. I noted in Part II that even Simple Fundamentalism can accept that we often have complex collectives of components only describable by non-linear dynamics. As I further outlined in Chapter 6, Fundamentalists like Bedau have plausibly shown that in such cases where we cannot deduce, and hence predict in Boogerd et al.'s terms, the laws and behaviors of the components in complex collectives from those in simpler systems alone, then we can often *still* explain by simulation using the laws holding in the simpler collective. This allows the Simple Fundamentalist to make a case that even when we have failure of derivability/deducibility the components are still determined solely by other components according to these laws. Hence the Fundamentalist can potentially justify determinative completeness of components even when we have the kind of emergence that Boogerd at al. focus upon. Consequently, the Simple Fundamentalist can plausibly run the Argument from Composition and Completeness even in such a case where we have the kind of emergence Boogerd et al. champion.

Boogerd et al.'s discussion plausibly suffers from the continuing struggle to divorce *ontological* notions of reduction and emergence from *semantic* ones. And Boogerd et al.'s theoretical framework for emergence only supplies a form of Weak emergence. The focus on unpredictability or undeducibility in Boogerd et al.'s notion of emergence does not preclude even Simple Fundamentalism from being true and hence clearly does not suffice to establish the existence of a composed property instance that is still determinative, i.e. S- or Strong emergence of the type I have examined at length. However, it is only fair to highlight two further points.

First, even given the latter conclusions, I should note that a distinctive form of Weak emergence has been established to exist by Boogerd et al. in the biochemical cases they examine and this is an important finding in itself that should not be lightly dismissed. For example, their work provides a powerful counterexample to greedy semantic reductionism. Second, and more interestingly for our purposes, one can read Boogerd et al. as seeking to articulate something close to S-emergence, but having mis-stepped in their focus on unpredictability and und-educibility in seeking to capture the intended ontological notion.

Stepping around Boogerd et al.'s own framework for emergence, it is therefore an important question whether using my framework for Mutualism the claims of Boogerd et al. about their biochemical case would suffice to establish a Mutualist hypothesis about the example. Let me therefore examine where their evidence leads when put into my Mutualist framework for S-emergence.

As I outlined in Chapter 8, it is a necessary condition of the applicability of a Mutualist hypothesis to some example of compositional explanation that we have differential powers holding of components in the relevant complex collective. One potential form of evidence for the existence of such differential powers is that we cannot use the laws holding of simpler systems alone, however we come to know them, to account for the powers of components in complex collectives. Against this background, it might be thought that Boogerd et al.'s claims about biochemical networks are of some moment.

Unfortunately, although the case may have other features that could help establish Mutualism, the simple focus of Boogerd et al. on undifferentiated unpredictability means that the evidence they have provided is still insufficient to establish S-emergence for two reasons. As I have just noted, unpredictability or undeducibility of the kind Boogerd et al. highlight does not even rule out Simple Fundamentalism. Thus, by itself, undifferentiated unpredictability or undeducibility of the behavior of components in complex collectives, from the laws in simpler collectives, is too blunt an instrument to establish that we have differential powers or emergent laws. For example, it is yet to be shown even that we need to posit emergent laws holding across both simpler and complex collectives, rather than simply accepting a Simple Fundamentalist account under which the laws in simpler systems hold in the complex collective but do not allow predictions, although explanations by simulation or something similar is available.

Second, even if the evidence that Boogerd at al. provide did suffice to show that we have a case of Conditioned aggregation involving differential powers, and/or laws, then this still would not establish a Mutualist hypothesis. Again, we need to remember the challenge of *Conditioned Fundamentalism* that presents a rival hypothesis also endorsing the existence of differential powers and which also entails that the laws holding of components in simpler collectives will fail to cover all the behaviors and powers of components in certain complex collectives, hence also entailing we have undeducibility/non-predictability of the type Boogerd et al. focus upon in their example.

To counter Conditioned Fundamentalism, a crucial further task for a proponent of Mutualism is to provide evidence that the differential powers of the components in the biological networks are machretically determined by composed properties, relations, processes, and/or individuals, rather than component-level properties, relations, processes, and/or individuals. Since undeducibility/non-predictability of the component's powers and behaviors in the complex collective is compatible with Conditioned Fundamentalism, the mere existence of such undifferentiated, bare undeducibility/non-predictability is too crude an instrument to resolve this key question for Mutualism. We therefore have a second reason why the evidence outlined by Boogerd et al. fails to establish a Mutualist hypothesis, since the evidence they adduce also fails to show a Mutualist account is better than its Conditioned Fundamentalist rival.

In their discussion of the foundations of systems biology, O'Malley and Dupré come closer to the evidence needed to establish a Mutualist hypothesis when they tell us:

Systems biology of any persuasion has to demonstrate that when single components come together and form a system, they engage in novel behavior and produce novel phenomena by the system itself constraining the components. (O'Malley and Dupré (2005), p. 1273)

This comes close to the relevant point, but we need to remember the lesson of Part II that even scientific reductionists such as Weinberg explicitly do, and can, accept "More is different" and allow that component individuals and properties bearing powerful relations to each other in collectives have novel powers resulting in novel behavior. Rather than simply having novel powers and behaviors of components in complex collectives my work has sharpened our understanding of the type of evidence actually needed to establish Mutualism. Contra the claims of Simple Fundamentalist rivals, it needs to be shown that the components have differential powers. And then against Conditioned Fundamentalist accounts it needs to be established that that contribution of such differential powers by components derives from the machretic determination of a composed entity.

Could Boogerd et al.'s work, and especially the suggestive features of biochemical networks that they highlight, be sharpened into a defense of Strong emergence using my theoretical accounts? Or is there other work in systems biology that could sustain such a defense? I have now framed the empirical evidence that is required, so this important issue is potentially resolvable. And Boogerd et al. offer intriguing evidence that suggests that such cases pass Threshold and have real promise for Mutualism, especially because the key question may be amenable to resolution using available research techniques. Contrary to the claims of Bickle, Rosenberg, and Schaffner, once we appreciate the actual terrain of our new debates, and more properly appreciate the rival positions doing battle, then we see that the molecular level is far from being an obvious bastion of scientific reductionism and Fundamentalism. Nonetheless, I have also shown here that Boogerd et al. are (as yet) unsuccessful in establishing a Mutualist hypothesis about biochemical networks.

10.2.3 Defending Mutualism (III): Laughlin on High-Energy Superconductivity

As my third formal example of a defense of Mutualism let me finally turn to the case of high-energy superconductivity that prominent scientific emergentists such as Anderson and Laughlin, who have been at the heart of work on the nature of this phenomenon, have argued to be of special moment. Laughlin claims that our understanding of superconductors, as well as the quantum mechanical laws holding of their components in simpler systems, is now so well developed that it confirms Mutualism holds at least sometimes in nature. In my terms, Laughlin's claim appears to be that high-energy superconductivity, and connected phenomena, are of such seminal importance because our scientific understanding in these cases passes conditions (a), (b), and (c) of Threshold, and Laughlin further claims that we can use this evidence to show that a Mutualist hypothesis is the best account of high-energy superconductivity.[14] Thus, for example, Laughlin (2005) stresses that electrons take preferred positions in superconductors that they would not take if the laws of quantum mechanics were the only

---

[14] Laughlin makes similar claims about von Klitzing's findings about the quantum Hall effect. Note that Laughlin's own work on the quantum hall effect eventually won him his Nobel prize (Laughlin (2005), p. 76).

determinative laws in such cases. But how does this, or other, features of the case support a Mutualist hypothesis?

To answer this question, I need to move beyond Laughlin's treatments that are either inaccessible, specialized accounts in articles, or the popular treatment in his (2005) book, which rarely provides much detail about particular examples. Fortunately, the physicist turned philosopher of science Sunny Auyang provides a more accessible account of high-energy superconductivity articulating the key features of the case that scientific emergentists such as Laughlin apparently use to defend their Mutualist hypotheses. I therefore quote Auyang at length, and she tells us that with high-energy superconductors:

> The superconducting current cannot be obtained by aggregating the contributions of individual electrons; similarly it cannot be destroyed by deflecting individual electrons. The microscopic mechanism ... baffled physicists for many years, and was finally explained in 1957 by the Nobel prize-winning BCS theory developed by John Bardeen, Leon Cooper, and Robert Schrieffer. An electron momentarily deforms the crystal lattice by pulling the ions towards it, creating in its wake a surplus of positive charges, which attract a second electron. Thus the two electrons form a pair, called the Cooper pair ... All the Cooper pairs interlock to form a pattern spanning the whole superconductor. Since the interlocking prevents the electrons from being individually scattered, the whole pattern of electrons does not decay but persists forever in moving coherently as a unit. The structure of the interlocked pattern as a whole, not the motion of the individual electrons, is the crux of superconductivity.
>
> ... No special ingredient is needed for [this] emergence; superconductors involve the same old electrons organized in different ways. The structured organization in superconductivity constrains the scattering of individual electrons and forces them to move with the whole, which can be viewed as a kind of downward causation. (Auyang (1998), pp. 180–1)

Here we confirm that every entity in the case at the higher level is fully composed and that we have a qualitative account of the relevant components underpinning something like compositional explanation. And the echoes of Dyke's "structured structuring structures" in this passage allows us to begin to appreciate the basis for the scientific emergentist's claims that the case is one where a Mutualist account is correct.

Crucially, the features of the example outlined by Auyang provide the scope for arguing that the composed entities are machretically determining, and reducing the degrees of freedom of, the component electrons in ways going beyond the laws holding in simpler collectives. Laughlin apparently argues that with high-energy superconductivity we know the quantum mechanical laws holding of components in simpler collectives and that exhaustively determine the components in those simpler collectives. But the extreme accuracy and precision of our understanding of the behaviors and powers of the components in high-energy superconductors shows that the laws holding of components in simpler collectives *cannot* be the only ones that hold of such components in superconductors. Laughlin claims we know that the quantum mechanical laws holding in simpler collectives would determine the component entities have *different* powers and behaviors than we actually find in superconductors. Therefore, concludes Laughlin, we see that further, emergent laws hold that indicate that "organizational" entities, such as composed entities, play a machretic role in determining the differential powers contributed to components such as electrons.

Providing a full assessment of Laughlin's claims about the case of high-energy superconductivity, and other cases from physics, requires a detailed engagement with his, and others', work on these phenomena in physics. However, the obvious question is whether the same kinds of concerns arise with Laughlin's defense of a Mutualist hypothesis about high-energy superconductivity. And, even without a more detailed engagement with Laughlin's arguments, I contend there are good reasons to think that such difficulties do arise.

First, let me note a concern pressed by Don Howard, one of the few philosophers of physics to critically engage Anderson and Laughlin, who contends that there is presently "no unified, general theory of condensed matter physics" (Howard (2007), p. 152). Consequently, Howard claims that at present we cannot draw any safe conclusions about the best hypothesis about the nature and implications of high-energy superconductivity. Until the latter issue is addressed, we consequently have a question concerning whether the empirical accounts in this case pass Threshold and hence about the success of Laughlin's defense of a Mutualist hypothesis.

Second, I need to note that we again have Conditioned Fundamentalist hypotheses about the phenomena of condensed matter physics.[15] And

---

[15] The Simple Fundamentalist can obviously also offer a hypothesis about this case too and I take no stand on whether Laughlin has successfully rebutted

the challenge of a Conditioned Fundamentalist hypothesis is especially pressing for Laughlin because this account also accepts that we find differential powers in the components in high-energy superconductors and potentially accepts that new fundamental laws about components hold in such examples, too. However, this Conditioned Fundamentalist rival claims that the only entities determining components like electrons have such differential powers are other components or entities at the component level. Thus, claims this rival, the electrons we find in Cooper pairs, and/or their properties and relations to each other, determine that other components have the relevant differential powers. And this Conditioned Fundamentalist account can even accept that quantum mechanical laws holding in simpler collectives do not exhaustively determine the behaviors and powers of electrons in superconductors. But this Fundamentalist hypothesis only accepts the existence of further *non-emergent* fundamental laws, solely involving component-level entities, that hold of the components in high-energy superconductors.

Such a Conditioned Fundamentalist hypothesis is important when assessing Laughlin's claims because it offers an important alternative account of the structure of the case endorsing many of the novel features Laughlin himself accepts. However, Laughlin has not attempted, so far as I know, to show that his Mutualist account of high-energy superconductivity provides a better account than this rival hypothesis. Given the comparative nature of theory appraisal, although Laughlin has outlined a promising Mutualist hypothesis about this case he therefore has yet to successfully establish this hypothesis as the best account, since he has not engaged the strongest relevant rival hypotheses.

At this point it is productive to return to the assessment of Hendry's important claims about the relations of quantum mechanical and chemical phenomena interpreted as a Mutualist hypothesis that I outlined in 10.1.3 above. For Hendry's Mutualist account faces the same issue as Laughlin's. Though skillfully engaging Simple Fundamentalist accounts, Hendry has not engaged rival Conditioned Fundamentalist hypotheses. And Hendry has not

that account either. The dialectical utility of the Conditioned Fundamentalist hypothesis for my purposes, as I outline below, is that it accepts many of Laughlin's key claims but Laughlin plausibly has not engaged it.

provided convincing Mutualist answers to questions Q2/Q2* or engaged the key issue of whether component or composed entities determine the contribution of differential powers. Consequently, Hendry has yet to establish a Mutualist view as the best hypothesis about the relation of chemical and quantum mechanical entities.[16] Nonetheless, Hendry's work again provides a promising basis for future work on a Mutualist account and in this case about phenomena that appeared to most writers to be a stronghold of Fundamentalism.

To conclude, I have shown that we are not, as yet, justified in accepting a Mutualist hypothesis about high-energy superconductivity or the relation of quantum mechanical and chemical phenomena. The lack of scientific consensus about which theories best explain the phenomenon is an issue. But, even if we accept their favored scientific framework, I have outlined how Laughlin, and also Hendry, have still failed to engage relevant rivals. However, despite this negative conclusion, what should again also be striking about the case of high-energy superconductivity, and even that of quantum mechanical and chemical phenomena, is that working scientists, in Laughlin and Woolley, have explicitly claimed Mutualism holds of this case; the scientific evidence may be sufficiently detailed to allow potential resolution of the key issues between this Mutualist hypothesis and its rivals; and it is an open question whether the Mutualist hypothesis provides the best account in the examples.

---

[16] As an aside, I should note that Hendry (2006) claims that, if successful, his arguments about such "approximate" derivations show that what he terms "physicalism" is false. Given my conclusions that a Mutualist scientific emergentism is a form of physicalism one immediately suspects there are theoretical confusions underlying this claim. And this is the case. Hendry (2006), (2010) accepts the explicit, but mistaken, claims of reductionist philosophers such as Kim that the truth of physicalism and the Completeness of Physics are inextricably intertwined. Having provided reasons to question the truth of the Completeness of Physics, through his examination of the practices of quantum chemistry, Hendry thus naturally uses his findings to question the truth of physicalism as well. However, my work in Chapter 7 shows Hendry's final conclusion is flawed, since I showed that claims about the necessary connection between the truth of physicalism and the Completeness of Physics are mistaken. Critiques such as those of Woolley, Laughlin, and Hendry thus do not challenge physicalism and only offer potential succor to Mutualist physicalists, rather than to Anti-Physicalist Ontological emergentists.

10.2.4 *Mutualism Has* Not *(Yet) Been Established Either Locally or Globally*

My survey of exemplar defenses of Mutualist hypotheses shows there are once more the same two kinds of problems with such justifications as we found with Fundamentalist defenses.[17] The first difficulty is that there are often real questions about whether our evidence actually passes Threshold to allow us to productively engage the issue of which local hypothesis provides the best account of the example's structure. The second kind of difficulty once more derives from the failure to appreciate that the strongest rival views. Extant defenses of Mutualism consequently fail to show that we should accept a Mutualist hypothesis over rival Simple or Conditioned fundamentalist accounts. Crucially, extant defenses do not offer the evidence necessary to resolve the disputes with these rivals, although we have now seen that these differences are empirically resolvable and that we may well already have findings that allow the resolution of the key disputes.

I want to emphasize that I have not surveyed all of the writers who appear to have defended Mutualism or even a fraction of the cases to which they point. There are many other scientists who look at further cases where they plausibly defend Strong emergence and Mutualism.[18] And there are also other philosophers than those I have considered

---

[17] At the start of the book, I put to one side the difficult cases of quantum mechanical phenomena and phenomenal consciousness as the focus of defenses of emergence or reduction in the sciences. And my work now supplies reasons to support this move and explains why I have not examined these examples in this chapter. Our scientific accounts of phenomenal consciousness, or quantum mechanical phenomena, plausibly fail condition (a), let alone conditions (b) or (c), of Threshold and hence these are simply not examples where it is presently productive to pursue the debates between Fundamentalism and Mutualism.

[18] For example, some of the other scientists, and the cases they focus on, which we could have considered instead include: Ilya Prigogine and his collaborators' work on thermodynamical systems (Prigogine (1968), (1997)); work on chemical signaling and other systems in slime mold and other organisms (Garfinkel (1987) and Goldbeter (1997)); Freeman's work on neural populations (Freeman (2000a), (2000b)); Couzin and Krause's research on the flocking in vertebrates (Couzin and Krause (2003)) and other examples of biological self-organization of which Camazine et al. (2001) provide a nice survey; and a host of writers from the sciences of complexity, including Kauffman, Prigogine, and his collaborators, and many others (Nicolis and Prigogine (1989), Kauffman (1993), (1995), and the work surveyed in Scott (2007)).

here who have mounted important defenses of Mutualism.[19] However, given the theoretical problems outlined in earlier chapters, I contend that in these extant defenses of Mutualist hypotheses the same two problems outlined with my examples arise once more.

Overall, although my own sympathies lie with Mutualism, my final conclusion is consequently that we are not presently justified in accepting a Mutualist hypothesis about a specific scientific case. Since establishing a Mutualist hypothesis in one scientific example is a necessary condition of establishing a global form of Mutualism, my conclusion is that we thus have yet to establish Mutualism either globally or locally. Although proponents of Mutualism may be disappointed by my negative conclusions, neutral observers will surely be surprised that we are even discussing fine-grained, empirical issues in relation to the success, or failure, of justifications of Mutualism, and the existence of Strong emergence, in concrete scientific cases. The broader message of my survey is thus actually a positive one for Mutualism, which, contrary to the "established" modes of thought, is amongst the live hypotheses about the structure of concrete cases of compositional explanation in the sciences.

10.3 Conclusion: Our Ongoing Battle of the Ages and the Road (Directly) Ahead

There have been far too many heroic solutions in philosophy; detailed work has too often been neglected; there has been too little patience. As was once the case in physics, a hypothesis is invented, and on top of this hypothesis a bizarre world is constructed, there is no effort to compare this world with the real world. The true method, in philosophy as in science, will be inductive, meticulous, and will not believe that it is the duty of every philosopher to solve every problem by himself.

– Bertrand Russell[20]

I began this book with a passage from E. O. Wilson highlighting the metaphysical excitement in recent scientific debates over reduction

---

[19] For instance, I could have looked at the case for Strong emergence made in Harald Atmanspacher and Robert Bishop's research about a number of examples from physics (Atmanspacher and Bishop (2006), Bishop (2005), (2006b)).

[20] Russell (1911), p. 61. I borrow this lovely passage of Russell's from Mulligan, Simons, and Smith (2006).

and emergence. In this final chapter, I have shown that Wilson is dead right. Contrary to the "established modes of thinking" on both sides of the debates, I have shown that our Battle of the Ages is ongoing both locally and globally. Neither side has, as yet, successfully established their favored hypothesis about their chosen scientific examples.

My work in this chapter also means that I can finally remove the clumsy appellation of a "candidate for a live position" that I have been using since Chapter 8. Whether in versions defended by scientists or philosophers, Simple Fundamentalism, Mutualism, and Conditioned Fundamentalism are the three live positions about the structure of cases of successful compositional explanation. And I have highlighted reasons to believe that in some cases we may already have empirical evidence that can be sharpened, using the theoretical machinery I have provided, so that it resolves which of these live views we should accept .

Traditional philosophers may be disappointed that, at the end of a long book, I have not produced a definitive conclusion. However, I take very seriously Russell's advice in the passage starting this section. I agree that naturalistic philosophers must often have patience and must also accept a division of labor. In various chapters, I have now given reasons supporting my starting contention that providing better theory was a necessary step before debates over reduction and emergence could move forward. And I have shown that existing theoretical frameworks for scientific composition, reduction, and emergence have, in one way or another, prevented us from clearly appreciating, and addressing, the nature of our ongoing debates.

Rather than what Russell would rightly characterize as a "heroic" attempt to accomplish all the necessary tasks here, I am therefore happy to have focused on providing the better theoretical frameworks needed to understand the deeper issues, appreciate the live views, and to connect these positions back up to their implications about concrete scientific cases and our findings about them. Perhaps most importantly, my theoretical frameworks provide a potential platform for a new phase of productive research by highlighting the empirical evidence needed to resolve the ongoing disputes. In concluding, let me therefore briefly provide a road map, using my findings to give some very broad guidance about the work that plausibly ought, and ought not, to be the primary focus in the next stretch of research.

In this regard, I suggest we have clear conclusions about debates over the live *global* positions. Philosophers have long been used to spending their time articulating various speculative global positions or theses about the structure of nature. And, as I have detailed, we do plausibly have evidence supporting certain global claims about nature. Thus, for example, there is a plausible case for physicalism understood as the view that everything in nature is either composed by, or identical to, the entities of microphysics. However, my work suggests that it is presently unproductive to pursue debates over which version of physicalism is true through discussions of whether we should accept a global species of Fundamentalism or Mutualism. The underlying difficulty is that I have shown that we cannot establish any of the live global positions until someone justifies our accepting one of the live positions in one or more scientific examples of compositional explanation. But given my findings about the ongoing nature of disputes over specific scientific cases, we therefore cannot presently resolve debates over the correct global position and form of physicalism. We need to patiently put such global debates to one side and focus our attention elsewhere.

Instead, the next stage of the debates over reduction and emergence needs to be focused *locally* just where the important work surveyed in the bulk of this chapter has been conducted – that is, in the careful examination of specific scientific cases of successful compositional explanation. Such work may consist in looking for new empirical evidence that can resolve the key issues between the live positions, or seeking to show that our existing evidence, when marshaled more carefully, can resolve which position provides the best account of some case. In particular, future discussions need to focus on whether differential powers exist in an example and, if they do, which entities determine the contribution of such powers. Such research on particular examples nicely fits Russell's characterization of being both "meticulous" and "inductive," since it involves the detailed task of looking, case by case, at concrete scientific examples. And it is important to emphasize that just this kind of meticulous work on concrete cases must plausibly also be at the center of the next stage of our theoretical discussions. And I have noted a number of places in our new debates where we desperately need more evidence from concrete examples to take our theoretical accounts forward.

My final conclusion is that our new debates over reduction and emergence in nature are very much ongoing. All of our three live views about the structure of nature, locally and globally, are to use Wilson's apt phrase "still walking on metaphysical thin ice" both theoretically and empirically. But it is still very hard not to join Wilson in feeling the excitement consequently gripping the sciences and, by extension, naturalistic philosophy. For in this situation, though the famous curse upon one's enemies is that they should live in interesting times, researchers focused on the structure of nature are far from cursed for they have an array of open empirical and theoretical questions to pursue in the next stage of the debates that beckon both scientists and philosophers alike.

Glossary

Anti-Physicalism – The position that there are entities in nature that are not ultimately composed by the entities of microphysics.

Argument from Composition – An ontological parsimony argument using the applicability of compositional notions to establish the conclusion that we should accept only component entities are either determinative or existent.

Argument from Composition and Completeness – An ontological parsimony argument using the applicability of compositional notions, *plus* the determinative completeness of component entities, to establish the conclusion that we should only accept component entities are either determinative or existent.

Argument from Exhaustion – Argument that in a case of compositional explanation composed entities can be determinative only if they are machretically determinative, where the reasoning works by showing composed entities cannot be determinative solely through other forms of determination than machresis.

Argument from Predicate Indispensability – Famous philosophical objection to Nagelian reductionism in the sciences showing that predicates of higher sciences are indispensable for expressing truths contrary to the implication of Nagelian reduction.

Coadunative, coadunativity – Entities are coadunative when they are necessarily found together.

Collective – A group of individuals usually bearing spatiotemporal, powerful, and/or productive relations to each other.

Collective propensity – The propensity of a group of interrelated individuals, i.e. a collective, to jointly behave in certain ways eventuating in certain results.

Collectivist Ontology – The ontology committed not only to component entities, but also to collective phenomena such as collectives of component individuals and their collective propensities.

Completeness of Physics – The thesis that microphysical entities are determined, in so far as they are determined, solely by other microphysical entities according to the laws or principles of composition covering their behavior in isolation or the simplest collectives.

Composition – See Scientific composition.

Compositional explanation – A scientific explanation of entity Y using entities X1–Xn that works by representing how X1–Xn compose Y.

Comprising – The compositional relation between powers posited in the sciences. See also Scientific composition.

Conditioned Fundamentalism – The position that in a case of compositional explanation the Conditioned view of aggregation is true, components are determinatively complete, and that we should accept that only the relevant component entities are determinative or exist. See Fundamentalism; Conditioned view of aggregation; Determinative completeness.

Conditioned view of aggregation – A picture of aggregation under which the contributions of powers by components in complex collectives is discontinuous with their contributions in simple collectives, components contribute differential powers, and the laws and/or principles of composition holding of components in simpler collectives do not exhaust the laws covering such components in the relevant complex collective.

Constitution – One name for the compositional relation between individuals posited in the sciences. See also Scientific composition.

Determination – Any species of relation between entities that makes a difference to the powers of individuals. See Scientific composition; Machresis; Production; Instantiation.

Determinative – An entity is determinative if it makes a difference to the powers of individuals.

Determinative completeness – The components in some case of compositional explanation are determinatively complete when these components are solely determined by other components (whether at the same or lower levels).

Differential powers – Powers contributed by components in complex collectives that are different from the powers that such components would contribute if the laws and/or principles of composition holding in simpler collectives solely accounted for such contributions of powers.

Emergence, Ontological – A property instance F is Ontologically emergent only if property instance F is a property of a higher-level individual, F is not realized, and F is determinative.

Emergence, Qualitative – A property instance F is Qualitatively emergent only if it is a property of a composed individual not had by any of the constituents of this individual.

Emergence, Strong – A property instance F is Strongly emergent only if F is a property of a composed individual, F is realized by the properties of the individual's constituents, and F is determinative. (One of the main contentions of the book is that a property instance F is Strongly emergent if and only if F is a property of a composed individual realized by the properties of the constituents of this individual and F machretically determines some of the powers contributed by its realizers.)

Emergence, Weak – A property instance F is Weakly emergent only if it is a property of a composed individual that is realized by the properties of the constituents of the individual but such that the laws/explanations/theories about F cannot be derived/predicted from the laws/explanations/theories about the realizers of F.

Entity – A catch-all term referring to any ontological category of thing, thus including powers, properties/relations, individuals, and processes.

Everyday reductionism – The methodology that supplies, and/or looks for, compositional explanations to explain higher-level entities using lower-level entities that compose them.

Exclusivism – The position that debates over reduction and emergence, and/or the structure of nature, can be resolved *solely* using the evidence of fundamental physics.

Fundamentalism – The position that in a case of compositional explanation, or the whole of nature, we should accept that only the relevant component entities are determinative or exist.

Fundamentalist Physicalism – A global view that takes all entities in nature to be identical to, or compositionally explained by, the entities of microphysics, but further claims that only the entities of microphysics are determinative or exist.

Implementation – The compositional relation between processes in the sciences and their compositional explanations. See also Scientific composition.

Inclusivism – The position that debates over reduction and emergence, and/or the structure of nature, can only be resolved using the evidence of an array of higher and lower sciences, rather than being resolvable simply using fundamental physics.

Instantiation –The relation between a powerful property or relation instance and an individual where the instance contributes powers to this individual.

Inter-level explanation – See Compositional explanation.

Joint role-filling – A many–one relation between working entities where none of the relata plays the role individuative of the other relata, but where these relata have roles that together fill the role of the other entity, thus entailing that the relata of joint role-filling are qualitatively distinct.

Machresis, machretic determination – The non-productive determinative relation between composed and component entities such that composed entities determine some of the powers of their components through role-shaping or role-constraining.

Manifest Image – Sellars's famous name for the ontological picture of nature given to us through our commonsense understanding and concepts.

Mass-energy neutral, mass-energy neutrality – A relation is mass-energy neutral when its relata have mass-energy, but the overall mass-energy of the relata equals the combined mass-energy of the entities on one side of the relation.

Mechanistic explanation – A species of compositional explanation explicitly positing compositional relations between processes and hence implementation relations. See also Compositional explanation.

Meta-Explanatory Adequacy Test (MEAT) – A condition of adequacy on theories about nature that they are acceptable only if the ontology of the theory accounts for the success and/or truth of relevant scientific explanations.

Metaphysics for science – An approach to ontological concepts, issues, arguments, and positions in the sciences that seeks to understand such scientific phenomena using theoretical frameworks developed for areas *other than* the sciences.

Metaphysics of science – An approach to understanding ontological concepts, issues, arguments, and positions in the sciences as they arise *within* or *from* the sciences and their explanations, theories, models, findings, etc.

Mutualism – The position that in a case of compositional explanation, or the whole of nature, the Conditioned view of aggregation is true and we have mutually determinative, and interdependent, composed entities that bear both machretic as well as compositional relations to each other.

Mutualist physicalism – A global view that takes all entities in nature to be identical to, or composed by, the entities of microphysics, but also claims that some composed and component entities are mutually determinative, and interdependent, entities that bear both machretic as well as compositional relations to each other.

Non-reductive physicalism – The ontological position that in a case of a compositional explanation, or the whole of nature, there are both higher-level determinative entities and the lower-level determinative entities that compose them.

Non-reductive physicalism, Enriched – A sub-species of non-reductive physicalism that offers an ontological reason to defend its commitment to both higher-level determinative entities and the lower-level determinative entities that compose such higher-level entities.

Non-reductive physicalism, Naked – A widely held sub-species of non-reductive physicalism that does *not* offer an *ontological* reason to defend its commitment to both higher-level determinative entities and the lower-level determinative entities that compose such higher-level entities.

Ontologically Unifying Power (OUP) – The feature of a successful compositional explanation that it shows the entities of its explanans and explanandum, found at different levels, are in some sense the same.

Physicalism – The global view that takes all entities in nature to be identical to, or composed by, the entities of microphysics.

Piercing Explanatory Power (PEP) – The feature of a compositional explanation that it explains entities of one kind at one level in terms of qualitatively distinct kinds of entities at some other level.

Process, productive process, production – A process, production, or productive process is the manifestation of the powers of an individual to result in an effect.

Realization – The compositional relation between property/relations instances posited in the sciences and their compositional explanations. See also Scientific composition.

Role-playing – A one–one relation where one of the relata plays the role individuative of the other relata, thus entailing that the relata are qualitatively similar or the same.

Role-filling – See Joint role-filling.

Scientifically Manifest Image – The account of a compositional explanation, and the whole of nature, that arises from a literal interpretation of intra-level productive and also inter-level compositional explanations in the sciences, hence accepting that there are both higher-level determinative entities and the lower-level determinative entities that compose them.

Scientific Composition – The compositional relations posited between various category of entity in compositional explanations in the sciences.

Scientific reductionism – The position that everyday reductionism and the compositional explanations it supplies show, on reflection, that we should accept only that component entities are either determinative or existent, but combined with the compromising claim that higher sciences and their predicates are indispensable for expressing truths about the collective phenomena in nature. See also Fundamentalism.

Simple Fundamentalism – The position that in a case of compositional explanation the Simple view of aggregation is true, components are determinatively complete, and that in this case we should accept that only the relevant component entities are determinative or exist. See Fundamentalism; Simple view of aggregation; Determinative completeness.

Simple view of aggregation – A picture of aggregation under which the contributions of powers by components in complex collectives is continuous with their contributions in simple collectives and such that the contribution of powers by components is determined solely by other components to the degree it is determined at all.

Working components – Entities involved in compositional relations such that these relations are centered, at least in part, upon the processes associated with the composed and component entities.

Working entities – Entities, whether individuals, powers, or properties, that are individuated, at least in part, by the processes with which they are associated and hence by roles.

Bibliography

Aizawa, K. 2007: "The Biochemistry of Memory Consolidation: A Model System for the Philosophy of Mind." *Synthese*, 155, pp. 65–98.

Aizawa, K., and Gillett, C. 2009a: "Levels, Individual Variation and Massive Multiple Realization in Neurobiology." In Bickle (2009), pp. 539–81.

2009b: "The (Multiple) Realization of Psychological and Other Properties in the Sciences." *Mind & Language*, 24(2), pp. 181–208.

2011: "The Autonomy of Psychology in the Age of Neuroscience." In P. McKay Illari, F. Russo, and J. Williamson (eds.), *Causality in the Sciences*. Oxford: Oxford University Press.

(eds.) 2016: *Scientific Composition and Metaphysical Ground*. London: Palgrave MacMillan.

Forthcoming: *The Parts of Sciences: Scientific Composition and Compositional Explanation*. London: Routledge.

Unpublished: "Multiple Realization and Methodology in Neuroscience and Psychology."

Alexander, S. 1920: *Space, Time and Deity*. 2 vols. The Gifford Lectures 1916–18. Toronto: Macmillan.

Anderson, P. 1972: "More is Different: Broken Symmetry and the Nature of the Hierarchical Structure of Science." *Science*, 177, pp. 393–6.

1990: "Solid-State Experimentalists: Theory Should Be on Tap, Not on Top." *Physics Today*, 43(9), pp. 9–11.

1995: "Historical Overview of the Twentieth Century in Physics." In Brown, Pais, and Pippard (1995), vol. 3, pp. 2017–33.

Andersen, P., Christiansen, P., Emmeche, C., and Finnemann, N. (eds.) 2000: *Downward Causation: Minds, Bodies and Matter*. Aarhus: Aarhus University Press.

Armstrong, D.M. 1997: *A World of States of Affairs*. New York: Cambridge University Press.

Atkins, P. 1995: "The Limitless Power of Science." In Cornwell (ed.) (1995), pp. 122–31.

Atmanspacher, H., and Bishop, R. 2006: "Contextual Emergence in the Description of Properties." *Foundations of Physics*, 36, pp. 1753–77.

Auyang, S. 1998: *Foundations of Complex-Systems Theories in Economics, Evolutionary Biology, and Statistical Physics*. Cambridge: Cambridge University Press.

Ayala, F., and Arp, R. (eds.) 2009: *Contemporary Debates in Philosophy of Biology*. Oxford: Wiley-Blackwell.

Ayala, F., and Dobzhansky, T. (eds.) 1974: *Studies in the Philosophy of Biology: Reduction and Related Problems*. New York: Macmillan.

Bechtel, W., and Mundale, J. 1999: "Multiple Realizability Revisited." *Philosophy of Science*, v.66, pp. 175–207.

Bechtel, W., and Richardson, R. 1993: *Discovering Complexity*. Princeton, NJ: Princeton University Press.

Beckermann, A., Flohr, H., and Kim, J. (eds.) 1992: *Emergence or Reduction?* New York: de Gruyter.

Bedau, M. 1997: "Weak Emergence." *Philosophical Perspectives*, 11, pp. 375–99.

2002: "Downward Causation and the Autonomy of Weak Emergence." *Principia*, 6, pp. 5–50.

2008: "Is Weak Emergence Just in the Mind?" *Minds and Machines*, 18, pp. 443–59.

Bennett, K. 2003: "Why the Exclusion Problem Seems Intractable, and How, Just Maybe, to Tract it." *Noûs*, 37(3), pp. 471–97.

2011: "Construction Area: No Hard Hat Required." *Philosophical Studies*, 154, pp. 79–104.

Bennett, K., and McLaughlin, B. 2005: "Supervenience." *The Stanford Encyclopedia of Philosophy*. <http://plato.stanford.edu/archives/win2003/entries/supervenience/>.

Bertalanffy, V. 1950: "An Outline of General System Theory." *British Journal for the Philosophy of Science*, 1, pp. 134–65.

Bickle, J. 1998: *Psychoneural Reduction: The New Wave*. Cambridge, MA: MIT Press.

2003: *Philosophy and Neuroscience: A Ruthlessly Reductive Account*. Boston: Kluwer.

2006: "Reducing Mind to Molecular Pathways." *Synthese*, v.151, pp. 411–34

2008: "Real Reduction in Real Neuroscience: Metascience, Not Philosophy of Science (and Certainly Not Metaphysics!)." In Hohwy and Kallestrup (2008), pp. 34–51.

(ed.) 2009: *The Oxford Handbook of Philosophy and Neuroscience*. Oxford: Oxford University Press.

Bishop, R. 2005: "Patching Physics and Chemistry Together." *Philosophy of Science*, 72, pp. 710–22.

2006a: "The Hidden Premise in the Causal Argument for Physicalism." *Analysis*, 66, pp. 44–52.

2006b: "Downward Causation in Fluid Convection." *Synthese*, 160, pp. 229–48.

Blitz, D. 1992: *Emergent Evolution*. New York: Kluwer.

Boogerd, F., Bruggeman, F., Hofmeyr, S., and Westerhoff, H. 2007: *Systems Biology: Philosophical Foundations*. Amsterdam: Elsevier.

Boogerd F., Bruggeman F., Richardson, R., Stephan, A., and Westerhoff, H. 2005: "Emergence and its Place in Nature: A Case Study of Biochemical Networks." *Synthese*, 145, pp. 131–64.

Borchert, D. (ed.) 2006: *Encyclopedia of Philosophy*. 2nd edition. New York: Gale Research.

Bowler, P. 2001: *Reconciling Science and Religion: The Debate in Early Twentieth-Century Britain*. Chicago, IL: University of Chicago Press.

Broad, C. D. 1925: *The Mind and its Place in Nature*. London: Routledge Kegan Paul.

Brown, J. 1991: *Laboratory of the Mind: Thought Experiments in the Natural Sciences*. London: Routledge.

2004: "Why Thought Experiments Transcend Experience." In Hitchcock (2004), pp. 23–43.

Brown, L., Pais, A., and Pippard, B. (eds.) 1995: *Twentieth Century Physics*. Second Edition. New York: CRC Press.

Brown, W., Murphy, N., and Malony, H. 1998: *Whatever Happened to the Soul?* Minneapolis, MN: Augsburg Fortress Press.

Camazine, S., Deneubourg, J., Franks, N., Sneyd, J., Theraulaz, G., and Bonabeau, E. 2001: *Self-Organization in Biological Systems*. Princeton, NJ: Princeton University Press.

Campbell, D. 1974: "'Downward Causation' in Hierarchically Organized Biological Systems." In Ayala and Dobzhansky (1974), pp. 179–86.

Cartwright, N. 1983: *How the Laws of Physics Lie*. Oxford: Clarendon Press.

1994: "Fundamentalism vs. the Patchwork of Laws." *Proceedings of the Aristotelian Society*, 103, pp. 279–92.

1999: *The Dappled World*. New York: Cambridge University Press.

Cat, J. 1998: "The Physicists' Debates on Unification in Physics at the End of the 20th Century." *Historical Studies in the Physical Sciences*, 28, pp. 253–99.

Chalmers, D. 1997: *The Conscious Mind*. New York: Oxford University Press.

2006: "Strong and Weak Emergence." In Clayton and Davies (2004, pp. 244–54.

Churchland, P.M. 1985: "Reduction, Qualia and the Direct Introspeciton of Brain States." *Journal of Philosophy*, v.82, pp. 8–28.

Churchland, P.S. 1986: *Neurophilosophy*: Toward a Unified Science of the Mind-Brain. Cambridge, MA: MIT Press.

Clapp, L. 2001: "Disjunctive Properties: Multiple Realizations." *Journal of Philosophy*, 98, pp. 111–36.

Clayton, P. 1997: *God and Contemporary Science*. Grand Rapids, MI: Eerdmans.

2004: *Mind and Emergence*. Oxford: Oxford University Press.

(ed.) 2006: *Oxford Handbook of Religion and Science*. Oxford: Oxford University Press.

Clayton, P., and Davies, P. (eds.) 2004: *The Re-Emergence of Emergence*. Oxford: Oxford University Press.

Corning, P. 2002: "The Re-Emergence of 'Emergence': A Venerable Concept in Search of a Theory." *Complexity*, 7, pp. 18–30.

Cornwell, J. (ed.) 1995: *Nature's Imagination*. New York Oxford University Press.

Corradini, A., and O'Connor, T. (eds.) 2011: *Emergence in Science and Philosophy*. New York: Routledge.

Couzin, I., and Krause, J. 2003: "Self-Organization and Collective Behavior in Vertebrates." *Advances in the Studies of Behavior*, 33, pp. 1–75.

Crane, T. 2001: *The Elements of Mind*. Oxford: Oxford University Press.

Craver, C. 2007: *Explaining the Brain*. New York: Oxford University Press.

Craver, C., and Bechtel, B. 2007: "Top-Down Causation without Top-Down Causes." *Biology and Philosophy*, 22, pp. 547–63.

Craver, C., and Darden, L. 2001: "Discovering Mechanisms in Neurobiology: The Case of Spatial Memory." In P. Machamer et al. (eds.), *Theory and Method in Neuroscience*. Pittsburgh, PA: University of Pittsburgh Press.

Crick, F. 1966: *Of Mice and Molecules*.   Seattle, WA: University of Washington.

1994: *The Astonishing Hypothesis: The Scientific Search for the Soul*. New York: Scribner.

Crook, S., and Gillett, C. 2001: "Why Physics Alone Cannot Define the 'Physical'." *Canadian Journal of Philosophy*, 31, pp. 333–60.

Crutchfield, J., Farmer, J., Packard, N., and Shaw, R. 1986: "Chaos." *Scientific American*, 255, pp. 46–57.

Cummins, R. 1975: "Functional Analysis." *Journal of Philosophy*, 72, pp. 741–65.

1983: *The Nature of Psychological Explanation*. Cambridge, MA: MIT Press.

Darden, L., and Maull, N. 1977: "Interfield Theories." *Philosophy of Science*, 44, pp. 43–64.

Dawkins, R. 1982: *The Extended Phenotype*. San Francisco, CA: W. H. Freeman.

1985: "Editorial." In Dawkins and Ridley (1985).

1987: *The Blind Watchmaker*. New York: Norton.

Dawkins, R., and Ridley, M. 1985: *Oxford Surveys in Evolutionary Biology*, *vol. 2*. New York: Oxford University Press.

Delehanty, M. 2005: "Emergent Properties and the Context Objection to Reduction." *Biology and Philosophy*, 20, pp. 715–34.

Dennett, D. 1969: *Content and Consciousness*. London: Routledge Kegan Paul.

1978: *Brainstorms*. Montgomery, VT: Bradford Books.

1991: "Real Patterns." *Journal of Philosophy*, 88, pp. 27–51.

1996: *Darwin's Dangerous Idea*. New York: Simon and Schuster.

Dupré, J. 1993: *The Disorder of Things: Metaphysical Foundations of the Disunity of Science*. Boston, MA: Harvard University Press.

2001: *Human Nature and the Limits of Science*. New York: Oxford University Press.

2008: *The Constituents of Life*. Assen: Van Gorcum.

2009: "It Is Not Possible to Reduce Biological Explanations to Explanations in Chemistry and/or Physics." In Varela and Arp (2009), pp. 32–47.

Dyke, C. 1988: *The Evolutionary Dynamics of Complex Systems: A Study in Biological Complexity*. New York: Oxford University Press.

Elkana, Y. 1974: *The Discovery of the Conservation of Energy*. London: Hutchinson.

Ellis, B. 2002: *The Philosophy of Nature*. Montreal: McGill University Press.

Endicott, R. 2006: "Multiple Realizability." In Borchert (2006).

Esfeld, M., and Sachse, C. 2011: *Conservative Reductionism*. New York: Routledge.

Fodor, J. 1968: *Psychological Explanation*. New York: Random House.

1974: "Special Sciences: Or, the Disunity of Science as a Working Hypothesis." *Synthese*, 28, pp. 97–115.

1975: *The Language of Thought*. New York: Crowell.

1997: "Special Sciences: Still Autonomous after All These Years." *Philosophical Perspectives*, 11.

1998: "Look! Review of E.O.Wilson's *Consilience*." *London Review of Books*, v.20, No.21, 29 October, pp. 3–6.

Foster, J. 1991: *The Immaterial Self: A Defence of the Cartesian Dualist Conception of the Mind*. London: Routledge.

Freeman, W. 1995: *Societies of Brains*. Mahwah, NJ: Erlbaum.

1999: "Consciousness, Intentionality and Causality." In Nunez and Freeman (1999), pp. 143–72.

2000a: *How Brains Make Up Their Minds*. New York: Columbia University Press.

2000b: *Neurodynamics: An Exploration of Mescopic Brain Dynamics*. London: Springer-Verlag.

Garfinkel, A. 1987: "The Slime Mold *Dictyostelium* as a Model of Self-Organization in Social Systems." In Yates (1987), pp. 181–212.

Gilbert, S., and Sarkar, S. 2000: "Embracing Complexity: Organicism for the 21st Century." *Developmental Dynamics*, 219, pp. 1–9.

Gillett, C. 2002a: "The Varieties of Emergence: Their Purposes, Obligations and Importance." *Grazer Philosophische Studien*, 65, pp. 89–115.

2002b: "The Dimensions of Realization: A Critique of the Standard View." *Analysis,* 62, pp. 316–23.

2003a: "Non-Reductive Realization and Non-Reductive Identity: What Physicalism Does Not Entail." In S. Walter and H.-D. Heckmann (eds.) *Physicalism and Mental Causation.* Bowling Green, OH: Imprint Academic, pp. 31–58.

2003b: "Strong Emergence as a Defense of Non-Reductive Physicalism." Special issue on emergence. *Principia*, 6, pp. 83–114.

2003c: "The Metaphysics of Realization, Multiple Realizability and the Special Sciences." *Journal of Philosophy*, 100, pp. 591–603.

2006a: "Samuel Alexander's Emergentism: Or, Higher Causation for Physicalists." *Synthese,* 153, pp. 261–96.

2006b: "The Hidden Battles over Emergence." In Clayton (1996).

2006c: "Special Sciences." In Borchert (2006).

2006d: "The Metaphysics of Mechanisms and the Challenge of the New Reductionism." In Schouten and de Joong (2006).

2007a: "Understanding the New Reductionism: The Metaphysics of Science and Compositional Reduction." *The Journal of Philosophy*, 104, pp. 193–216.

2007b: "A Mechanist Manifesto for the Philosophy of Mind: A Third Way for Functionalists." *Journal of Philosophical Research*, 32, pp. 21–42.

2007c: "Hyper-Extending the Mind? Setting Boundaries in the Special Sciences." *Philosophical Topics*, 351, pp. 161–88.

2010: "Moving Beyond the Subset Model of Realization." *Synthese*, 177, pp. 165–92.

2011a: "Multiply Realizing Scientific Properties and their Instances: A Response to Polger and Shapiro." *Philosophical Psychology*, v.24, pp. 727–38.

2011b: "On the Implications of Scientific Composition and Completeness: Or, the Troubles, and *Troubles*, of Non-Reductive Physicalism." In Corradini and O'Connor (2011).

2012: "Understanding the Sciences through the Fog of 'Functionalism(s)'." In Hunneman (2012), pp. 159–81.

2013: "Constitution, and Multiple Constitution, in the Sciences: Using the Neuron as a Guide." *Minds and Machines*, v.23, pp. 309–37.

2016: "The Metaphysics of Nature, Science and the Rules of Engagement" in Aizawa and Gillett (2016).

Unpublished: "Making Sense of Levels in the Sciences."

Gillett, C., and Loewer, B. (eds.) 2001: *Physicalism and its Discontents*. Cambridge: Cambridge University Press.

Gillett, C., and Rives, B. 2001: "Does the Argument from Realization Generalize? Responses to Kim." *Southern Journal of Philosophy*, 39, pp. 79–98.

Gillett, C., and Rives, B. 2005: "The Non-Existence of Determinables: Or, a World of Absolute Determinates as Default Hypothesis." *Nous*, 39, pp. 483–504.

Glennan, S. 1996: "Mechanisms and the Nature of Causation." *Erkenntnis*, 44, pp. 49–71.

Goldbeter, A. 1997: *Biochemical Oscillations and Cellular Rhythms: The Molecular Basis of Periodic and Chaotic Behavior*. Cambridge: Cambridge University Press.

Goldbeter, A., Gonze, D., Houart, G., Leloup, J., Halloy, J., and Dupont, G. 2001: "From Simple to Complex Oscillatory Behavior in Metabolic and Genetic Control Network." *Chaos*, 11, pp. 247–60.

Goldstein, J. 1999: "Emergence as a Construct: History and Issues." *Emergence*, 1, pp. 49–72.

Goodwin, B. 2001: *How the Leopard Changed its Spots: The Evolution of Complexity*. Princeton, NJ: Princeton University Press.

Gregory R., 1994: "DNA in the Mind's Eye". *Nature*, 368, pp. 359–60.

Groff, R., and Greco, J. 2012: *Powers and Capacities in Philosophy: The New Aristotelianism*. New York: Routledge.

Hall, N. 2004: "Two Concepts of Causation." In J. Collins, N. Hall and L. Paul (eds.) *Causation and Counterfactuals*, pp. 225–76. Cambridge, MA: The MIT Press.

Haraway, D. 1976: *Crystals, Fabrics and Fields*. New Haven, CT: Yale University Press.

Harold, F. 2001: *The Way of the Cell*. New York: Oxford University Press.

Hasker, W. 1999: *The Emergent Self*. Ithaca, NY: Cornell University Press.

Healey, R. 2013: "Physical Composition." *Studies in the History and Philosophy of Science Part B*, 44, pp. 48–62.

Heil, J. 1999: "Multiple Realizability." *American Philosophical Quarterly*, 36, pp. 189–208.

2003: *From an Ontological Point of View*. New York: Oxford University Press.

Hempel, C. 1965: *Aspects of Scientific Explanation and Other Essays in the Philosophy of Science*. New York: Free Press.

Hempel, C., and Oppenheim, P. 1965: "On the Idea of Emergence." In Hempel (1965), pp. 258–64.

Hendry, R. 1998: "Models and Approximations in Quantum Chemistry." In N. Shanks (ed.), *Idealization in Contemporary Physics*. Amsterdam: Rodopoi.

2006: "Is There Downward Causation in Chemistry?" In Baird et al. (eds.) *Philosophy of Chemistry*, pp. 173–89. Dordrecht: Springer.

2010: "Emergence vs. Reduction in Chemistry." In MacDonald and MacDonald (2010a), pp. 205–21.

Hendry, R., and Needham, P. 2007: "Le Poidevin on the Reduction of Chemistry." *British Journal for the Philosophy of Science*, 58, pp. 339–53.

Hitchcock, C. (ed.) 2004: *Contemporary Debates in the Philosophy of Science*. Malden, MA: Blackwell.

Hohwy, J., and Kallestrup, J. (eds.) 2008: *Being Reduced: New Essays on Reduction, Explanation and Causation*. New York: Oxford University Press.

Holland, J. 1999: *Emergence: From Chaos to Order*. New York: Basic Books.

Holldobler, B., and Wilson, E. 1990: *The Ants*. Cambridge, MA: Harvard University Press.

2009: *The Superorganism: The Beauty, Elegance and Strangeness of Insect Societies*. New York: Norton

Hooker, C. A. 1981: "Towards a General Theory of Reduction. Part I: Historical and Scientific Settings. Part II: Identity. Part III: Cross Categorial Reduction." *Dialogue*, 20, pp. 38–59, 201–36, 496–529.

Horgan, J. 1997: *The End of Science*. New York: Broadway Books.

Horgan, T. 1993a: "From Supervenience to Superdupervenience: Meeting the Demands of a Material World." *Mind*, 102, pp. 555–86.

1993b: "Nonreductive Materialism and the Explanatory Autonomy of Psychology." In Wagner, S. and Warner, R. (eds.) *Naturalism*. South Bend: Notre Dame University Press.

Horst, S. 2007: *Beyond Reduction: Philosophy of Mind and Post-Reductionist Philosophy of Science*. New York: Oxford University Press.

Howard, D. 2007: "Reduction and Emergence in the Physical Sciences: Some Lessons from the Particle Physics and Condensed Matter Debate." In Murphy and Stoeger (2007).

Huemer, M. 2009: "When Is Parsimony a Virtue?" *Philosophical Quarterly*, 59, pp. 216–36.

Hull, D. 2002: "Varieities of Reductionism." In Regenmortel and Hull (2002), pp. 161–72.

Humphreys, P. 1997: "How Properties Emerge". *Philosophy of Science*, 64, pp. 1–17.

2008a: "Synchronic and Diachronic Emergence." *Minds and Machines,* 18, pp. 431–42.

2008b: "Computational and Conceptual Emergence." *Philosophy of Science,* 75, pp. 584–94.

Hunneman, P. (ed.) 2012: *Functions: Selection and Mechanisms.* Dordrecht: Kluwer.

Imbert, C. 2007: "Why Diachronically Emergent Properties Must Also Be Salient." In Gershenson, Aerts, and Edmonds (eds.), *Worldviews, Science and Us,* pp. 99–116. Hackensack, NJ: World Scientific Publishing Company.

Juarrero, A. 1999: *Dynamics in Action: Intentional Behavior as a Complex System.* Cambridge, MA: MIT Press.

Kandel, E. 2006: *In Search of Memory.* New York: Norton.

Kauffman, S. 1993: *The Origins of Order: Self-Organization and Selection in Evolution.* New York: Oxford University Press.

1995: *At Home in the Universe.* New York: Oxford University Press.

Kauffman, S., and Clayton, P. 2006: "On Emergence, Agency and Organization." *Biology and Philosophy,* 21, pp. 501–21.

Kellert, S. 1993: *In the Wake of Chaos.* Chicago, IL: University of Chicago Press.

Kim, J. 1972: "Phenomenal Properties, Psychological Laws, and Identity Theory." *Monist,* 56, pp. 177–92.

1992a: "'Downward Causation' in Emergentism and Nonreductive Physicalism." In Beckermann, Flohr, and Kim (1992).

1992b: "Multiple Realization and the Metaphysics of Reduction." *Philosophy and Phenomenological Research,* 52, pp. 1–26.

1993a: *Supervenience and Mind.* Cambridge: Cambridge University Press.

1993b: "The Nonreductionist's Troubles with Mental Causation." In Kim (1993a).

1993c: "Dretske on How Reasons Explain Behaviour." In Kim (1993a).

1993d: "Postscripts on Supervenient Causation." In Kim (1993a).

1993e: "Mechanism, Purpose and Explanatory Exclusion." In Kim (1993a).

1997: "The Mind–Body Problem: Taking Stock after Forty Years." *Philosophical Perspectives,* 11.

1998: *Mind in a Physical World.* Cambridge, MA: MIT Press.

1999: "Making Sense of Emergence." *Philosophical Studies,* 95, pp. 3–44.

2003: "Blocking Causal Drainage and Other Maintenance Chores with Mental Causation." *Philosophy and Phenomenological Research,* 67, pp. 151–76.

2006: "Being Realistic about Emergence." In Clayton and Davies (2004), pp. 189–202

Kistler, M. 2010: "Mechanisms and Downward Causation." *Philosophical Psychology*, 22, pp. 595–609.

Kitcher, P. 1984: "1953 and All That: A Tale of Two Sciences." *Philosophical Review*, 93, pp. 335–73. Reprinted in his (2003). All references are to the reprint.

    2003: *In Mendel's Mirror*. New York: Oxford University Press.

Kuhn, T. 1970: *The Structure of Scientific Revolutions*. Chicago, IL: University of Chicago Press.

Ladyman, J., and Ross, D. 2007: *Everything Must Go*. New York: Oxford University Press.

Laughlin, R. 2005: *A Different Universe: Reinventing Physics from the Bottom Down*. New York: Basic Books.

Laughlin, R., and Pines, D. 2000: "The Theory of Everything." *Proceedings of the National Academy of Science*, 97, pp. 28–31.

Laughlin, R., Pines, D., Schmalien, J., Stojkovic, B., and Wolynes, P. 2000: "The Middle Way," *Proceedings of the National Academy of Science*, 97, pp. 32–7.

Levins, R., and Lewontin, R. 1987: *The Dalectical Biologist*. Cambridge, MA: Harvard University Press.

Lewin, R. 1992: *Complexity: Life at the Edge of Chaos*. New York: Macmillan.

Lewis, David 1972: "Psychophysical and Theoretical Identifications." *Australasian Journal of Philosophy*, v.50, pp. 249–58.

    1983: "New Work for a Theory of Universals." *Australasian Journal of Philosophy*, v.61, pp. 343–77.

Loewer, B. 1995: "An Argument for Strong Supervenience." In Savellos and Yalcin (1995).

    2001: "From Physics to Physicalism." In Gillett and Loewer (eds.) (2001).

    2008: "Why There *Is* Anything Except Physics." In Hohwy and Kallestrup (2008).

Lowe, J. 1996: *The Subjects of Experience*. Cambridge: Cambridge University Press.

    2008: *Personal Agency: The Metaphysics of Mind and Action*. New York: Oxford University Press.

    2009: "Dualism." In McLaughlin, Beckermann, and Walter (2009), pp. 66–84.

Machamer, P., Darden, L., and Craver, C. 2000: "Thinking about Mechanisms." *Philosophy of Science*, 67, pp. 1–25.

Maturana, H., and Varela, F. 1973: "Autopoiesis and Cognition." In R. S. Cohen and M. W. Wartofsky (eds.), *Boston Studies in the Philosophy of Science*, vol. 10, p. 42.

Mayr, E. 1988a: "The Limits of Reductionism." *Nature*, 33.

1988b: *Toward a New Philosophy of Biology*. Cambridge, MA: Harvard University Press.

2002: *What Makes Biology Unique?* Cambridge: Cambridge University Press.

MacDonald, C., and MacDonald, G. (eds.) 2010a: *Emergence in Mind*. New York: Oxford University Press.

2010b: "Emergence and Downward Causation." In MacDonald and MacDonald (2010a), pp. 139–68.

McGinn, C. 1993: *The Problem of Consciousness: Essays Towards a Resolution*. Oxford: Blackwell.

McLaughlin, B. 1992: "The Rise and Fall of British Emergentism." In Beckermann, Flohr, and Kim (1992).

1995: "The Varieties of Supervenience." In Savellos and Yalcin (1995).

McLaughlin, B., Beckermann, A., and Walter, S. 2009: *The Oxford Handbook of Philosophy of Mind*. New York: Oxford University Press.

Mellor, H. 2008: "Micro-Composition". Royal Institute of Philosophy Supplements, v.83, pp.81-106.

Melnyk, A. 1995: "Two Cheers for Reductionism; Or, The Dim Prospects for Nonreductive Materialism." *Philosophy of Science*, 62, pp. 370–88.

1997: "How to Keep the 'Physical' in Physicalism." *Journal of Philosophy*, 94, pp. 622–37.

2003: *A Physicalist Manifesto*. New York: Cambridge University Press.

Merricks, T. 2001: *Objects and Persons*. New York: Oxford University Press.

Midgley, M. 1995: "Reductive Megalomania." In Cornwell (1995), pp. 133–47.

Mitchell, S. 2003: *Biological Complexity and Integrative Pluralism*. Cambridge: Cambridge University Press.

2009: *Unsimple Truths: Science, Complexity and Policy*. Chicago, IL: University of Chicago Press.

2012: "Emergence: Logical, Functional and Dynamical." *Synthese*, v.185, pp. 171–86.

Morowitz, H. 2002: *The Emergence of Everything: How the World Became Complex*. New York: Oxford University Press.

Mulligan, K., Simons, P., and Smith, B. 2006: "What's Wrong with Contemporary Philosophy?" *Topoi*, 25, pp. 63–7.

Murphy, N., and Brown, W. 2007: *Did My Neurons Make Me Do It?* New York: Oxford University Press.

Murphy, N., and Stoeger, W. 2007: *Evolution and Emergence: Systems, Organisms, Persons*. New York: Oxford University Press.

Nagel, E. 1952: "Wholes, Sums and Organic Unities." *Philosophical Studies*, 2, pp. 16–32.

1961: *The Structure of Science*. New York: Harcourt Brace.

Newman, D. 1996: "Emergence and Strange Attractors." *Philosophy of Science*, 63, pp. 245–61.

Newton-Smith, W. 1981: *The Rationality of Science*. London: Routledge & Kegan Paul.

Nicolis, G., and Prigogine, I. 1989: *Exploring Complexity*. New York: Freeman.

Norton, J., 2004: "Why Thought Experiments Do Not Transcend Empiricism." In Hitchcock (2004), pp. 44–66.

Nunez, R., and Freeman, W. (eds.) 1999: *Reclaiming Cognition: The Primacy of Action, Intention and Emotion*. Bowling Green, OH: Imprint Academic.

O'Connor, T. 1994: "Emergent Properties." *American Philosophical Quarterly*, 31, pp. 91–104.

O'Connor, T., and Churchill, J. 2010: "Is Non-Reductive Physicalism Viable within a Causal Powers Metaphysics?" In MacDonald and MacDonald (2010a), pp. 43–60.

O'Connor, T., and Jacobs, J. 2003: "Emergent Individuals." *Philosophical Quarterly*, 53, pp. 540–55.

O'Connor, T., and Wong, H. 2003: "The Metaphysics of Emergence." *Noûs*, 39, pp. 659–79.

O'Malley, M., and Dupré, J. 2005: "Fundamental Issues in Systems Biology." *BioEssays*, 27, pp. 1270–6.

Papineau, D. 1993: *Philosophical Naturalism*. Oxford: Basil Blackwell.

1995: "Arguments for Supervenience and Physical Realization." In Savellos and Yalcin (1995).

2001: "The Rise of Physicalism." In Gillett and Loewer (2001).

2002: *Thinking about Consciousness*. New York: Oxford University Press.

Pattee, H. 1970: "The Problem of Biological Hierarchy." In Waddington (1970), vol. 3, pp. 117–36.

1973a: "The Physical Basis and Origin of Hierarchical Control." In Pattee (1973b).

(ed.) 1973b: *Hierarchy Theory*. New York: George Braziller.

Peacocke, A. 1990: *Theology for a Scientific Age*. Oxford: Basil Blackwell.

1995: "God's Interaction with the World." In Russell et al. (1995).

1999: "The Sound of Sheer Silence: How Does God Communicate with Humanity?" In Russell et al. (1999).

Pereboom, D. 2011: *Consciousness and the Prospects of Physicalism*. New York: Oxford University Press

Pfeifer, J., and Sarkar, S. 2006: *Philosophy of Science: An Encyclopedia*. New York: Routledge.

Piccinini, G. 2004: "Functionalism, Computationalism, and Mental States." *Studies in the History and Philosophy of Science*, 35, pp. 811–33.

Poland, J. 1994: *Physicalism*. Oxford: Clarendon Press.

Polanyi, M. 1968: "Life's Irreducible Structure." *Science*, 160, pp. 1308–12.

Polger, T. 2007: "Realization and the Metaphysics of Mind." *Australasian Journal of Philosophy*, 85, pp. 233–59.

Prigogine, I. 1968: *Introduction to Thermodynamics of Irreversible Processes*. New York: Wiley.

1997: *End of Certainty*. New York: The Free Press.

Prigogine, I., and Stengers, I. 1984: *Order out of Chaos: Man's New Dialogue with Nature*. New York: Bantam Books.

Primas, H. 1983: *Chemistry, Quantum Mechanics, and Reductionism*. Berlin: Springer- Verlag.

Putnam, H. 1967: "Psychological Predicates." In W. Capitan and D. Merrill (eds.), *Art, Mind and Religion*. Pittsburgh: Pittsburgh University Press. Reprinted as "The Nature of Mental States" in Putnam (1975b). (All references are to the reprint.)

1975a: "Philosophy and Our Mental Life." In Putnam (1975b).

1975b: *Mind, Language and Reality: Philosophical Papers, vol. 2*. Cambridge: Cambridge University Press.

Regenmortel, M., and Hull, D. 2002: *Promises and Limits of Reductionism in the Biomedical Sciences*. Chichester: John Wiley and Sons.

Richardson, R., and Stephan, A. 2007: "Mechanisms and Mechanical Explanation in Systems Biology." In Boogerd et al. (2007), pp. 123–44.

Rosenberg, A. 2006: *Darwinian Reductionism: Or, How to Stop Worrying and Love Molecular Biology*. Chicago, IL: Chicago University Press.

Ross, D., and Spurrett, D. 2004: "What to Say to a Skeptical Metaphysician? A Defense Manual for Cognitive and Behavioral Scientists." *Behavioral and Brain Sciences*, 27, pp. 603–27.

Rueger, A. 2000: "Robust Supervenience and Emergence." *Philosophy of Science*, 67, pp. 466–91.

Russell, B. 1911: "Le Réalisme Analytique." *Bulletin de la Société Française de Philosophie*, 11, pp. 53–61.

Russell, J., Murphy, N., Meyering, T., and Arbib, A. (eds.) 1999: *Neuroscience and the Person: Scientific Perspectives on Divine Action*. Vatican City State: Vatican Observatory Publications.

Russell, J., Murphy, N., and Peacocke, A. (ed.) 1995: *Chaos and Complexity: Scientific Perspectives on Divine Action*. Vatican City State: Vatican Observatory Publications.

Salthe, S. 1985: *Evolving Hierarchical Systems: Their Representation and Structure*. New York: Columbia University Press.

Sarkar, S. 2005: *Molecular Models of Life*. Cambridge, MA: MIT Press.

Savellos, E., and Yalcin, U. (eds.) 1995: *Supervenience: New Essays.* Cambridge: Cambridge University Press.

Scerri, E. 1994: "Has Chemistry Been at Least Approximately Reduced to Quantum Mechanics?" In D. Hull, M. Forbes, and R. M. Burian (eds.), *PSA* vol. 1, pp. 160–70.

1997: "Has the Periodic Table Been Successfully Axiomitized?" *Erkentnis,* 47, pp. 229–43.

1998a: "Popper's Naturalized Approach to the Reduction of Chemistry." *International Studies in the Philosophy of Science,* 12, pp. 33–44.

1998b: "How Good is the Quantum Mechanical Explanation of the Periodic Table?" *Journal of Chemical Education,* 75, pp. 1384–5.

2007: "Reduction and Emergence in Chemistry: Two Recent Approaches." *Philosophy of Science,* 74, pp. 920–31.

Schaffer, J. 2007: "From Nihilism to Monism." *Australasian Journal of Philosophy,* 85, pp. 175–91.

2010: "Monism: The Priority of the Whole." *Philosophical Review,* 119, pp. 31–76.

Schaffner, K. 1967: "Approaches to Reduction". *Philosophy of Science,* v.34, pp. 137–47.

1993: *Discovery and Explanation in Biology and Medicine.* Chicago, IL: University of Chicago Press.

2002: "Reductionism, Complexity and Molecular Medicine: Genetic Chips and the 'Globalization' of the Genome." In Regenmortel and Hull (2002), pp. 323–51.

2006: "Reduction: The Cheshire Cat Problem and a Return to Roots." *Synthese,* 151, pp. 377–402.

Schouten, M., and de Joong, H. 2006: *The Matter of Mind: Philosophical Essays on Psychology, Neuroscience and Reduction.* Oxford: Blackwell.

Scott, A. 2007: *The Non-Linear Universe: Chaos, Emergence, Life.* New York: Springer.

Seeley, T. 1989: "Social Foraging in Honey Bees: How Nectar Foragers Assess Their Colony's Nutritional Status." *Behavioral Ecology and Sociobiology,* 24, pp. 181–99.

1995: *The Wisdom of the Hive: The Social Physiology of Honey Bee Colonies.* Cambridge, MA: Harvard University Press.

Sellars, W. 1963: "Philosophy and the Scientific Image of Man." In R. Colodny (ed.), pp. 35–78, *Science, Perception, and Reality.* New York: Humanities Press/Ridgeview.

Shapiro, L. 2000: "Multiple Realizations". *Journal of Philosophy,* 97, pp. 635–54.

2004: *The Mind Incarnate.* Cambridge, MA: MIT Press.

2010: "Lessons from Causal Exclusion." *Philosophy and Phenomenological Research*, 88, pp. 594–604.

2011: "Mental Manipulations and the Problem of Causal Exclusion." *Australasian Journal of Philosophy*, 90, pp. 1–18.

Shoemaker, S. 1980: "Causality and Properties." In Van Inwagen (1980).

2001: "Realization and Mental Causation." In Gillett and Loewer (2001).

2003a: "Realization, Micro-Realization and Coincidence." *Philosophy and Phenomenological Research*, 67, pp. 1–23.

2003b: "Kim on Emergence." *Philosophical Studies*, 108, pp. 53–63.

2007: *Physical Realization*. New York: Oxford University Press.

Siderits, M. 2007: *Buddhism as Philosophy: An Introduction*. Cambridge: Hackett.

Silberstein, M., and McGeever, J. 1999: "The Search for Ontological Emergence." *The Philosophical Quarterly*, 49, pp. 183–200.

Simon, H. 1969: *The Sciences of the Artificial*. Cambridge, MA: MIT Press.

Sklar, L. 1967: "Types of Inter-Theoretic Reduction." *The British Journal for the Philosophy of Science*, 18, pp. 109–24.

1995: *Physics and Chance: Philosophical Issues in the Foundations of Statistical Mechanics*. New York: Cambridge University Press.

Sorenson, A., Busch, T., and Vinson, S. 1985: "Control of Food Influx by Temporal Subcastes in the Fire Ant, Solenopsis invicta." *Behavioral Ecology and Sociobiology*, 17, pp. 191–8.

Sorger, P. 2005: "A Reductionist's Systems Biology." *Current Opinion in Cell Biology*, 17, pp. 9–11.

Sperry, R. 1986: "Macro-Determinism vs. Microdeterminism." *Philosophy of Science*, 53, pp. 265–70.

1992: "Turnabout on Consciousness: A Mentalist View." *Journal of Mind and Behaviour*, 13, pp. 259–80.

Spurrett, D. 1999: *The Completeness of Physics*. Phd thesis: University of Natal.

Stephan, A. 1999: "Varieties of Emergentism." *Evolution and Cognition*, 5, pp. 49–59.

Stump, E. 2012: "Emergence, Causal Powers, and Aristotelianism in Metaphysics." In Groff and Greco (2012).

Teller, P. 1992: "A Contemporary Look at Emergence." In Beckermann et al. (1992), pp. 139–53.

Van Fraassen, B. 1980: *The Scientific Image*. Oxford: Clarendon Press.

Van Gulick, R. 1993: "Who's in Charge Here? And Who's Doing All the Work?" In Heil and Mele (eds.) *Mental Causation*. New York: Clarendon Press.

2001: "Reduction, Emergence and Other Options on the Mind/Body Problem." *Journal of Consciousness Studies,* 8, pp. 1–34.

Van Inwagen, P. (ed.) 1980: *Time and Cause.* Dordrecht: Reidel.

1990: *Material Beings.* Ithaca, NY: Cornell University Press.

Waddington, C. 1970: *Towards a Theoretical Biology.* 4 vols. Chicago, IL: Aldine Publishing.

Weinberg, S. 1992: *Dreams of a Final Theory.* New York: Random House.

2001: *Facing Up: Science and its Cultural Adversaries.* Cambridge, MA: Harvard University Press.

Weiner, N. 1948: *Cybernetics or Control and Communication in the Animal and the Machine.* Cambridge, MA: MIT Press.

Whitehead, A. N. 1925: *Science and the Modern World.* London: Free Press.

Wilkes, K. 1988: *Real People: Personal Identity without Thought Experiments.* Oxford: Clarendon Press.

Williams, G. C. 1985: "A Defense of Reductionism in Evolutionary Biology." In Dawkins and Ridley (1985), pp. 1–27.

Wilson, E. 1998: *Consilience: The Unity of Knowledge.* New York: Knopf.

Wilson, E., and Holldobler, B. 1988: "Dense Heterarchies and Mass Communication as the Basis of Organization in Ant Colonies." *Trends in Ecology and Evolution,* 3, pp. 65–84.

Wilson, J. 1999: "How Superduper Does a Physicalist Supervenience Need to Be?" *Philosophical Quarterly,* 49, pp. 33–52.

2005: "Supervenience-Based Formulations of Physicalism." *Nous,* 29, pp. 426–59.

2010: "Non-Reductive Physicalism and Degrees of Freedom." *British Journal for the Philosophy of Science,* 61, pp. 279–311.

2014: "No Work for a Theory of Grounding." *Inquiry,* 57, pp. 1–45.

Forthcoming: "Metaphysical Emergence: Weak and Strong." In T. Bigaj and C. Wuthrich (eds.) *Metaphysics in Contemporary Physics.* Poznan Studies in the Philosophy of the Sciences and the Humanities.

Wimsatt, W. 1976: "Reductionism, Levels of Organization and the Mind–Body Problem." In *Conciousness and the Brain,* ed. by G. Globus, G. Maxwell, and I. Savodnik. New York: Plenum Press.

1997: "Aggregativity: Reductive Heuristics for Finding Emergence." *Philosophy of Science,* 64, pp. S372–S384.

2000: "Emergence as Nonaggregativity and the Biases of Reductionisms." *Foundations of Science,* 5, pp. 269–97.

2007: *Re-Engineering Philosophy for Limited Beings.* Cambridge, MA: Harvard University Press.

Wimsatt, W., and Sarkar, S. 2006: "Reductionism." In Pfeifer and Sarkar (2006), vol. 2, pp. 696–703.

Woolley, R. 1985 "The Molecular Structure Conundrum". *Journal of Chemical Education*, v.62.

Woolley, R and Sutcliffe, B. 1977: "Molecular Structure and the Born-Oppenheimer Approximation". *Chemical Physics Letters*, v.45.

Yates, F. (ed.) 1987: *Self-Organizing Systems: The Emergence of Order.* New York: Plenum Press.

Index